Guide to Cultivating Complex Analysis

Working the Complex Field

Jiří Lebl

November 5, 2025
(version 1.8)

2

Acknowledgments:
I would like to thank Adam Cartisano, Josiah Ireland, Hoai Dao, Haridas Das, Uddhaba Pandey, Amanullah Nabavi, Rajan Adhikari, Abdullah Al Helal, Preston Kelley, Kevin Fernández, Jonathan Hunt, Sunday Oluwafemi Gbodogbe, and generally all students in my classes for pointing out typos/errors and helpful suggestions.

More information:
See https://www.jirka.org/ca/ for more information (including contacts).

Contents

Introduction

If you cannot prove a man wrong, don't panic. You can always call him names.
—Oscar Wilde

The purpose of this book is to teach a one-semester graduate course in complex analysis for incoming graduate students.[*] The first seven chapters are a natural first semester in a two-semester sequence where the second semester could be several complex variables (e.g., [L3]) or perhaps harmonic analysis. It could perhaps be used for a more elementary two-semester sequence if the appendix is covered first, and all the optional bits of the main text are also covered. We assume basic knowledge of undergraduate analysis in the real variable, called "advanced calculus" in some schools. The text assumes knowledge of metric spaces and differential calculus in several variables, but if the reader is not confident in these topics or has not yet seen them, the useful results are presented (with proofs) in the appendices. With that, a basic prerequisite for the course would be at least a single semester of undergraduate analysis if the appendices are also covered or read, and if the student has seen metric spaces and mappings in $\mathbb{R}^2$, then the course can just start in Chapter 1. Very basic undergraduate linear and abstract algebra is also useful.

The analysis prerequisites can be mostly found in [L1, L2, R1]. Further recommended reading on complex analysis is [B, C1, C2, R2, U]. See the aptly named Further Reading chapter.

This book takes the view that we do not need to redefine and reprove things that we have done in a basic undergraduate real analysis course, especially with regard to mappings of the plane. We can quite quickly jump to holomorphic functions as solutions of the Cauchy–Riemann equations, for instance. The connection is to understand both the derivative of a planar mapping and multiplication by a complex number as a 2×2 real matrix. When we introduce line integrals, we connect them to the line integrals the student has seen in calculus. The holomorphic inverse function theorem can be introduced early as a consequence of the standard inverse function theorem in $\mathbb{R}^2$. An outline of a pure complex analysis proof is left for later as an exercise. These are not simply time-saving measures. The point is to stress that we are not defining some totally new and different world.

---

[*]I wrote it specifically to teach Math 5283 at Oklahoma State University.

We also try to introduce the $z, \bar{z}$ approach instead of just the purely x, y approach. For example, we introduce and use the Wirtinger operators. It is really a better way to think about complex variables.

We try not to define any terminology or notation conflicting with what the reader has learned before. Mainly, the term "differentiable" is generally left for the real derivative and we use "complex differentiable" when needed. Although to be sure, we often write "(real) differentiable" or "differentiable (in the real sense)" to make it clear when we mean real differentiability.

Finally, some sections in the first seven chapters are marked with a $\star$ and those can be easily skipped on first reading (though it does not mean they are not important, just not necessary for what follows). Skipping some may make it possible to cover other later topics.

The general dependence of the non-appendix chapters is the following diagram. The way I usually run my semester course is to go through chapters 1–5, skipping the homotopy versions of Cauchy, to get through basic theory of holomorphic functions, then getting to 6 (Montel and Riemann mapping), and some bits of 7 (harmonic functions). There are some extra topics for a different plan such as 8 (Weierstrass factorization), 9 (Runge), and 10 (analytic continuation).

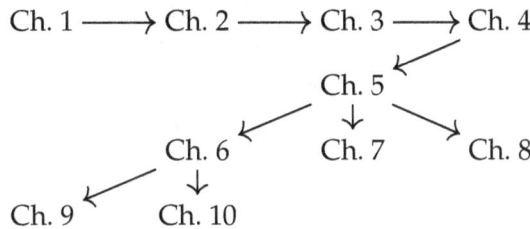

The only reason why 9 (Runge) depends on 6 (Montel and Riemann mapping) is that we prove Lemma 6.3.7 (around every compact there exists a cycle homologous to zero) as an example application of Riemann mapping.

1*i* The Complex Plane

It's clearly a budget. It's got a lot of numbers in it.

—George W. Bush

1.1*i* Complex numbers

Modern* mathematics is taking a false statement such as "all polynomials have a root" and redefining what a "root" could be, that is, redefining "number," so that the statement is true. In this instance, we arrive at the complex numbers. Although this technique (moving the goalposts) feels like cheating, it gave us essentially all the mathematics we know, both pure and applied. This same technique starts with the natural numbers $\mathbb{N} = \{1, 2, 3, \ldots\}$, the only numbers obvious from nature, and gives us zero and negative numbers producing the integers $\mathbb{Z}$, so that we can solve equations such as $n + 2 = 1$. From $\mathbb{Z}$, we define the rational numbers $\mathbb{Q}$ to solve[†] equations such as $2x = 1$. We extend $\mathbb{Q}$ to the real numbers $\mathbb{R}$ to solve equations such as $x^2 = 2$. Actually, our definition of the real numbers is such that we get theorems like the intermediate value theorem, Bolzano–Weierstrass, etc. It is then not much of a stretch to do the same thing when trying to solve $z^2 + 1 = 0$. Just as with the real numbers, the consequences of adding $\sqrt{-1}$ to the mix are much more profound than just finding roots of polynomials.

Curiously, while the step into analysis with real numbers is a step into the abyss, the step into analysis with complex numbers is a step into a fairytale wonderland. A first-year real analysis course crushes the student's hopes and dreams. Reasonable statements are false and bizarre counterexamples abound. In contrast, a complex analysis course fills the student with unrealistic optimism. It is replete with naïve and silly statements that only a bad calculus student could entertain.[‡] The two are the good cop/bad cop, the yin/yang, of contemporary analysis.

---

*In this context, *modern* means "later than the middle ages."

[†]Do we really solve $2x = 1$ by writing $x = 1/2$? After all, $1/2$ is just a placeholder for an object that we can't describe in any way other than "whatever 1 divided by 2 would be if it existed."

[‡]E.g.: If you can differentiate once, you can differentiate twice. Every function acts sort of like a linear function. If all derivatives are zero at a point, the function is constant. Etc.

1.1.1i The complex numbers as the plane

You have surely seen the complex number field, but let us review its definition anyway. As a set, let the *complex number field* $\mathbb{C}$ be the set $\mathbb{R}^2 = \mathbb{R} \times \mathbb{R}$. The set is a plane, so we call it the *complex plane**. To make it a field, we define addition and product:

$$(a,b) + (c,d) \overset{\text{def}}{=} (a+c, b+d),$$
$$(a,b)(c,d) \overset{\text{def}}{=} (ac - bd, bc + da).$$

> *Exercise* **1.1.1:** *Check that $\mathbb{C}$ is a field, where the additive identity is $0 = (0,0)$ and the multiplicative identity is $1 = (1,0)$. That is, $\mathbb{C}$ is an abelian group under addition, the nonzero complex numbers are an abelian group under multiplication, and the distributive law holds. Hint: The multiplicative inverse of (a,b) is $\left(\frac{a}{a^2+b^2}, \frac{-b}{a^2+b^2}\right)$.*

When we write a real number x, we identify it with the complex number $(x,0)$. With this identification $\mathbb{R} \subset \mathbb{C}$. We also define the *imaginary unit*[†]

$$i \overset{\text{def}}{=} (0,1).$$

With this notation, $(x,y) = x + iy$. From now on, $x + iy$ is the only way we will write complex numbers in terms of the coordinates x and y. We call $x + iy$ the *cartesian form* of the complex number. The number i has the magical property that

$$i^2 = -1.$$

For this reason we sometimes[‡] write $i = \sqrt{-1}$. Note that there is another square root of -1, that is, $-i$. The numbers i and $-i$ are the solutions to $z^2 + 1 = 0$. We will prove later that every polynomial has roots over the complex numbers.

Given a complex number $z = x + iy$, its "evil twin" is the *complex conjugate* of z:

$$\bar{z} \overset{\text{def}}{=} x - iy.$$

The number x is called the *real part* and y is called the *imaginary part*. We write

$$\operatorname{Re} z = \operatorname{Re}(x + iy) = \frac{z + \bar{z}}{2} = x, \qquad \operatorname{Im} z = \operatorname{Im}(x + iy) = \frac{z - \bar{z}}{2i} = y.$$

A particularly useful observation is that we wrote the real part and the imaginary part in terms of z and $\bar{z}$. Any expression we write in terms of the real and imaginary parts of z, we can equally well write in terms of z and $\bar{z}$. And vice versa. For example,

$$x^3 + y^3 + 3ixy = \left(\frac{z + \bar{z}}{2}\right)^3 + \left(\frac{z - \bar{z}}{2i}\right)^3 + 3i\left(\frac{z + \bar{z}}{2}\right)\left(\frac{z - \bar{z}}{2i}\right),$$

---

*Although there is that odd mathematician out there who thinks that the *complex plane* is $\mathbb{C}^2 = \mathbb{C} \times \mathbb{C}$. If you hear someone say that, politely whack them over the head for me.

[†]Beware of engineers; they think it is called j, despite there being no "j" in "imaginary."

[‡]There are those who *always* write $\sqrt{-1}$ instead of i. Those people also deserve a good whack.

or

$$z^2 - i\bar{z}^2 + z\bar{z} = (x + iy)^2 - i(x - iy)^2 + (x + iy)(x - iy).$$

It may seem that an expression in terms of z and $\bar{z}$ is more complicated. Namely, z and $\bar{z}$ are not "independent variables." However, it is particularly powerful to think in terms of z and $\bar{z}$ instead of x and y, and to pretend in many contexts as if z and $\bar{z}$ were actually independent variables.

1.1.2*i*   The geometry and topology of the plane

The size of z is measured by the so-called *modulus*, which is just the *euclidean distance* from the origin to z:

$$|z| \overset{\text{def}}{=} \sqrt{z\bar{z}} = \sqrt{x^2 + y^2}.$$

More simply, $|z|^2 = z\bar{z}$. Notice $|z| \geq 0$, and $|z| = 0$ if and only if $z = 0$.

Proposition 1.1.1 (Cauchy–Schwarz and the triangle inequality). *If $z, w \in \mathbb{C}$, then*

(i) $|\operatorname{Re} z\bar{w}| \leq |z||w|$ *(Cauchy–Schwarz inequality*, note: $\operatorname{Re} z\bar{w}$ is the $\mathbb{R}^2$ dot product),*

(ii) $|z + w| \leq |z| + |w|$ *(Triangle inequality).*

Proof. The modulus squared of a complex number is always nonnegative. Thus,

$$\begin{aligned}
0 &\leq |z\bar{w} - \bar{z}w|^2 \\
&= (z\bar{w} - \bar{z}w)(\bar{z}w - z\bar{w}) \\
&= 2z\bar{z}w\bar{w} - z^2\bar{w}^2 - \bar{z}^2w^2 \\
&= 4z\bar{z}w\bar{w} - (z\bar{w} + \bar{z}w)^2 \\
&= (2|z||w|)^2 - (2\operatorname{Re} z\bar{w})^2.
\end{aligned}$$

This proves Cauchy–Schwarz. We prove the triangle inequality via Cauchy–Schwarz:

$$\begin{aligned}
|z + w|^2 &= (z + w)(\bar{z} + \bar{w}) \\
&= z\bar{z} + w\bar{w} + z\bar{w} + \bar{z}w \\
&\leq z\bar{z} + w\bar{w} + 2|z||w| \\
&= (|z| + |w|)^2. \qquad \qquad \square
\end{aligned}$$

Exercise 1.1.2: *Prove the* polarization identity $4z\bar{w} = |z + w|^2 - |z - w|^2 + i(|z + iw|^2 - |z - iw|^2)$.

---

*The name is wrong. Some (wrongly) say it should be Cauchy–Bunyakovsky–Schwarz. Bunyakovsky and Schwarz proved the infinite-dimensional version. This version ought to be called Cauchy inequality, but lamentably that name could refer to a different inequality, the Cauchy estimates.

The distance between two numbers z and w is measured by

$$|z - w|.$$

This distance makes $\mathbb{C}$ into a complete metric space. By complete, we mean that Cauchy sequences have limits. See Appendix A for an introduction to metric spaces.

Proposition 1.1.2. *Complex addition, multiplication, division, and conjugation are continuous: Suppose $\{a_n\}$ and $\{b_n\}$ are two convergent sequences of complex numbers. Then,*

(i) $\displaystyle\lim_{n\to\infty}(a_n + b_n) = \left(\lim_{n\to\infty} a_n\right) + \left(\lim_{n\to\infty} b_n\right),$

(ii) $\displaystyle\lim_{n\to\infty} a_n b_n = \left(\lim_{n\to\infty} a_n\right)\left(\lim_{n\to\infty} b_n\right),$

(iii) $\displaystyle\lim_{n\to\infty} \frac{1}{a_n} = \frac{1}{\lim\limits_{n\to\infty} a_n},$ *as long as* $\displaystyle\lim_{n\to\infty} a_n \neq 0,$

(iv) $\displaystyle\lim_{n\to\infty} \bar{a}_n = \overline{\lim_{n\to\infty} a_n}.$

Exercise **1.1.3:** *Prove the proposition.*

The basic neighborhood (that is, an open ball) in $\mathbb{C}$ is called a *disc*. Given $p \in \mathbb{C}$ and $r > 0$, define the disc of radius r around p as

$$\Delta_r(p) \overset{\text{def}}{=} \{z \in \mathbb{C} : |z - p| < r\}.$$

A disc centered at the origin of radius 1 is called the *unit disc*

$$\mathbb{D} \overset{\text{def}}{=} \Delta_1(0) = \{z \in \mathbb{C} : |z| < 1\}.$$

The unit disc will come up often in this course, as it turns out that a lot of complex analysis can be done by looking at just the unit disc.

A useful "version" of the unit disc is the *upper half-plane*:

$$\mathbb{H} \overset{\text{def}}{=} \{z \in \mathbb{C} : \operatorname{Im} z > 0\}.$$

We will see in a moment that $\mathbb{D}$ and $\mathbb{H}$ are equivalent in a very nice way. Things done on the unit disc can just as well be done on the upper half-plane.

The following definition is perhaps somewhat unnecessary, but it is easier to write and say than *open and connected*, and it is commonly used in complex analysis.[*]

Definition 1.1.3. An open and connected set $U \subset \mathbb{C}$ is called a *domain*.[†]

---

[*]We generally consider our sets also nonempty, but usually the statements of results for empty open sets or domains are simply vacuous.

[†]Perhaps "domain" is a tad unfortunate since we also call the X in $f: X \to Y$ a "domain" of the function, even if X is not a domain in the sense of topology.

1.1.3i   Complex-valued functions

It is possible that the analysis you have seen so far in your mathematical career has been for real-valued functions $f \colon X \to \mathbb{R}$. In this book, we are concerned with complex-valued functions $f \colon X \to \mathbb{C}$. The results for real-valued functions are then applied by thinking of either the components of f separately or by thinking of $\mathbb{C}$ as the real vector space $\mathbb{R}^2$.

When we find ourselves in the possession of a complex-valued function $f \colon X \to \mathbb{C}$, we write $u = \operatorname{Re} f$ and $v = \operatorname{Im} f$ for real-valued functions $u, v \colon X \to \mathbb{R}$, and then

$$f = u + iv.$$

If $X \subset \mathbb{C}$, we think of $X \subset \mathbb{R}^2$. A derivative in x or y (where $z = x + iy$) is then applied to the components (just as if f was valued in $\mathbb{R}^2$):

$$\frac{\partial f}{\partial x} = \frac{\partial u}{\partial x} + i\frac{\partial v}{\partial x} \qquad \text{and} \qquad \frac{\partial f}{\partial y} = \frac{\partial u}{\partial y} + i\frac{\partial v}{\partial y}.$$

If $X \subset \mathbb{R}$, that is, if f is a complex-valued function of one real variable, then $f' = u' + iv'$. Equivalently, we treat f as a function from $\mathbb{R}$ to $\mathbb{R}^2$ and hence f' is a 2×1 matrix—a column vector, or in other words f' represents a complex number if we are identifying $\mathbb{C}$ and $\mathbb{R}^2$.

Matters are similar for integration. For $f \colon [a, b] \to \mathbb{C}$, we say f is (Riemann) integrable if u and v are, and then

$$\int_a^b f(t)\,dt = \int_a^b u(t)\,dt + i\int_a^b v(t)\,dt.$$

Indeed, that is the way one integrates vector-valued functions for any vector space, and $\mathbb{C} = \mathbb{R}^2$ is a vector space. Basic analysis tells us that if given a Riemann integrable real-valued function $u \colon [a, b] \to \mathbb{R}$, then $|u|$ is Riemann integrable and $\left| \int_a^b u(t)\,dt \right| \leq \int_a^b |u(t)|\,dt$. A similar result holds for complex-valued functions.

Proposition 1.1.4. *Suppose $f \colon [a, b] \to \mathbb{C}$ is (Riemann) integrable. Then $|f|$ is (Riemann) integrable and*

$$\left| \int_a^b f(t)\,dt \right| \leq \int_a^b |f(t)|\,dt.$$

Exercise 1.1.4: *Prove the proposition. Hint: After you know integrability, consider a Riemann sum and the regular triangle inequality.*

1.1.4*i*   Matrix representation of complex numbers

As $\mathbb{C}$ is $\mathbb{R}^2$, we can think of $\mathbb{C}$ as a real two-dimensional vector space by forgetting about the complex multiplication. The standard basis is 1 and i. To put multiplication back into the picture, we think of linear operators on $\mathbb{R}^2$. Given a complex number ξ, the map $z \mapsto \xi z$ is a real-linear operator*. A real-linear operator on $\mathbb{R}^2$ is given by a 2×2 real matrix. Namely, the complex number $\xi = a + ib$ can be represented by the 2×2 matrix

$$\begin{bmatrix} a & -b \\ b & a \end{bmatrix}. \tag{1.1}$$

Let us check. If we think of a complex number $a + ib$ as a matrix and $c + id$ as a column vector, then complex multiplication makes sense as matrices:

$$\begin{bmatrix} a & -b \\ b & a \end{bmatrix} \begin{bmatrix} c \\ d \end{bmatrix} = \begin{bmatrix} ac - bd \\ bc + ad \end{bmatrix}.$$

The matrices

$$1 \quad " = " \quad \begin{bmatrix} 1 & 0 \\ 0 & 1 \end{bmatrix} \quad \text{and} \quad i \quad " = " \quad \begin{bmatrix} 0 & -1 \\ 1 & 0 \end{bmatrix}$$

are the identity and the rotation counterclockwise by 90 degrees respectively—precisely what we expect multiplication by 1 and i to do.

Complex conjugation is also a real-linear operator and can be represented by the matrix $\begin{bmatrix} 1 & 0 \\ 0 & -1 \end{bmatrix}$. Notice that complex conjugation is not a multiplication by a complex number. Below you will prove, however, that multiplications by complex numbers together with conjugation do in fact "generate" all the real-linear operators.

For those matrices representing complex numbers, we can also multiply the 2×2 matrices themselves, and this matrix multiplication is the same as multiplication of the complex numbers. Similarly with addition. That is, we can view the field of complex numbers as a subring of $M_2(\mathbb{R})$ (exercise below).

Exercise **1.1.5:** *Prove that a) the matrix multiplication on matrices of the form (1.1) is commutative and b) reproduces the complex number multiplication, and that these matrices form a subring of $M_2(\mathbb{R})$. c) Prove that nonzero matrices of this form are invertible (the subring is a field). Specifically, notice the determinant appearing in the denominator for the multiplicative inverse.*

Exercise **1.1.6:** *Prove that if the 2×2 matrix M represents a complex number $a + ib$, then M has two eigenvalues: $a \pm ib$ with the corresponding eigenvectors $\begin{bmatrix} 1 \\ \mp i \end{bmatrix}$.*

Exercise **1.1.7:** *Prove that every real-linear operator on $\mathbb{C}$ (that is, every 2×2 real matrix M) can be represented by two complex numbers ξ and ζ, and the formula $z \mapsto \xi z + \zeta \bar{z}$.*

---

*$L: \mathbb{C} \to \mathbb{C}$ is real-linear if $L(az + bw) = aL(z) + bL(w)$ for all $a, b \in \mathbb{R}$ and $z, w \in \mathbb{C}$.

Exercise **1.1.8:**

  a) *Suppose a 2×2 real matrix M represents multiplication by $\xi \in \mathbb{C}$. Show that $\det M = |\xi|^2$.*

  b) *Suppose a 2×2 real matrix M is represented by $z \mapsto \xi z + \zeta \bar{z}$ (see previous exercise). Show that $\det M = |\xi|^2 - |\zeta|^2$.*

This representation of complex numbers comes up quite often in applications. For instance, an $m \times n$ complex matrix can be represented by a $2m \times 2n$ real matrix by replacing each entry by a 2×2 matrix. So software set up for working with real matrices can easily be duped into working with complex matrices.

For us, the main application will be to understand the derivative of a complex-valued function of a complex variable $f \colon \mathbb{C} \to \mathbb{C}$. Thinking of the function as a mapping $f \colon \mathbb{R}^2 \to \mathbb{R}^2$, the real derivative of f is a 2×2 matrix. The object of study of complex analysis, the holomorphic (or analytic) functions, are those functions whose real derivative matrix corresponds to a multiplication by a complex number.

1.2i Polar form and the exponential

1.2.1i The exponential

The exponential is the most fundamental and useful function in complex analysis. Assume we know e^x for real numbers. Define e^z for complex numbers (see Figure 1.1):

$$\exp(z) = e^z = e^{x+iy} \overset{\text{def}}{=} e^x e^{iy} = e^x \cos y + i e^x \sin y.$$

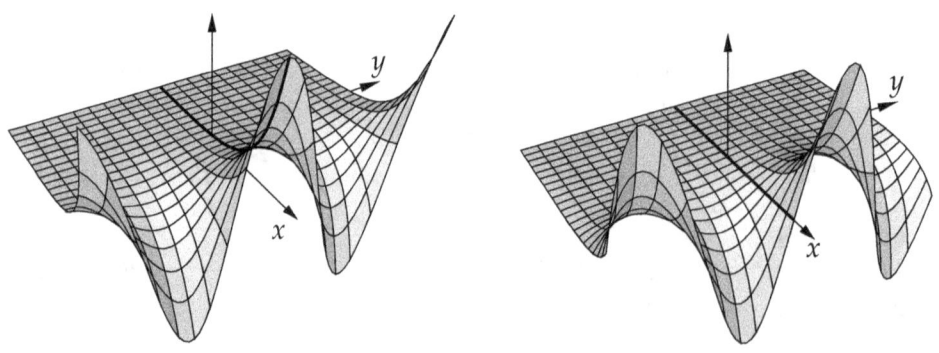

Figure 1.1: Graphs of the real part (left) and imaginary part (right) of the complex exponential $e^z = e^{x+iy}$. The plot of the real exponential ($y = 0$) is marked in a bold line.

The definition agrees with the standard exponential for real numbers (when $y = 0$). Furthermore,

$$e^{\bar{z}} = e^{x-iy} = e^x \cos y - i e^x \sin y = \overline{e^z},$$

and

$$|e^z| = |e^{x+iy}| = e^x.$$

It is possible to define the complex exponential without resorting to the real exponential, sine, and cosine, and we will do so in due course. But we are impatient and we want something to play around with now, without waiting.

Proposition 1.2.1. *For any two complex numbers $z, w \in \mathbb{C}$,*

$$e^{z+w} = e^z e^w.$$

> *Exercise* **1.2.1:** *Prove the proposition using the definition and trigonometric identities.*

For real θ, the definition of the exponential gives the so-called *Euler's formula*:

$$e^{i\theta} = \cos\theta + i\sin\theta.$$

In other words, $e^{i\theta}$ parametrizes the unit circle. The formula says that for real θ,

$$\cos\theta = \operatorname{Re} e^{i\theta} = \frac{e^{i\theta} + e^{-i\theta}}{2}, \qquad \sin\theta = \operatorname{Im} e^{i\theta} = \frac{e^{i\theta} - e^{-i\theta}}{2i}.$$

We define cosine and sine for complex numbers by plugging those numbers into the formulas above, now that we know how to evaluate the exponential:

$$\cos z \overset{\text{def}}{=} \frac{e^{iz} + e^{-iz}}{2}, \qquad \sin z \overset{\text{def}}{=} \frac{e^{iz} - e^{-iz}}{2i}.$$

1.2.2i Polar coordinates

As complex numbers are just the plane, we can use *polar coordinates* to represent complex numbers. That is, $x = r\cos\theta$ and $y = r\sin\theta$. We write z in *polar form* as

$$z = x + iy = re^{i\theta}.$$

Here, $r = |z| = \sqrt{x^2 + y^2}$ is the modulus, and θ is the angle that $x + iy$ makes with the real axis (the x-axis). The θ is called the *argument*. See Figure 1.2. The reason for the notation is the Euler's formula, so

$$z = re^{i\theta} = r\cos\theta + ir\sin\theta = x + iy.$$

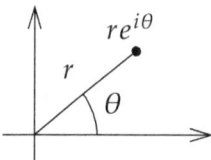

Figure 1.2: Polar coordinates.

Polar form is particularly nice for multiplication and for powers. Suppose $z = re^{i\theta}$ and $w = se^{i\psi}$, then

$$zw = re^{i\theta}se^{i\psi} = rse^{i(\theta+\psi)}, \qquad \frac{1}{z} = \frac{1}{re^{i\theta}} = \frac{1}{r}e^{-i\theta}, \qquad z^n = \left(re^{i\theta}\right)^n = r^n e^{in\theta}.$$

Multiplication rotates by the argument and scales by the modulus. Namely, we see again that multiplication by $i = e^{i\pi/2}$ is rotation counterclockwise by 90 degrees. The downside is that the polar form is particularly terrible for addition. You win some, you lose some.

Exercise **1.2.2:** *Let z, w be two nonzero complex numbers and let ℓ_z and ℓ_w be the lines through the origin and z and w respectively. Write θ for the angle between ℓ_z and ℓ_w. Then prove that $\operatorname{Re} z\bar{w} = |z||w|\cos\theta$. Note that $\operatorname{Re} z\bar{w}$ is the standard real dot product in $\mathbb{R}^2$, and so you are asked to prove the formula for the dot product from calculus.*

1.2.3*i*   The argument

We attempt to define the argument of $z = re^{i\theta}$ as

$$\arg z \overset{\text{def?}}{=} \theta,$$

but we run up against the problem that if θ is an argument of z, then so is $\theta + 2\pi$, $\theta - 2\pi$, or $\theta + k2\pi$ for any integer k. In other words, $\arg z$ is not a function in the classical sense, but a *multivalued function**. The correct definition is

$$\arg z \overset{\text{def}}{=} \ldots, \theta - 4\pi, \theta - 2\pi, \theta, \theta + 2\pi, \theta + 4\pi, \ldots$$

One more minor issue remains. If $z = 0$, then $z = 0 = 0e^{i\theta}$ for any θ whatsoever. Therefore, we only define the argument for nonzero z.

---

*Non-complex analysts will sometimes claim that a multivalued function is nonsense, but you can safely ignore those troublemakers.

It may at times be useful to nail down a particular number for the argument. We define the *principal branch of* arg as

$$\operatorname{Arg} z \stackrel{\text{def}}{=} \theta, \qquad \text{where} - \pi < \theta \leq \pi.$$

It may seem like a good solution to the multivaluedness of arg, but one's hopes are dashed by the cruel reality of Arg not being continuous on the negative real axis. See Figure 1.3. The principal branch is somewhat less useful than one might think. There is also the issue that not everyone agrees on what "principal branch" means; some mathematicians sacrifice the positive real axis and let θ be in the range $[0, 2\pi)$.

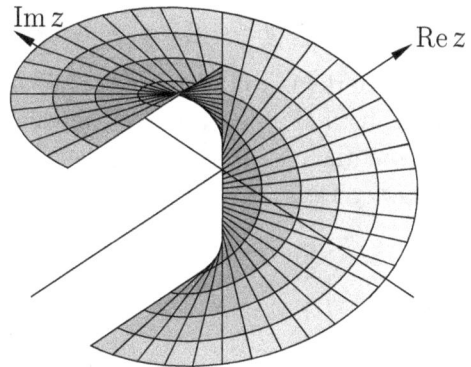

Figure 1.3: Graph of the principal branch of argument.

Exercise **1.2.3:** *Show that* Arg *as defined above is not a continuous function on* $\mathbb{C} \setminus \{0\}$.

1.2.4i   Mapping properties of the exponential

Let us see what the exponential does to the complex plane. The identity $e^{z+w} = e^z e^w$ implies that the exponential is never zero (exercise). From the known properties of polar coordinates and the real exponential, it follows that the complex exponential is onto $\mathbb{C} \setminus \{0\}$. The complex exponential is not one-to-one, it is infinitely-many-to-one. For any integer k,

$$e^{z+ik2\pi} = e^z e^{ik2\pi} = e^z. \tag{1.2}$$

Exercise **1.2.4:** *Prove that (1.2) are the only such identities by showing that if $2k\pi < \operatorname{Im} z \leq 2(k+1)\pi$ and $2k\pi < \operatorname{Im} w \leq 2(k+1)\pi$, then $e^z = e^w$ implies $z = w$.*

Exercise **1.2.5:** *Use $e^{z+w} = e^z e^w$ and $e^0 = 1 \neq 0$ to show that $e^z \neq 0$ for all $z \in \mathbb{C}$. In other words, show that if a function f satisfies $f(z+w) = f(z)f(w)$ and $f(0) = 1$, then $f(z) \neq 0$ for all z.*

Consider a vertical line given by $x = c$. As

$$e^z = e^{x+iy} = e^x e^{iy},$$

and $|e^{iy}| = 1$, the exponential takes the vertical line $x = c$ to a circle of radius e^c. See Figure 1.4. Thus, the exponential $w = e^z$ takes the strip $a < x < b$ to the annulus

$$\{w \in \mathbb{C} : e^a < |w| < e^b\}.$$

On the other hand, the horizontal line $y = c$ is taken to the ray from the origin to infinity where $\theta = c$ in polar coordinates. Again see Figure 1.4. Hence, the exponential takes the strip $a < y < b$ to the sector

$$\{w \in \mathbb{C} : \text{“}a < \arg w < b\text{”}\}.$$

The reason for the quotation marks is that the inequality makes no sense without interpreting it properly. It means that at least one of the values of $\arg w$ is between a and b. In particular, the exponential e^z takes the set given by $2k\pi < \operatorname{Im} z \leq 2(k+1)\pi$ in a one-to-one fashion (see Exercise 1.2.4) onto $\mathbb{C} \setminus \{0\}$.

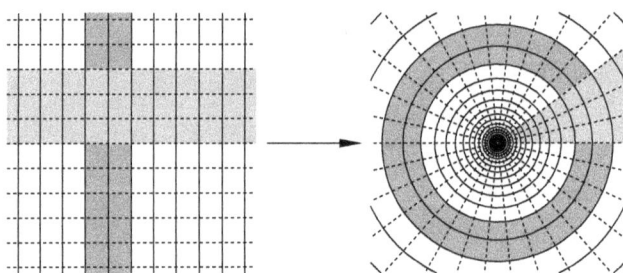

Figure 1.4: Horizontal and vertical lines, and a horizontal and a vertical strip, mapped by the exponential. Note that each horizontal line only goes to a ray from the origin.

1.3i \ The Riemann sphere

It is sometimes useful to extend the real numbers by adding $\pm\infty$. A similar concept exists for the complex plane, although we only add one infinity. We write

$$\mathbb{C}_\infty = \mathbb{C} \cup \{\infty\},$$

and we call $\mathbb{C}_\infty$ the *Riemann sphere*. We want the topology of $\mathbb{C}_\infty$ to be the same as that of $\mathbb{C}$ when we are away from infinity. Define the function $g \colon \mathbb{C}_\infty \to \mathbb{C}_\infty$ by

$$g(z) = \begin{cases} 1/z & \text{if } z \neq 0 \text{ and } z \neq \infty, \\ \infty & \text{if } z = 0, \\ 0 & \text{if } z = \infty. \end{cases}$$

The function g is bijective (one-to-one and onto). Any neighborhood of the origin in $\mathbb{C}$ is taken to a set that includes infinity and we call those the neighborhoods of ∞. When talking about a neighborhood of infinity in $\mathbb{C}_\infty$, then we really want to think of this map, and think of the corresponding neighborhood of the origin.

More concretely, we can give $\mathbb{C}_\infty$ a metric space structure. Let S^2 be the unit sphere in $\mathbb{R}^3$, that is, the set described by $x^2 + y^2 + z^2 = 1$ if (x, y, z) are the coordinates.* The plane $\mathbb{R}^2$ with coordinates (x, y) can be identified with $\mathbb{C}$ with coordinate ξ by taking $x + iy = \xi$. Given any point $p \in S^2$ that is not the north pole $(0, 0, 1) \in S^2$, there is a unique line in $\mathbb{R}^3$ through the point $(0, 0, 1)$ and p. It is not difficult to prove that this line is never parallel to the xy-plane, that is, to $\mathbb{C}$, and hence it must intersect $\mathbb{C}$ in a unique point $\xi \in \mathbb{C}$ (a plane and a line intersect at a unique point unless the two are parallel). Define

$$\Phi(p) = \xi.$$

Let $\Phi\big((0, 0, 1)\big) = \infty$, so that we have a map $\Phi \colon S^2 \to \mathbb{C}_\infty$. This map is called the *stereographic projection*. See Figure 1.5. The map is bijective (exercise below), and so define a metric on $\mathbb{C}_\infty$ by using a metric on S^2, which can be the subspace metric coming from the euclidean metric on $\mathbb{R}^3$. Another possibility could be the great circle distance, Example A.1.7. Both distances would lead to the same topology and so the same limits.

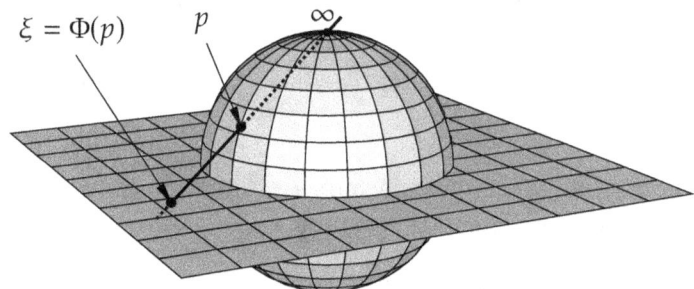

Figure 1.5: Stereographic projection of the sphere S^2 to the complex plane.

Exercise 1.3.1: Show that Φ is bijective.

Exercise 1.3.2: Suppose (ϕ, θ) are spherical coordinates on S^2, where $0 \leq \phi \leq \pi$ is the zenith (angle made with the z-axis) and $-\pi < \theta \leq \pi$ the azimuth, and we write points in $\mathbb{C}$ using polar coordinates re^θ. Then prove that Φ takes (ϕ, θ) to $\cot(\phi/2)\, e^{i\theta}$.

Exercise 1.3.3: Show that the topology induced on $\mathbb{C}$ by the topology of S^2 using Φ as above is equivalent to the standard one. That is, show that a set $U \subset \mathbb{C}$ is open in the euclidean metric on $\mathbb{C}$ if and only if it is open using the metric coming from S^2.

---

*This paragraph is possibly the only instance in this book where z is a real number.

The point of the Riemann sphere is to give the value ∞ to certain limits and to allow limits as z tends to ∞. For a function $f \colon U \subset \mathbb{C}_\infty \to \mathbb{C}_\infty$, we define

$$\lim_{z \to z_0} f(z) = L$$

as the limit using the topology on the Riemann sphere, including the cases when $z_0 = \infty$, $L = \infty$, or both. Conveniently the Riemann sphere is compact. Indeed, topologically, it is the same as the "one-point compactification" of $\mathbb{C}$.

Exercise 1.3.4: *Suppose $L \in \mathbb{C}$ and $z_0 \in \mathbb{C}$. Show that $\lim_{z \to z_0} f(z) = L$ in the sense of the Riemann sphere if and only if $\lim_{z \to z_0} f(z) = L$ in the usual sense (the euclidean metric on $\mathbb{C}$).*

Exercise 1.3.5: *Suppose $L \in \mathbb{C}$. Show that $\lim_{z \to \infty} f(z) = L$ in the sense of the Riemann sphere if and only if for every $\epsilon > 0$ there exists an M such that $|f(z) - L| < \epsilon$ whenever $|z| > M$.*

Exercise 1.3.6: *Suppose $z_0 \in \mathbb{C}$. Show that $\lim_{z \to z_0} f(z) = \infty$ in the sense of the Riemann sphere if and only if for every $M > 0$ there exists a $\delta > 0$ such that $|f(z)| > M$ whenever $|z - z_0| < \delta$.*

Exercise 1.3.7: *Suppose $L \in \mathbb{C}_\infty$. Show that $\lim\limits_{z \to \infty} f(z) = L$ if and only if $\lim\limits_{z \to 0} f(1/z) = L$.*

Exercise 1.3.8: *Suppose $z_0 \in \mathbb{C}_\infty$. Show that $\lim\limits_{z \to z_0} f(z) = \infty$ if and only if $\lim\limits_{z \to z_0} \frac{1}{f(z)} = 0$.*

It then makes sense to talk about the value of $1/z$ at the origin as ∞, and the value of z at ∞ as ∞. In fact, every nonconstant polynomial is ∞ at ∞.

Exercise 1.3.9: *Suppose $P(z) = a_d z^d + a_{d-1} z^{d-1} + \cdots + a_1 z + a_0$ is a polynomial where $a_0, \ldots, a_d \in \mathbb{C}$. Prove that if $d \geq 1$ and $a_d \neq 0$ (P is nonconstant), then $\lim_{z \to \infty} P(z) = \infty$. Hint: If $a_d = 1$, then using $\left| \frac{a_{d-1} z^{d-1} + \cdots + a_1 z + a_0}{z^d} \right|$, one finds $|P(z)| \geq \frac{1}{2} |z|^d$ for large z.*

In calculus and in basic real analysis, you likely encountered infinite limits in the sense of the extended reals. Despite that the two types of infinite limits, either in the sense of the extended reals or in the sense of the Riemann sphere, look similar, and despite using essentially the same notation, they are different. For example,

$$\lim_{x \to 0} \frac{1}{x} \quad \text{does not exist,} \quad \text{but} \quad \lim_{z \to 0} \frac{1}{z} = \infty.$$

Here on the left-hand side we tacitly use the extended real sense (x is real, no?) and on the right-hand side we tacitly use the Riemann sphere sense (z seems complex).

Such implicit assumptions could, obviously, cause confusion. In this book, limits are going to be in the Riemann sphere sense unless either otherwise noted or obvious. If confusion could arise, we will write the extended real ∞ as $+\infty$.

The (partial) arithmetic that one can reasonably define with the ∞ of the Riemann sphere is quite different from the ∞ of the extended reals. For $c \neq \infty$, we could define $c + \infty = \infty$, but neither $\infty + \infty$ nor $\infty - \infty$ makes sense. For instance, if $f(z) = z$ and $g(z) = c - z$, then $\lim_{z \to \infty} f(z) = \infty$, $\lim_{z \to \infty} g(z) = \infty$, but $\lim_{z \to \infty} (f(z) + g(z)) = c$, so $\infty + \infty$ does not make sense. Unlike for the extended reals, it is reasonable (and at times useful) to define $c/0 = \infty$ for $c \neq 0$, $c/\infty = 0$ for $c \neq \infty$, and $c \cdot \infty = \infty$ for $c \neq 0$. The expressions $0 \cdot \infty$, $0/0$, and ∞/∞ had better be left undefined.

1.4i \ Linear fractional transformations

A convenient set of transformations of the complex plane or the Riemann sphere are the *linear fractional transformations* (LFT) (sometimes called the *Möbius transformations*). A function

$$f(z) = \frac{az + b}{cz + d}$$

is a linear fractional transformation if $ad \neq bc$. The requirement on a, b, c, d guarantees that the ratio does not simplify and that the function is nonconstant.

If $c \neq 0$, the expression is really defined only on $\mathbb{C} \setminus \{-d/c\}$; however, as in the last section, write

$$f\left(\frac{-d}{c}\right) = \infty, \qquad \text{and} \qquad f(\infty) = \frac{a}{c}.$$

If $c = 0$, then set $f(\infty) = \infty$. In either case, f is a map of the Riemann sphere to itself.

> **Exercise 1.4.1:** *Prove that an LFT is a bijective mapping of the Riemann sphere to itself.*
>
> **Exercise 1.4.2:** *Prove that an LFT extended to the Riemann sphere as above is continuous.*

Any LFT is a composition of *translations*

$$T_a(z) = z + a, \qquad a \in \mathbb{C},$$

complex dilations

$$D_a(z) = az, \qquad a \in \mathbb{C} \setminus \{0\},$$

and *inversions*

$$I(z) = \frac{1}{z}.$$

Consider an LFT $f(z) = \frac{az+b}{cz+d}$. Without loss of generality, assume that either $c = 1$ or $c = 0$. Suppose $c = 1$ first,

$$f(z) = \frac{az + b}{z + d} = \frac{b - ad}{z + d} + a = T_a\left(D_{b-ad}\left(I\left(T_d(z)\right)\right)\right).$$

If $c = 0$, then we can also assume that $d = 1$ and so $f(z) = az + b$. So

$$f(z) = az + b = T_b\big(D_a(z)\big).$$

Translations are easy to understand, they move everything in $\mathbb{C}$ in one direction. Complex dilation D_a is the traditional euclidean plane geometry dilation by $|a|$ and rotation by $\arg a$. The inversion is the euclidean plane geometry inversion across the unit circle and complex conjugation, see Figure 1.6. The euclidean inversion across the circle simply inverts the distance to the origin:

$$\frac{1}{|z|}e^{i\arg z} = \frac{|z|}{|z|^2}e^{i\arg z} = \frac{1}{\bar{z}}.$$

To get our complex inversion $I(z)$ we also conjugate. Translations and dilations both take ∞ to ∞. The inversion swaps ∞ and 0.

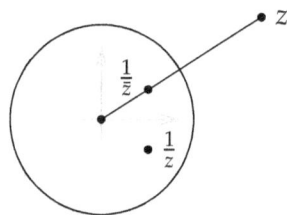

Figure 1.6: Complex inversion.

This sort of decomposition is quite useful in proving statements about LFTs that are preserved under composition; one only needs to prove them for T_a, D_a, and I. This technique will come in handy in the very next exercise. Let us include straight lines in the set of circles. After all, a straight line is just a circle through infinity: Think about the circle of radius r centered at ir as $r \to +\infty$. Then an LFT takes circles to circles. We leave this fact as an exercise.

Exercise 1.4.3: Prove that if we include straight lines in the set of "circles," then an LFT takes circles to circles.

Exercise 1.4.4: Prove that given any circle or a straight line, there exists an LFT that takes that circle or line to the real line.

One way to view an LFT is as a 2×2 complex matrix. For this purpose, we need to view the Riemann sphere as the so-called one-dimensional *projective space*. Define the equivalence relation $\sim$ on $\mathbb{C}^2 \setminus \{0\} = \mathbb{C} \times \mathbb{C} \setminus \{(0,0)\}$ by $u \sim v$ if and only if $u = \lambda v$ for some $\lambda \in \mathbb{C}$. The *one-dimensional complex projective* space is then defined as the quotient

$$\mathbb{CP}^1 \stackrel{\text{def}}{=} \mathbb{C}^2 \setminus \{0\}\big/_{\sim}.$$

In other words, $\mathbb{CP}^1$ is the set of "complex lines through the origin" in $\mathbb{C}^2$, or yet in other words, it is the set of one-dimensional vector subspaces of $\mathbb{C}^2$.

We identify $\mathbb{CP}^1$ with $\mathbb{C}_\infty$ in the following way. Denote by $[z : w] \in \mathbb{CP}^1$ the equivalence class (under $\sim$) of vectors in $\mathbb{C}^2 \setminus \{0\}$ that contains $(z, w) \in \mathbb{C}^2 \setminus \{0\}$. Define the bijection $\Psi \colon \mathbb{C}_\infty \to \mathbb{CP}^1$ as

$$\Psi(z) = \begin{cases} [z : 1] & \text{if } z \in \mathbb{C}, \\ [1 : 0] & \text{if } z = \infty. \end{cases}$$

Exercise 1.4.5: *Prove that the Ψ defined above is bijective.*

Let us check that an LFT

$$f(z) = \frac{az + b}{cz + d}$$

corresponds to an invertible linear map given by the matrix

$$M = \begin{bmatrix} a & b \\ c & d \end{bmatrix}.$$

An invertible M takes one-dimensional subspaces to one-dimensional subspaces, so it sounds plausible.

First, $\Psi \circ f$ for $z \in \mathbb{C} \setminus \{-d/c\}$ (or $z \in \mathbb{C}$ if $c = 0$) is equal to

$$\Psi \circ f(z) = \left[\frac{az + b}{cz + d} : 1 \right] = [az + b : cz + d],$$

where the second equality follows by definition of $\sim$. When $z = -d/c$, then $cz + d = 0$, and $\Psi \circ f(z) = \Psi(\infty) = [1 : 0] = [az + b : cz + d]$ as well.

Let us consider $\Psi \circ f \circ \Psi^{-1}$. If $w \neq 0$, then $[z : w] = [z/w : 1]$. So

$$\Psi \circ f \circ \Psi^{-1}([z : w]) = \Psi \circ f \left(\frac{z}{w} \right) = \left[a\frac{z}{w} + b : c\frac{z}{w} + d \right] = [az + bw : cz + dw].$$

And one checks that the same equality holds if $w = 0$. As $M \left[\begin{smallmatrix} z \\ w \end{smallmatrix} \right] = \left[\begin{smallmatrix} az+bw \\ cz+dw \end{smallmatrix} \right]$, the function f corresponds to the linear map $v \mapsto Mv$ on $\mathbb{C}^2$. The requirement $ad \neq bc$ implies $\det M \neq 0$, or in other words, M is invertible. So every LFT is represented by an invertible 2×2 matrix M (not uniquely), and conversely every invertible 2×2 matrix M corresponds to an LFT.

An invertible 2×2 matrix M gives a map from $\mathbb{C}^2 \setminus \{0\}$ to $\mathbb{C}^2 \setminus \{0\}$. Let $\pi \colon \mathbb{C}^2 \setminus \{0\} \to \mathbb{CP}^1$ be the map* $\pi((z, w)) = [z : w]$. The following commutative diagram[†] may

---

*The "quotient map" or the "natural projection."

[†]A diagram is commutative if taking two different routes in the picture gives the same map.

illustrate the entire situation better:

$$\mathbb{C}^2 \setminus \{0\} \xrightarrow{\ M\ } \mathbb{C}^2 \setminus \{0\}$$

$$\begin{array}{ccc}
\mathbb{C}^2 \setminus \{0\} & \xrightarrow{\ M\ } & \mathbb{C}^2 \setminus \{0\} \\
\downarrow{\scriptstyle\pi} & & \downarrow{\scriptstyle\pi} \\
\mathbb{CP}^1 & \xrightarrow{\Psi \circ f \circ \Psi^{-1}} & \mathbb{CP}^1 \\
{\scriptstyle\Psi}\upharpoonright\downharpoonright{\scriptstyle\Psi^{-1}} & & {\scriptstyle\Psi}\upharpoonright\downharpoonright{\scriptstyle\Psi^{-1}} \\
\mathbb{C}_\infty & \xrightarrow{\ f\ } & \mathbb{C}_\infty
\end{array}$$

Example 1.4.1: A handy LFT is the *Cayley map*:

$$C(z) = \frac{z - i}{z + i}.$$

The map is clearly an LFT, and it takes the upper half-plane $\mathbb{H} = \{z \in \mathbb{C} : \operatorname{Im} z > 0\}$ to the unit disc $\mathbb{D}$. Let us see why. The map takes $z \in \mathbb{C}$ to the unit disc if

$$1 > \left|\frac{z - i}{z + i}\right| = \frac{|z - i|}{|z + i|}.$$

In other words, $|z + i| > |z - i|$: The distance of z to $-i$ is larger than the distance of z to i. It is straightforward plane geometry to see that $z \in \mathbb{H}$. See Figure 1.7.

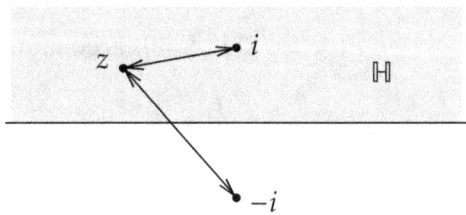

Figure 1.7: Why does the Cayley map take $\mathbb{H}$ to $\mathbb{D}$.

Any LFT is bijective if thought of as a map from $\mathbb{C}_\infty$ to itself, so C^{-1} exists, and it is also a useful map. We leave it as an exercise to figure out the inverse.

Exercise 1.4.6: *Figure out what C^{-1} (inverse of Cayley) is. Hint: Think of $\mathbb{C}_\infty$ as $\mathbb{CP}^1$ and C as a matrix. It is really easy to invert 2×2 matrices.*

Exercise 1.4.7: *For every LFT $f(z) = \frac{az+b}{cz+d}$, find f^{-1}. Hint: Same hint as above.*

In the exercises, you have essentially just shown (or at least finished showing) that LFTs form a group under composition, called the *Möbius group*. This group is generated by the elements T_a, D_a, and I for $a \in \mathbb{C}$.

1.5*i* Cross ratio ★

There is a certain quantity that is preserved by LFTs, the *cross ratio*:

$$(z_1, z_2; z_3, z_4) = \frac{(z_3 - z_1)(z_4 - z_2)}{(z_3 - z_2)(z_4 - z_1)} = \frac{z_3 - z_1}{z_3 - z_2} : \frac{z_4 - z_1}{z_4 - z_2},$$

where z_1, z_2, z_3, z_4 are complex numbers. The cross ratio was already described by the ancient Greeks[*] and plays a key role in projective geometry. The definition is extended to when one of the numbers is ∞ by simply erasing the affected terms from the ratios. For instance, if $z_1 = \infty$, then pretend that $z_3 - \infty$ is really equal to $z_4 - \infty$ and thus they cancel—which makes sense as $\lim_{z \to \infty} \frac{z_3 - z}{z_4 - z} = 1$. Consequently,

$$(\infty, z_2; z_3, z_4) = \frac{z_4 - z_2}{z_3 - z_2}, \qquad\qquad (z_1, \infty; z_3, z_4) = \frac{z_3 - z_1}{z_4 - z_1},$$

$$(z_1, z_2; \infty, z_4) = \frac{z_4 - z_2}{z_4 - z_1}, \qquad\qquad (z_1, z_2; z_3, \infty) = \frac{z_3 - z_1}{z_3 - z_2}.$$

By "preserved by LFTs" we mean:

Proposition 1.5.1. *Suppose that f is an LFT, then*

$$(z_1, z_2; z_3, z_4) = \big(f(z_1), f(z_2); f(z_3), f(z_4)\big).$$

Exercise 1.5.1: *Prove Proposition 1.5.1.*

Exercise 1.5.2: *Prove that four distinct points are on a line or a circle if and only if the cross ratio is real. Hint: See also Exercise 1.4.4.*

Cross ratios give a convenient way to describe LFTs. For three distinct numbers z_2, z_3, z_4, the function

$$f(z) = (z, z_2; z_3, z_4)$$

is an LFT such that $f(z_2) = 1$, $f(z_3) = 0$ and $f(z_4) = \infty$. In other words, $(z, z_2; z_3, z_4) = \big(f(z), 1; 0, \infty\big)$.

Exercise 1.5.3: *Given two sets of distinct points $z_1, z_2, z_3 \in \mathbb{C}_\infty$ and $w_1, w_2, w_3 \in \mathbb{C}_\infty$, explicitly find an LFT f, such that $f(z_1) = w_1$, $f(z_2) = w_2$, and $f(z_3) = w_3$.*

Exercise 1.5.4: *Given distinct points $z_1, z_2, z_3 \in \mathbb{C}$ and using the cross ratio definition of an LFT, explicitly find the equation of a circle (or the straight line) through the three points. Hint: Inverse image of the real line is a circle or a straight line.*

---

[*]By Pappus of Alexandria who lived in the early 4th century AD. So it's not as impressively ancient as say Thales's theorem from about a thousand years earlier. Also, isn't Alexandria in Egypt?

2*i*   Holomorphic and Analytic Functions

If this is coffee, please bring me some tea; but if this is tea, please bring me some coffee.

—Abraham Lincoln

2.1*i*   Holomorphic functions and Cauchy–Riemann

2.1.1*i*   Holomorphic functions

The functions we wish to study are those that in some sense generalize polynomials in $z \in \mathbb{C}$; we wish to study functions that, at least locally, behave like $P(z) = a_n z^n + a_{n-1} z^{n-1} + \cdots + a_1 z + a_0$. Polynomials are easy to understand and easy to work with. Alas, there aren't that many of them. For instance, there is no nonzero polynomial that solves the most basic of differential equations: $f' = f$. We must enlarge our horizons a bit.

Consider a polynomial $P(z)$ and expand it near some $z_0 \in \mathbb{C}$:

$$P(z) = c_0 + c_1(z - z_0) + c_2(z - z_0)^2 + \cdots + c_n(z - z_0)^n.$$

In other words, $P(z_0 + h) = c_0 + c_1 h + c_2 h^2 + \cdots + c_n h^n$. Then

$$\lim_{h \to 0} \frac{P(z_0 + h) - P(z_0)}{h} = \lim_{h \to 0} \frac{P(z_0 + h) - c_0}{h} = c_1.$$

So $P(z_0 + h)$ is approximated (locally) by $c_0 + c_1 h$ up to an error that vanishes faster than h. We should emphasize that the limits are "as a *complex h* goes to 0."

Accordingly, we wish to study functions that are locally approximated by $c_0 + c_1 h$ in the same way. More formally, we want functions such that

$$f(z_0 + h) = \underbrace{f(z_0)}_{c_0} + \underbrace{\xi h}_{c_1 h} + o(|h|)$$

for some $\xi \in \mathbb{C}$, where $o(|h|)$ means any function $\varphi(h)$ such that $\lim_{h \to 0} \frac{\varphi(h)}{|h|} = 0$.

Definition 2.1.1. Suppose $U \subset \mathbb{C}$ is open. Given $f : U \to \mathbb{C}$ and $z_0 \in U$, we say f is *complex differentiable* at z_0 if the limit

$$f'(z_0) \overset{\text{def}}{=} \lim_{h \to 0} \frac{f(z_0 + h) - f(z_0)}{h} \qquad (= \xi) \tag{2.1}$$

exists. We call $f'(z_0)$ the *complex derivative* of f at z_0. The notation $\frac{df}{dz}$ is also useful.

A function $f : U \to \mathbb{C}$ is *holomorphic* if it is complex differentiable at every point. That is, if (2.1) exists for all $z_0 \in U$.

Above, we proved the following proposition, which justifies our motivation for the complex derivative.

Proposition 2.1.2. *If $P(z)$ is a polynomial, then $P : \mathbb{C} \to \mathbb{C}$ is holomorphic.*

The most basic result about holomorphic functions is that they are continuous. This fact can be proved either directly from the definition (exercise), or using that a holomorphic function is (real) differentiable as we will observe shortly.

Proposition 2.1.3. *If $U \subset \mathbb{C}$ is open and $f : U \to \mathbb{C}$ is holomorphic, then f is continuous.*

Exercise **2.1.1:** *Directly from the definition of the complex derivative, show that a holomorphic function is continuous (prove Proposition 2.1.3).*

Exercise **2.1.2:** *Show that $f(z) = \bar{z}$ is not complex differentiable at any point.*

Exercise **2.1.3:** *Show that $f(z) = z\bar{z} = |z|^2$ is complex differentiable at the origin, but nowhere else.*

2.1.2*i*   Cauchy–Riemann equations

Suppose $U \subset \mathbb{C}$ is open, and let $f : U \to \mathbb{C}$ be a differentiable (in the real sense and as a function of two real variables, see section B.3 in the appendix) function. If we think of $\mathbb{C}$ as $\mathbb{R}^2$, then the real derivative of f is a 2×2 real matrix Df that approximates f locally. That is, f is (real) differentiable at z_0 if there exists a 2×2 real matrix $Df|_{z_0}$ such that

$$\lim_{h \to 0} \frac{|f(z_0 + h) - f(z_0) - (Df|_{z_0})h|}{|h|} = 0.$$

We think of h as a column vector in $\mathbb{R}^2$ to be able to apply it to the 2×2 real matrix $Df|_{z_0}$. A key point here is that the limit is taken as h moves in $\mathbb{C}$ (or $\mathbb{R}^2$ if you wish). So $f(z_0 + h) - f(z_0) - (Df|_{z_0})h$ is $o(|h|)$ as needed. The trick is to see when $(Df|_{z_0})h$ corresponds to ξh for some $\xi \in \mathbb{C}$.

Write $f = u + iv$, that is, as a mapping into $\mathbb{R}^2$ it is $f = (u, v)$. Then

$$Df|_{z_0} = \begin{bmatrix} \frac{\partial u}{\partial x}\big|_{z_0} & \frac{\partial u}{\partial y}\big|_{z_0} \\ \frac{\partial v}{\partial x}\big|_{z_0} & \frac{\partial v}{\partial y}\big|_{z_0} \end{bmatrix}.$$

We have seen in the previous chapter that only a matrix of the form $\left[\begin{smallmatrix} a & -b \\ b & a \end{smallmatrix}\right]$ corresponds to multiplication by a complex number $a + ib$. Ergo, the derivative $Df|_{z_0}$ corresponds to multiplication by a complex number if and only if it is of the form $\left[\begin{smallmatrix} a & -b \\ b & a \end{smallmatrix}\right]$, namely if

$$\left.\frac{\partial u}{\partial x}\right|_{z_0} = \left.\frac{\partial v}{\partial y}\right|_{z_0}, \qquad \left.\frac{\partial v}{\partial x}\right|_{z_0} = -\left.\frac{\partial u}{\partial y}\right|_{z_0}.$$

In that event, $Df|_{z_0}$ corresponds to multiplication by the number $\xi = \left.\frac{\partial u}{\partial x}\right|_{z_0} + i\left.\frac{\partial v}{\partial x}\right|_{z_0}$, which is equal to $\left.\frac{\partial v}{\partial y}\right|_{z_0} - i\left.\frac{\partial u}{\partial y}\right|_{z_0}$. Consequently,

$$0 = \lim_{h \to 0} \frac{|f(z_0 + h) - f(z_0) - \xi h|}{|h|} = \lim_{h \to 0} \left| \frac{f(z_0 + h) - f(z_0)}{h} - \xi \right|,$$

that is to say,

$$\lim_{h \to 0} \frac{f(z_0 + h) - f(z_0)}{h} = \xi.$$

So f is complex differentiable at z_0 and $f'(z_0) = \xi$.

We proved that if f is differentiable at z_0 (in the real sense), with $\left.\frac{\partial u}{\partial x}\right|_{z_0} = \left.\frac{\partial v}{\partial y}\right|_{z_0}$ and $\left.\frac{\partial v}{\partial x}\right|_{z_0} = -\left.\frac{\partial u}{\partial y}\right|_{z_0}$, then the complex derivative exists at z_0. Conversely, if f is complex differentiable at z_0, then it is real differentiable at z_0, as the complex derivative $f'(z_0)$ gives the $Df|_{z_0}$, and the two equations $\left.\frac{\partial u}{\partial x}\right|_{z_0} = \left.\frac{\partial v}{\partial y}\right|_{z_0}$ and $\left.\frac{\partial v}{\partial x}\right|_{z_0} = -\left.\frac{\partial u}{\partial y}\right|_{z_0}$ must also hold. Let us formalize what we just proved.

Proposition 2.1.4. *Let $U \subset \mathbb{C}$ be open and $f = u + iv \colon U \to \mathbb{C}$ be a function. Then f is complex differentiable at $z_0 \in U$ if and only if f is (real) differentiable at $z_0 \in U$ with $\left.\frac{\partial u}{\partial x}\right|_{z_0} = \left.\frac{\partial v}{\partial y}\right|_{z_0}$ and $\left.\frac{\partial v}{\partial x}\right|_{z_0} = -\left.\frac{\partial u}{\partial y}\right|_{z_0}$. In this case, $f'(z_0) = \left.\frac{\partial u}{\partial x}\right|_{z_0} + i\left.\frac{\partial v}{\partial x}\right|_{z_0} = \left.\frac{\partial v}{\partial y}\right|_{z_0} - i\left.\frac{\partial u}{\partial y}\right|_{z_0}$.*

If the partial derivatives exist and are continuous, then f is (real) differentiable (see section B.3 again). Consequently, we have the following, perhaps easier to apply, result for continuously differentiable functions.

Corollary 2.1.5. *Let $U \subset \mathbb{C}$ be open and let $f = u + iv \colon U \to \mathbb{C}$ be a function such that $\frac{\partial u}{\partial x}, \frac{\partial u}{\partial y}, \frac{\partial v}{\partial x},$ and $\frac{\partial v}{\partial y}$ exist and are continuous (that is, f is continuously differentiable). Then*

$$\frac{\partial u}{\partial x} = \frac{\partial v}{\partial y}, \qquad \frac{\partial v}{\partial x} = -\frac{\partial u}{\partial y} \tag{2.2}$$

if and only if f is holomorphic (complex differentiable at all $z \in U$), or in other words,

$$f'(z) = \lim_{h \to 0} \frac{f(z + h) - f(z)}{h} \qquad \text{exists for all } z \in U.$$

The equations (2.2) are called the *Cauchy–Riemann equations**. Complex analysis, then, is the study of their solutions.

---

*Interestingly, the equations first appeared in the work of d'Alembert, and it was Euler who first connected them to analytic functions. Perhaps they had better be called the French-guy–German-guy equations, except that Euler was really Swiss—he only lived in Germany for a long time.

Hence, continuously differentiable functions satisfying the Cauchy–Riemann equations are holomorphic. On the other hand, a holomorphic function is differentiable in the real sense, and thus the partial derivatives exist and satisfy the Cauchy–Riemann equations. We will show later that holomorphic functions are continuously differentiable—and not just that, they are infinitely differentiable.

As an application of the Corollary, make sure to do the next exercise: The complex exponential, sine, and cosine are all holomorphic functions.

Exercise 2.1.4: Show that the complex exponential function, and hence also sine and cosine, is holomorphic, and show that $\exp' = \exp$.

Exercise 2.1.5: Let $U \subset \mathbb{C}$ be a domain (open and connected), and $f : U \to \mathbb{R}$ be a real-valued function that is holomorphic. Prove that f is constant.

Exercise 2.1.6: Let $U \subset \mathbb{C}$ be open and $f : U \to \mathbb{C}$ holomorphic. Write $f = u + iv$. Show that u and v are harmonic, *that is, $\frac{\partial^2 u}{\partial x^2} + \frac{\partial^2 u}{\partial y^2} = 0$ and $\frac{\partial^2 v}{\partial x^2} + \frac{\partial^2 v}{\partial y^2} = 0$. Feel free to assume that both u and v are twice continuously differentiable.*

Exercise 2.1.7: Let $U \subset \mathbb{C}$ be open and $f : U \to \mathbb{C}$ holomorphic. Write $f = u + iv$. Show that whenever the second derivative test applies to u or v, you get a saddle, that is, prove $\frac{\partial^2 u}{\partial x^2} \frac{\partial^2 u}{\partial y^2} - \left(\frac{\partial^2 u}{\partial x \partial y}\right)^2 \leq 0$ and $\frac{\partial^2 v}{\partial x^2} \frac{\partial^2 v}{\partial y^2} - \left(\frac{\partial^2 v}{\partial x \partial y}\right)^2 \leq 0$. Feel free to assume that both u and v are twice continuously differentiable.

Exercise 2.1.8: Consider the polar coordinates $z = re^{i\theta}$.
  a) Show that the Cauchy–Riemann equations (outside the origin) on $f = u + iv$ are

$$\frac{\partial u}{\partial r} = \frac{1}{r}\frac{\partial v}{\partial \theta}, \qquad \frac{\partial v}{\partial r} = \frac{-1}{r}\frac{\partial u}{\partial \theta}.$$

  b) Use the computation to (locally) find the form of all solutions to the Cauchy–Riemann equations where $\mathrm{Re}\, f = u$ does not depend on the argument θ. By "locally" we mean only in some neighborhood U of every point $p \neq 0$.

2.2*i* Basic properties of holomorphic functions

2.2.1*i* Elementary calculus

Let us solve a differential equation. A common technique in analysis to prove an equality is to differentiate and then show that the derivative is zero.

Proposition 2.2.1. *Let $U \subset \mathbb{C}$ be a domain (open and connected), $f : U \to \mathbb{C}$ holomorphic, and $f'(z) = 0$ for all $z \in U$. Then f is a constant.*

Proof. Follows from the standard real result, see Theorem B.3.10 in the appendix. □

Next, what would calculus be without the chain rule.

Proposition 2.2.2 (Chain rule). *Let $U \subset \mathbb{C}$ and $V \subset \mathbb{C}$ be open, $f \colon U \to V$ complex differentiable at $z \in U$, and $g \colon V \to \mathbb{C}$ complex differentiable at $f(z)$. Then the composition $g \circ f$ is complex differentiable at z and $(g \circ f)'(z) = g'(f(z))f'(z)$.*

Proof. We offer two proofs. The first works only for holomorphic functions, and the second allows a generalization to nonholomorphic functions with the Wirtinger operators in the next section (an exercise).

Let $h \neq 0$, and let $k = f(z + h) - f(z)$. Assume first that $k \neq 0$. Then

$$\frac{(g \circ f)(z + h) - (g \circ f)(z)}{h} = \frac{g(f(z + h)) - g(f(z))}{f(z + h) - f(z)} \frac{f(z + h) - f(z)}{h}$$

$$= \frac{g(f(z) + k) - g(f(z))}{k} \frac{f(z + h) - f(z)}{h}.$$

A differentiable function is continuous, so $k \to 0$ as $h \to 0$. If $k = 0$, then the difference quotient for $g \circ f$ is zero, but $k = 0$ only happens for arbitrarily small $h \neq 0$ if $f'(z) = 0$. So if $f'(z) = 0$, the difference quotient of $g \circ f$ goes to zero whether or not k is ever 0. The proof then follows by continuity of complex multiplication and taking the limit as $h \to 0$.

Let's see the second proof. Complex differentiable functions are real differentiable, so we apply the standard real chain rule (Theorem B.3.7). Let $w = f(z) \in V$. Then

$$D(g \circ f)|_z = Dg|_w Df|_z.$$

The 2×2 matrices $Dg|_w$ and $Df|_z$ correspond to complex numbers $g'(w)$ and $f'(z)$ as g and f are complex differentiable at w and z respectively. A product $Dg|_w Df|_z$ of two such matrices again corresponds to a complex number, the product of the two. So $D(g \circ f)|_z$ corresponds to the pertinent complex number. Hence $g \circ f$ is complex differentiable at z and the given equality holds. $\qquad\square$

This simple statement of the chain rule still holds if we plug a real differentiable function of one variable into a complex differentiable one. If $\gamma \colon (a, b) \to \mathbb{C}$ is a (real) differentiable function, where $\gamma = \alpha + i\beta$, then write $\gamma' = \alpha' + i\beta'$, which can also be interpreted as a 2×1 matrix (column vector) $\begin{bmatrix} \alpha' \\ \beta' \end{bmatrix}$.

Proposition 2.2.3 (Chain rule). *Let $U \subset \mathbb{C}$ be open, $\gamma \colon (a, b) \to U$ (real) differentiable at $t \in (a, b)$, and $f \colon U \to \mathbb{C}$ complex differentiable at $\gamma(t)$. Then the composition $f \circ \gamma$ is (real) differentiable at t and $(f \circ \gamma)'(t) = f'(\gamma(t))\gamma'(t)$.*

Proof. The first proof follows almost in the same way. But it is useful to see how we think of it in terms of real derivatives. Let $z = \gamma(t)$. Then

$$D(f \circ \gamma)|_t = Df|_z D\gamma|_t.$$

That is an equation of real linear operators. Now $Df|_z$ corresponds to multiplication by the complex number $f'(z)$, and $D\gamma|_t$ is the 2×1 matrix (column vector) represented by $\gamma'(t)$. The result follows. $\qquad\square$

Finally, every calculus student needs linearity, product and quotient rules, and the power rule.

Proposition 2.2.4. *Let $U \subset \mathbb{C}$ be open, and $f : U \to \mathbb{C}$ and $g : U \to \mathbb{C}$ holomorphic.*

(i) $f + g$ *is holomorphic and* $\frac{d}{dz}\big[f(z) + g(z)\big] = f'(z) + g'(z)$.

(ii) fg *is holomorphic and* $\frac{d}{dz}\big[f(z)g(z)\big] = f'(z)g(z) + f(z)g'(z)$.

(iii) $1/g$ *is holomorphic on* $\big\{z \in U : g(z) \neq 0\big\}$ *and* $\frac{d}{dz}\big[\frac{1}{g(z)}\big] = \frac{-g'(z)}{\big(g(z)\big)^2}$.

The proof is left as an exercise below. There are again several ways to do it. One way is almost identical to the proof for functions of one real variable. Note that a holomorphic function is continuous, and so the set $\big\{z \in U : g(z) \neq 0\big\}$ is open.

Proposition 2.2.5 (Power rule and its consequences)**.**

(i) *For every integer n, the function $z \mapsto z^n$ is holomorphic where defined (outside the origin if n negative) and $\frac{d}{dz}\big[z^n\big] = nz^{n-1}$ if $n \neq 0$ and $\frac{d}{dz}\big[z^0\big] = 0$.*

(ii) *A polynomial $P(z) = \sum_{n=0}^{d} c_n z^n$ is holomorphic and $P'(z) = \sum_{n=0}^{d-1}(n+1)c_{n+1}z^n$.*

(iii) *Rational functions $\frac{P(z)}{Q(z)}$ are holomorphic on the set where Q is not zero.*

The proof is again left as an exercise.

Exercise 2.2.1: *Prove Proposition 2.2.4. Hint for product: $f(z + h)g(z + h) - f(z)g(z) = f(z + h)g(z + h) - f(z)g(z + h) + f(z)g(z + h) - f(z)g(z)$.*

Exercise 2.2.2: *Prove the first item of Proposition 2.2.5, the power rule. Hint: First prove that z is holomorphic, then prove z^n is holomorphic for positive n (use product rule and induction), and finally prove that z^n is holomorphic for negative n. Note: You are proving both that the complex derivative exists and computing it.*

Exercise 2.2.3: *Prove the last two items of Proposition 2.2.5: Polynomials $P(z)$ are holomorphic on $\mathbb{C}$ (and compute their derivative), and rational functions $\frac{P(z)}{Q(z)}$ are holomorphic on the set where Q is nonzero.*

Perhaps the reader may ask: Is every solution to the Cauchy–Riemann equations holomorphic? Above, we saw that the answer is affirmative for continuously differentiable functions, or at least functions differentiable as functions of two real variables. Surprisingly, the answer is false* if we only assume the existence of partial derivatives, as the following exercise shows.

---

*A good thorough account of this problem is: J. D. Gray and S. A. Morris, *When is a Function that Satisfies the Cauchy-Riemann Equations Analytic?* The American Mathematical Monthly, Vol. 85, No. 4 (Apr., 1978), pp. 246–256.

Exercise 2.2.4: *Let $f(0) = 0$ and $f(z) = e^{-z^{-4}}$ for $z \neq 0$. Prove that partial derivatives exist at every point (including the origin) and f satisfies the Cauchy–Riemann equations at every point, but f is not complex differentiable at the origin (f is not even continuous).*

2.2.2*i*   Wirtinger operators

Suppose $z = x + iy$. The so-called *Wirtinger operators*,

$$\frac{\partial}{\partial z} \overset{\text{def}}{=} \frac{1}{2}\left(\frac{\partial}{\partial x} - i\frac{\partial}{\partial y} \right), \qquad \frac{\partial}{\partial \bar{z}} \overset{\text{def}}{=} \frac{1}{2}\left(\frac{\partial}{\partial x} + i\frac{\partial}{\partial y} \right),$$

provide a way to understand the Cauchy–Riemann equations.* These operators are determined by insisting

$$\frac{\partial}{\partial z}z = 1, \quad \frac{\partial}{\partial z}\bar{z} = 0, \quad \frac{\partial}{\partial \bar{z}}z = 0, \quad \frac{\partial}{\partial \bar{z}}\bar{z} = 1.$$

The Cauchy–Riemann equations are then expressed as

$$\frac{\partial f}{\partial \bar{z}} = 0. \tag{2.3}$$

That seems a far nicer statement of the equations than (2.2), and it is just one complex equation. It says a function is holomorphic if and only if it depends on z but not on $\bar{z}$. That statement had better make no sense at first glance. After all, the Wirtinger operators are not really derivatives with respect to actual variables, they are simply formal operators. Also, and more importantly, how could something possibly depend on z but not on $\bar{z}$. But let us humor ourselves and check what (2.3) means:

$$\frac{\partial f}{\partial \bar{z}} = \frac{1}{2}\left(\frac{\partial f}{\partial x} + i\frac{\partial f}{\partial y} \right) = \frac{1}{2}\left(\frac{\partial u}{\partial x} + i\frac{\partial v}{\partial x} + i\frac{\partial u}{\partial y} - \frac{\partial v}{\partial y} \right) = \frac{1}{2}\left(\frac{\partial u}{\partial x} - \frac{\partial v}{\partial y} \right) + \frac{i}{2}\left(\frac{\partial v}{\partial x} + \frac{\partial u}{\partial y} \right).$$

This expression is zero if and only if the real and imaginary parts are zero. Namely,

$$\frac{\partial u}{\partial x} - \frac{\partial v}{\partial y} = 0 \qquad \text{and} \qquad \frac{\partial v}{\partial x} + \frac{\partial u}{\partial y} = 0.$$

That is, the Cauchy–Riemann equations are satisfied. For emphasis, we state this result as a proposition.

Proposition 2.2.6. *Let $U \subset \mathbb{C}$ be open. Then $f \colon U \to \mathbb{C}$ is holomorphic if and only if f is (real) differentiable and*

$$\frac{\partial f}{\partial \bar{z}} \equiv 0.$$

---

*Despite the notation, these are *not* partial derivatives in z and $\bar{z}$ (whatever that would mean).

The Wirtinger derivative in z computes the holomorphic derivative if f is holomorphic. We can write the z derivative in two different ways:

$$\frac{\partial f}{\partial z} = \frac{1}{2}\left(\frac{\partial u}{\partial x} + \frac{\partial v}{\partial y}\right) + \frac{i}{2}\left(\frac{\partial v}{\partial x} - \frac{\partial u}{\partial y}\right) = \frac{\partial u}{\partial x} + i\frac{\partial v}{\partial x} = \frac{\partial f}{\partial x}$$

$$= \frac{1}{i}\left(\frac{\partial u}{\partial y} + i\frac{\partial v}{\partial y}\right) = \frac{1}{i}\frac{\partial f}{\partial y}.$$

In the second form, we want to think of the derivative in the imaginary direction as a derivative in iy and not the partial derivative in y. That is why the $1/i$ is there. If f is complex differentiable, h can approach zero from every direction:

$$f'(z) = \lim_{\substack{h \to 0 \\ h \in \mathbb{C}}} \frac{f(z+h) - f(z)}{h} = \lim_{\substack{t \to 0 \\ t \in \mathbb{R}}} \frac{f(z+t) - f(z)}{t} = \frac{\partial u}{\partial x}\Big|_z + i\frac{\partial v}{\partial x}\Big|_z = \frac{\partial f}{\partial x}\Big|_z,$$

and

$$f'(z) = \lim_{\substack{h \to 0 \\ h \in \mathbb{C}}} \frac{f(z+h) - f(z)}{h} = \lim_{\substack{t \to 0 \\ t \in \mathbb{R}}} \frac{f(z+it) - f(z)}{it} = \frac{1}{i}\left(\frac{\partial u}{\partial y}\Big|_z + i\frac{\partial v}{\partial y}\Big|_z\right) = \frac{1}{i}\frac{\partial f}{\partial y}\Big|_z.$$

So for a holomorphic function,

$$f' = \frac{\partial f}{\partial z}.$$

The complex derivative f', sometimes written as $\frac{df}{dz}$, only exists for holomorphic functions. The Wirtinger operators $\frac{\partial f}{\partial z}$ and $\frac{\partial f}{\partial \bar{z}}$ make sense for every real differentiable function. Do not confuse the notation even though $\frac{df}{dz}$ and $\frac{\partial f}{\partial z}$ look similar. Consider a polynomial P in x and y, or equivalently in z and $\bar{z}$.* The Wirtinger operators apply and work as if z and $\bar{z}$ really were independent variables. For example:

$$\frac{\partial}{\partial z}\left[z^2\bar{z}^3 + z^{10}\right] = 2z\bar{z}^3 + 10z^9 \quad \text{and} \quad \frac{\partial}{\partial \bar{z}}\left[z^2\bar{z}^3 + z^{10}\right] = z^2(3\bar{z}^2) + 0.$$

So at least for polynomials, a function is holomorphic if it does not depend on $\bar{z}$. Note that the function $z^2\bar{z}^3 + z^{10}$ is not holomorphic and $\frac{d}{dz}\left[z^2\bar{z}^3 + z^{10}\right]$ does not exist.

Exercise **2.2.5:** *Justify the statement about Wirtinger operators: Consider the function $z^m\bar{z}^n$ for any nonnegative integers m and n. Compute $\frac{\partial}{\partial z}\left[z^m\bar{z}^n\right]$ and $\frac{\partial}{\partial \bar{z}}\left[z^m\bar{z}^n\right]$.*

Exercise **2.2.6:** *Let $f\colon U \subset \mathbb{C} \to \mathbb{C}$ be real differentiable at $p \in U$. The derivative $Df|_p$ can be represented by two numbers ξ and ζ: It is the real-linear map $h \mapsto \xi h + \zeta \bar{h}$ (see Exercise 1.1.7). Show that $\frac{\partial f}{\partial z}\big|_p = \xi$ and $\frac{\partial f}{\partial \bar{z}}\big|_p = \zeta$.*

---

*Usually when we say "polynomial" in this book, we mean a polynomial in z, so we will always explicitly mention if we mean a polynomial in x and y, or z and $\bar{z}$.

Exercise 2.2.7: Prove that $ix^2 - 2xy - iy^2 + 3x + 3iy + i$ is a holomorphic function of $z = x + iy$, not by differentiating, but by writing it as a polynomial in z and not $\bar{z}$. That is, write x and y in terms of z and $\bar{z}$, and then show that $\bar{z}$ cancels.

Exercise 2.2.8: Suppose $f : U \to \mathbb{C}$ is real differentiable and let $\bar{f}$ denote the complex conjugate of f. Show

$$\overline{\left(\frac{\partial f}{\partial z} \right)} = \frac{\partial \bar{f}}{\partial \bar{z}} \qquad \text{and} \qquad \overline{\left(\frac{\partial f}{\partial \bar{z}} \right)} = \frac{\partial \bar{f}}{\partial z}.$$

Exercise 2.2.9: Let $U \subset \mathbb{C}$ be a domain and $f : U \to \mathbb{C}$ is such that both f and its conjugate $\bar{f}$ are holomorphic. Show that f is constant.

Exercise 2.2.10: Prove a Wirtinger operator version of the chain rule for real differentiable functions: Let $U \subset \mathbb{C}$ and $V \subset \mathbb{C}$ be open, and $f : U \to V$ (real) differentiable at $p \in U$, $g : V \to \mathbb{C}$ (real) differentiable at $f(p) \in V$. Write $\bar{f}$ for the function that is the complex conjugate of f. Then the composition $g \circ f$ is (real) differentiable at p and

$$\left. \frac{\partial(g \circ f)}{\partial z} \right|_p = \left. \frac{\partial g}{\partial z} \right|_{f(p)} \left. \frac{\partial f}{\partial z} \right|_p + \left. \frac{\partial g}{\partial \bar{z}} \right|_{f(p)} \left. \frac{\partial \bar{f}}{\partial z} \right|_p,$$

and

$$\left. \frac{\partial(g \circ f)}{\partial \bar{z}} \right|_p = \left. \frac{\partial g}{\partial z} \right|_{f(p)} \left. \frac{\partial f}{\partial \bar{z}} \right|_p + \left. \frac{\partial g}{\partial \bar{z}} \right|_{f(p)} \left. \frac{\partial \bar{f}}{\partial \bar{z}} \right|_p.$$

Remark: This chain rule almost makes it seem like a nonholomorphic function is a function of not just z, but two "independent" variables z and $\bar{z}$.

Exercise 2.2.11: A function satisfying $\frac{\partial f}{\partial z} = 0$ is called **antiholomorphic**. Suppose $U \subset \mathbb{C}$ is open and $f : U \to \mathbb{C}$. Prove that if the following limit exists

$$g(z) = \lim_{h \to 0} \frac{f(z + h) - f(z)}{\bar{h}}$$

for all $z \in U$ (note the bar on the h), then f is real differentiable, and satisfies

$$\frac{\partial f}{\partial z} \equiv 0, \qquad \text{and} \qquad \frac{\partial f}{\partial \bar{z}} \equiv g.$$

Exercise 2.2.12:
  a) Suppose $U \subset \mathbb{C}$ is open and $f : U \to \mathbb{C}$ is holomorphic. Show that if $\frac{\partial f}{\partial z}$ is continuous, then $\frac{\partial f}{\partial x}$ and $\frac{\partial f}{\partial y}$ are continuous.
  b) Find an example of a function $f : \mathbb{C} \to \mathbb{C}$ for which $\frac{\partial f}{\partial x}$ and $\frac{\partial f}{\partial y}$ exist at all points, $\frac{\partial f}{\partial z}$ is continuous, but $\frac{\partial f}{\partial x}$ and $\frac{\partial f}{\partial y}$ are discontinuous. Hint: Consider the conjugate of the function from Exercise 2.2.4.

2.2.3*i*   Inverse function theorem and automorphisms

To work with a new category of functions, one should always ask what are the right changes of variables.

Definition 2.2.7. Let $U, V \subset \mathbb{C}$ be open sets. A holomorphic function $f \colon U \to V$ that is bijective and such that the inverse f^{-1} is also holomorphic* is called a *biholomorphism*. If there exists a biholomorphism $f \colon U \to V$, we say that U and V are *biholomorphic*. If $U = V$, then a biholomorphism f is called an *automorphism*[†]. Let Aut(U) denote the set of all automorphisms of U. Traditionally, a biholomorphism $f \colon U \to V$ is called a *conformal mapping* and then U and V are said to be *conformally equivalent*.[‡]

For example, the Cayley map $C(z) = \frac{z-i}{z+i}$ takes the upper half-plane $\mathbb{H} = \{z \in \mathbb{C} : \operatorname{Im} z > 0\}$ to the unit disc $\mathbb{D}$ and has a holomorphic inverse. In other words, $C|_{\mathbb{H}} \colon \mathbb{H} \to \mathbb{D}$ is a biholomorphism making $\mathbb{H}$ and $\mathbb{D}$ biholomorphic.

The reader can check that for a nonempty open $U \subset \mathbb{C}$, the set Aut(U) is a group under composition, although we will not be too worried about the group structure.

> *Exercise* 2.2.13: *Check that* Aut(U) *is a group under composition: Composition of two automorphisms is an automorphism, there is an identity element, composition is associative, and there exists an inverse for every element.*

> *Exercise* 2.2.14: *Show that for any constants $a, b \in \mathbb{C}$, $a \neq 0$, the function $az + b$ is an automorphism of $\mathbb{C}$.*

A biholomorphism f has the property that $f'(z) \neq 0$ for all z. Indeed, f and f^{-1} are holomorphic, so differentiate the equality $f^{-1}(f(z)) = z$ using the chain rule to find $(f^{-1})'(f(z))f'(z) = 1$. Hence, $f'(z)$ cannot be zero. If $w = f(z)$, then

$$(f^{-1})'(w) = \frac{1}{f'(z)} \qquad \text{or} \qquad f'(z) = \frac{1}{(f^{-1})'(w)}.$$

Locally, the relationship between nonzero derivative and invertibility is the inverse function theorem. Consider a holomorphic function $f = u + iv$, its real derivative Df, and its complex derivative f'. The real derivative is, as a matrix,

$$Df = \begin{bmatrix} \frac{\partial u}{\partial x} & \frac{\partial u}{\partial y} \\ \frac{\partial v}{\partial x} & \frac{\partial v}{\partial y} \end{bmatrix}.$$

---

*Surprisingly, we will (later) show that this condition is superfluous.

[†]The word automorphism is used in other contexts as well. It always means that it maps the set to itself and is the right sort of equivalence in the context you are in. In topology it means a homeomorphism, in differential geometry a diffeomorphism, in group theory an isomorphism.

[‡]In one complex variable only! In higher dimensions the definitions differ.

Using the Cauchy–Riemann equations, we compute the Jacobian determinant,

$$\det Df = \frac{\partial u}{\partial x}\frac{\partial v}{\partial y} - \frac{\partial u}{\partial y}\frac{\partial v}{\partial x} = \left(\frac{\partial u}{\partial x}\right)^2 + \left(\frac{\partial v}{\partial x}\right)^2 = |f'(z)|^2.$$

The Jacobian determinant is nonzero (positive) and Df is invertible whenever $f'(z)$ is nonzero. The computation also implies that the determinant of Df is always nonnegative, so a holomorphic function preserves orientation. Why is it the modulus of f' squared? The determinant of Df measures how area changes, and if we multiply a piece of the plane by $re^{i\theta}$, the area gets multiplied by r^2.

The real inverse function theorem (Theorem B.3.16) for continuously differentiable functions of $\mathbb{R}^2$ to $\mathbb{R}^2$ says that if Df is invertible at some point p, then f takes a neighborhood V of p bijectively to a neighborhood $f(V)$ of $f(p)$ and the inverse on that neighborhood is continuously differentiable with $D(f^{-1})|_{f(p)} = (Df|_p)^{-1}$.

An inverse of a 2×2 matrix that represents a complex number also represents a complex number (the reciprocal). Consequently, if f satisfies the Cauchy–Riemann equations, so does its inverse. We obtain the holomorphic inverse function theorem.

Theorem 2.2.8 (Inverse function theorem for holomorphic functions). *Suppose $U \subset \mathbb{C}$ is open, $f : U \to \mathbb{C}$ is holomorphic, $p \in U$, and $f'(p) \neq 0$. Suppose further that f is continuously differentiable. Then there exist open sets $V, W \subset \mathbb{C}$ such that $p \in V \subset U$, $f(V) = W$, the restriction $f|_V$ is injective (one-to-one), and hence a $g : W \to V$ exists such that $g(w) = (f|_V)^{-1}(w)$ for all $w \in W$. Furthermore, g is holomorphic and*

$$g'(w) = \frac{1}{f'(g(w))} \qquad \text{for all } w \in W.$$

The hypothesis that f is continuously differentiable is completely superfluous.[*] Every holomorphic function is continuously differentiable, although you will have to wait till around Theorem 3.3.3 for why that is true.

A holomorphic function whose derivative is nonzero everywhere need not be globally invertible. The exponential e^z is never zero, and thus neither is its derivative. However, $e^z = e^{z+2\pi i}$, so the exponential is not injective. That the inverse of the exponential, the logarithm, has infinitely many values at each point is fundamental to complex analysis. So much so that we've named a whole chapter after the logarithm.

Another interesting remark about biholomorphisms is that generally there are very few biholomorphisms for specific open sets U and V. We will compute later the automorphism group of a few sets such as the disc or the complex plane, and it is in fact rather small. For instance, automorphisms of $\mathbb{C}$ are simply the affine maps $az + b$. On the other hand, at each point, there are a huge number of holomorphic functions with nonzero derivative. So there are lots of local coordinate changes, but few global coordinate changes.

---

[*]Cauchy (early 1800s) assumed continuity of the derivative for his work. It was Goursat more than half a century later who showed that continuity of the derivative came for free.

In the following exercises, you may want to apply the inverse function theorem to show that the inverse is holomorphic.

Exercise 2.2.15:
   a) *Find a biholomorphism from the horizontal strip* $S = \{z \in \mathbb{C} : 0 < \text{Im}\, z < \pi\}$ *to the upper half-plane* $\mathbb{H} = \{z \in \mathbb{C} : \text{Im}\, z > 0\}$.
   b) *Find a biholomorphism from the horizontal strip* S *to the unit disc* $\mathbb{D}$.

Exercise 2.2.16:
   a) *Show that* z^2 *is 2-to-1 on* $\mathbb{C} \setminus \{0\}$ *while its derivative is nonzero.*
   b) *Show that* z^2 *is a biholomorphism of the right half-plane* $\{z \in \mathbb{C} : \text{Re}\, z > 0\}$ *and the slit plane* $\mathbb{C} \setminus (-\infty, 0] = \{z \in \mathbb{C} : \text{Im}\, z \neq 0 \text{ or } \text{Re}\, z > 0\}$.

Exercise 2.2.17: *Consider* $f(z) = z + 1/z$. *Show that* f *takes* $\mathbb{C} \setminus \overline{\mathbb{D}}$ *biholomorphically to* $\mathbb{C} \setminus [-2, 2]$, *and it also takes* $\mathbb{D} \setminus \{0\}$ *biholomorphically to* $\mathbb{C} \setminus [-2, 2]$.

Exercise 2.2.18: *Consider* Δ_1 *a closed disc of radius 1 centered at* i *and* Δ_2 *a closed disc of radius 1 centered* $-i$. *Find a biholomorphism of* $\mathbb{C} \setminus (\Delta_1 \cup \Delta_2)$ *onto the punctured unit disc* $\mathbb{D} \setminus \{0\}$. *Hint: Figure out what* $1/z$ *does to the two circles.*

Exercise 2.2.19: *Let* $f(z) = \frac{1-z^4}{1+z^4}$ *and* $g(z) = i\left(\frac{1-z^2}{1+z^2}\right)^2$. *Let* $S = \{z \in \mathbb{C} : |z| < 1, \text{Re}\, z > 0, \text{Im}\, z > 0\}$. *Find* $f(S)$ *and* $g(S)$. *Then show that they are both biholomorphisms onto their image. Think about the functions as composition.*

Exercise 2.2.20:
   a) *Show that if* $\Delta \subset \mathbb{C}$ *is a disc such that* $0 \notin \Delta$, *then there exist two distinct holomorphic functions* $f : \Delta \to \mathbb{C}$ *such that* $(f(z))^2 = z$. *In other words,* $f(z) = \pm\sqrt{z}$ *and the square root and its negative is holomorphic on* Δ.
   b) *Show that there does not exist a continuous* $f : \mathbb{C} \setminus \{0\} \to \mathbb{C}$ *such that* $(f(z))^2 = z$. *That is, we cannot choose a continuous square root in the punctured plane. Hint: Just consider the unit circle.*

Exercise 2.2.21:
   a) *Suppose* f *is antiholomorphic, that is, assume* f *is (real) differentiable and* $\frac{\partial f}{\partial z} = 0$. *Show that* $\det Df\big|_p = -\left|\frac{\partial f}{\partial \bar{z}}(p)\right|^2$. *In other words, the Jacobian determinant is nonpositive, and* f *flips orientation.*
   b) *More generally, if* f *is (real) differentiable, then* $\det Df\big|_p = \left|\frac{\partial f}{\partial z}(p)\right|^2 - \left|\frac{\partial f}{\partial \bar{z}}(p)\right|^2$.

2.2.4i   Conformality ★

The actual definition of "conformal mapping" is a (real) differentiable bijective mapping $f : U \to V$ of open $U, V \subset \mathbb{R}^2$ that preserves a) orientation, and b) angles.

Both of these are taken in the infinitesimal sense, that is, they are statements about what the linear mapping Df does to vectors. Consider two continuously differentiable curves $\gamma\colon (-\epsilon, \epsilon) \to \mathbb{R}^2$ and $\alpha\colon (-\epsilon, \epsilon) \to \mathbb{R}^2$, such that $\gamma(0) = \alpha(0) = p \in \mathbb{R}^2$. By preserving angles, we mean that the curves $f \circ \gamma$ and $f \circ \alpha$ meet at the same angle at $f(p)$, see Figure 2.1. In other words,

angle between $Df|_p\gamma'(0)$ and $Df|_p\alpha'(0)$ $\quad = \quad$ angle between $\gamma'(0)$ and $\alpha'(0)$.

As we are preserving orientation, we can take the angle to be the signed angle starting at one vector and ending at the other vector.

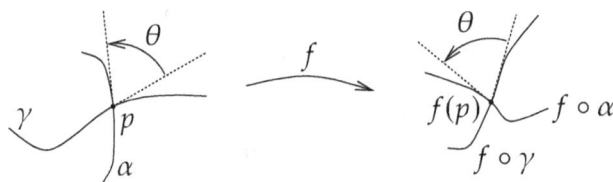

Figure 2.1: Preserving angles (and orientation).

By preserving angles we also mean that no vector can be taken to zero, as zero does not make any well-defined angle with anything else. Thus Df must be invertible at every point for a conformal map. We are really just doing linear algebra, so we start with the relevant linear algebra statement.

Proposition 2.2.9. *A 2×2 real matrix M preserves orientation and angles between vectors if and only if M corresponds to the multiplication by a nonzero complex number.*

Proof. Suppose that M preserves orientation and angles. As we said M must be nonsingular. Let $M = \begin{bmatrix} a & b \\ c & d \end{bmatrix}$. As M preserves angles, given two vectors v and w in $\mathbb{R}^2$, the angle between Mv and Mw is the same as the angle between v and w. The vectors $v = \begin{bmatrix} 1 \\ 0 \end{bmatrix}$ and $w = \begin{bmatrix} 0 \\ 1 \end{bmatrix}$ are orthogonal, and so Mv and Mw are orthogonal:

$$0 = Mv \cdot Mw = \begin{bmatrix} a \\ c \end{bmatrix} \cdot \begin{bmatrix} b \\ d \end{bmatrix} = ab + cd.$$

As M is nonsingular, either a or c is nonzero. In either case, there must exist some nonzero $r \in \mathbb{R}$ such that

$$M = \begin{bmatrix} a & -rc \\ c & ra \end{bmatrix}.$$

A similar calculation using $v = \begin{bmatrix} 1 \\ 1 \end{bmatrix}$ and $w = \begin{bmatrix} 1 \\ -1 \end{bmatrix}$ results in $a^2 + c^2 = b^2 + d^2 = r^2(a^2 + c^2)$. Or in other words, $r = \pm 1$. That M preserves orientation means $\det M > 0$. As $\det M = r(a^2 + c^2)$, we find $r = 1$. Hence,

$$M = \begin{bmatrix} a & -c \\ c & a \end{bmatrix}.$$

That is, M corresponds to multiplication by the complex number $a + ic$.

For the converse, assume M is multiplication by the nonzero complex number ξ. Let $z = |z|e^{i\theta}$ and $w = |w|e^{i\psi}$ be two nonzero complex numbers, thinking of them as vectors in $\mathbb{R}^2$. The (signed) angle between them, $\theta - \psi$, can be computed using

$$\frac{z\overline{w}}{|z||w|} = e^{i(\theta - \psi)}.$$

Similarly, the angle between ξz and ξw can be computed using

$$\frac{\xi z \overline{\xi w}}{|\xi z||\xi w|} = \frac{|\xi|^2 z \overline{w}}{|\xi|^2 |z||w|} = \frac{z\overline{w}}{|z||w|} = e^{i(\theta - \psi)}.$$

The (signed) angle is the same, so $z \mapsto \xi z$ preserves orientation and angles. □

Per Exercise 1.1.7, an arbitrary M is given by $Mz = \xi z + \zeta \bar{z}$ for some $\xi, \zeta \in \mathbb{C}$. The proposition says that M preserves orientation and angles if and only if $\zeta = 0$. Another way to see that if $Mz = \xi z$, then M preserves orientation is to note that $\det M = |\xi|^2 > 0$.

Applying the proposition to Df, we find:

Corollary 2.2.10. *Let $U \subset \mathbb{C}$ be open. A real differentiable function $f : U \to \mathbb{C}$ preserves orientation and angles if and only if f is holomorphic and f' never vanishes.*

In other words, conformal maps are holomorphic, and holomorphic maps with nonzero derivative preserve angles and orientation. Once we prove later that holomorphic maps are continuously differentiable and we will be able to apply the inverse function theorem we have just presented in the previous subsection, then we will see that conformal maps are also biholomorphic.

Exercise 2.2.22: *Prove that a 2×2 matrix M preserves angles and reverses orientation if and only if M corresponds to the mapping $h \mapsto \xi \bar{h}$ for some $\xi \in \mathbb{C}$.*

Exercise 2.2.23: *Let $U \subset \mathbb{C}$ be open. Prove that real differentiable function $f : U \to \mathbb{C}$ preserves angles and reverses orientation if and only if the conjugate $\bar{f}$ is holomorphic and its derivative never vanishes.*

2.3*i* Power series

2.3.1*i* The function z^n

To understand holomorphic functions locally, it is sufficient to understand $z \mapsto z^n$. We will prove that holomorphic functions are just power series and so we can always factor a z^n for some n out of a power series that vanishes at the origin, which is, after all, just a sum of such terms. Consequently, every holomorphic function really behaves sort of like z^n behaves near the origin for some n.

Let's first see what z^n does to angles. If $z = re^{i\theta}$, then

$$z^n = r^n e^{in\theta}.$$

So z^2 takes sectors with vertex at the origin and doubles their angle. See Figure 2.2. It takes the first quadrant $\{z \in \mathbb{C} : \operatorname{Re} z \geq 0, \operatorname{Im} z \geq 0\}$ to the closed upper half-plane $\{z \in \mathbb{C} : \operatorname{Im} z \geq 0\}$. Similarly, it takes the second quadrant $\{z \in \mathbb{C} : \operatorname{Re} z \leq 0, \operatorname{Im} z \geq 0\}$ to the closed lower half-plane $\{z \in \mathbb{C} : \operatorname{Im} z \leq 0\}$.

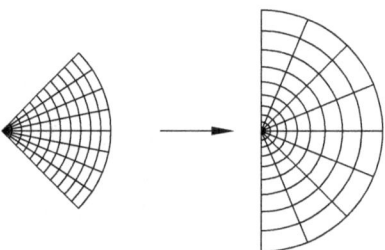

Figure 2.2: What z^2 does to the sector $\frac{-\pi}{4} \leq \operatorname{Arg} z \leq \frac{\pi}{4}, |z| < 1.1$.

Another key point about the function $z \mapsto z^n$ is that it is n-to-1. That is, there are n distinct n^{th} roots of every complex number except 0, which in some sense also has n roots, but they are all 0. For a nonzero number w, write $w = re^{i\theta}$. It is easy to verify that the n n^{th} roots of w are (using the polar form)

$$r^{1/n}e^{i\theta/n}, \quad r^{1/n}e^{i\theta/n+2\pi i/n}, \quad \ldots, \quad r^{1/n}e^{i\theta/n+2\pi i(n-1)/n}.$$

Those are the n different zs such that $z^n = w$. They are equally spaced out on a circle of radius $r^{1/n}$, see Figure 2.3. The roots of $w = 1$ are called the *roots of unity*.

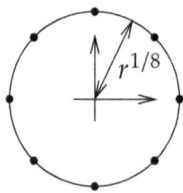

Figure 2.3: The eight 8th roots of a positive number r: $r^{1/8}, r^{1/8}e^{i\pi/4}, r^{1/8}e^{i\pi/2}$, etc.

Exercise 2.3.1: *Prove that*
  a) If $|z| < 1$, then $\lim\limits_{n \to \infty} z^n = 0$.
  b) If $|z| > 1$, then $\lim\limits_{n \to \infty} z^n = \infty$.
  c) If $z \neq 1$ is such that $|z| = 1$, then z^n diverges as $n \to \infty$.

> **Exercise 2.3.2** (Easy): *On the unit circle parametrized by the angle θ, write $\sin(n\theta)$ and $\cos(n\theta)$ as a linear combination of powers (including negative) of $z = e^{i\theta}$.*

2.3.2i   Power series and radius of convergence

A *power series* around $p \in \mathbb{C}$ is simply the series

$$\sum_{n=0}^{\infty} c_n(z - p)^n,$$

where c_n are some complex numbers. Where it converges, it defines a function of z. As the series clearly converges at $z = p$, we worry about convergence at other points. We say the series is *convergent* if it converges at some $z \neq p$.

The most important series, and in some sense the only one that we really know how to sum, is the *geometric series*.

Proposition 2.3.1 (Geometric series).

(i) *For $z \in \mathbb{D}$,*

$$\frac{1}{1 - z} = \sum_{n=0}^{\infty} z^n.$$

(ii) *For $z \notin \mathbb{D}$,*

$$\sum_{n=0}^{\infty} z^n \quad diverges.$$

(iii) *Given $0 < r < 1$, then for all $z \in \overline{\Delta_r(0)}$ (that is, $|z| \leq r$)*

$$\left| \frac{1}{1 - z} - \sum_{n=0}^{m} z^n \right| \leq \frac{r^{m+1}}{1 - r}.$$

Consequently, as $\frac{r^{m+1}}{1-r} \to 0$, the geometric series converges uniformly on $\overline{\Delta_r(0)}$.

Proof. All three items follow (details left as exercise) from

$$1 + z + z^2 + \cdots + z^m = \frac{1 - z^{m+1}}{1 - z},$$

for all $z \neq 1$, which follows by expanding $(1 - z)(1 + z + z^2 + \cdots + z^m)$. $\square$

> **Exercise 2.3.3:** *Fill in the details of the proof of Proposition 2.3.1. Do not forget about the boundary of the disc.*

A power series *converges absolutely* if the following series converges:

$$\sum_{n=0}^{\infty} |c_n||z - p|^n.$$

For $N < M$,

$$\left| \sum_{n=N+1}^{M} c_n(z - p)^n \right| \leq \sum_{n=N+1}^{M} |c_n||z - p|^n.$$

Hence, if the sequence of partial sums of $\sum |c_n||z - p|^n$ is Cauchy, so is the sequence of partial sums of $\sum c_n(z - p)^n$. Thus, an absolutely convergent series actually converges.

Let $r = |z - p|$, and consider the real series $\sum |c_n|r^n$. Define

$$R = \frac{1}{\limsup\limits_{n \to \infty} \sqrt[n]{|c_n|}}, \qquad (2.4)$$

where we interpret $1/\infty = 0$ and $1/0 = \infty$, so $R = \infty$ is allowed.[*] By the standard root test, the series $\sum |c_n|r^n$ converges if

$$\limsup_{n \to \infty} \sqrt[n]{|c_n|r^n} = r \limsup_{n \to \infty} \sqrt[n]{|c_n|} = r\frac{1}{R} < 1.$$

So the power series converges absolutely when $r < R$. If $r\frac{1}{R} > 1$, then for infinitely many n, $|c_n(z - p)^n| > 1$. So the series diverges if $r > R$. See Figure 2.4. We proved:

Proposition 2.3.2 (Cauchy–Hadamard theorem[†]). *A power series $\sum c_n(z - p)^n$ converges absolutely if $|z - p| < R$ and diverges if $|z - p| > R$, where R is defined by (2.4).*

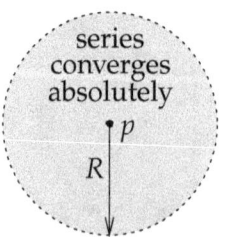

series
converges
absolutely

$\bullet\, p$

R

series
does not converge

Figure 2.4: Radius of convergence.

The number R is called the *radius of convergence*. The power series converges absolutely in the disc $\Delta_R(p)$, and diverges in the complement of the closure $\overline{\Delta_R(p)}$. Convergence (or divergence) on the boundary circle $\partial \Delta_R(p)$ is a tricky matter.

---

[*]We are not using the "extended reals sense" here, we are really extending just the nonnegative reals, and so something like $1/0 = \infty$ makes sense, for the same reason as on the Riemann sphere.

[†]Cauchy published this result in 1821, and Hadamard, despite also being French, didn't know about it and published it in his thesis in 1888.

A useful criterion for convergence is that the sequence $\{|c_n|r^n\}$ is bounded whenever $0 < r < R$.

Proposition 2.3.3. *The series $\sum c_n(z - p)^n$ converges in $\Delta_R(p)$ for some $R > 0$ if and only if for every r with $0 < r < R$, there exists an $M > 0$ such that*

$$|c_n| \leq \frac{M}{r^n} \qquad \text{for all } n.$$

It is not necessarily true that $\{|c_n|R^n\}$ is bounded if R is the radius of convergence. The two series $\sum z^n$ and $\sum nz^n$ both have radius of convergence 1, while the sequence of coefficients is bounded in the first case and not in the second. However, $\{nr^n\}$ is bounded for every $r < 1$.

Proof. Suppose the series converges in $\Delta_R(p)$ and $0 < r < R$. Then (by Cauchy–Hadamard) $\sum |c_n|r^n$ converges, and the terms of that series are bounded.

Conversely, fix r with $0 < r < R$, suppose $|c_n|r^n \leq M$ for all n, and suppose $0 < s < r$. Then

$$\sqrt[n]{|c_n|s^n} = \frac{s}{r}\sqrt[n]{|c_n|r^n} \leq \frac{s}{r}\sqrt[n]{M}.$$

The limsup of the right-hand side is strictly less than 1 as $s/r < 1$. So the series converges absolutely in $\overline{\Delta_s(p)}$ by the root test again. As s and r with $0 < s < r < R$ were arbitrary, the series converges (absolutely) in $\Delta_R(p)$. $\qquad \square$

The proof is fairly typical for convergence results of power series. Convergence in $\Delta_R(p)$, means boundedness of $\{|c_n|r^n\}$ for a smaller $\Delta_r(p)$, which only gets us convergence in $\Delta_s(p)$. See Figure 2.5. But since s and r are arbitrary, we get convergence in $\Delta_R(p)$.

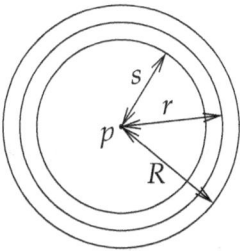

Figure 2.5: The three discs from the convergence proof.

Exercise 2.3.4: Prove the triangle inequality for series. If $\sum_{n=0}^{\infty} c_n$ converges, then $\left|\sum_{n=0}^{\infty} c_n\right| \leq \sum_{n=0}^{\infty}|c_n|$ (the right-hand side is possibly ∞).

The convergence within the radius of convergence is even nicer than just absolute. Let $K \subset \mathbb{C}$ be a set. A power series $\sum c_n(z - p)^n$ converges *uniformly absolutely* for $z \in K$ when $\sum |c_n||z - p|^n$ converges uniformly for $z \in K$. Suppose a series converges uniformly absolutely on K. It converges absolutely, so it converges, and

$$\left| \sum_{n=0}^{\infty} c_n(z - p)^n - \sum_{n=0}^{m} c_n(z - p)^n \right| = \left| \sum_{n=m+1}^{\infty} c_n(z - p)^n \right| \leq \sum_{n=m+1}^{\infty} |c_n||z - p|^n.$$

The right-hand side goes to zero uniformly in $z \in K$ as $m \to \infty$, and so a uniformly absolutely convergent series also converges uniformly. So the name fits the crime.

Proposition 2.3.4. *Let $\sum c_n(z - p)^n$ be a power series with radius of convergence $R > 0$. If $0 < r < R$, then the power series converges uniformly absolutely on $\overline{\Delta_r(p)}$. Furthermore, let $U = \Delta_R(p)$ if $R < \infty$ and $U = \mathbb{C}$ if $R = \infty$, and let $K \subset U$ be compact. Then the series converges uniformly absolutely on K.*

Less formally, a power series converges uniformly (absolutely) on compact subsets of its domain of convergence.

Proof. Without loss of generality suppose $R < \infty$. Suppose $0 < r < R$. As $\sum c_n(z - p)^n$ converges absolutely on $\Delta_R(p)$, the series $\sum |c_n|r^n$ converges (and in particular any tail of that series converges). Thus for $z \in \overline{\Delta_r(p)}$,

$$\left| \sum_{n=0}^{\infty} |c_n||z - p|^n - \sum_{n=0}^{m} |c_n||z - p|^n \right| \leq \sum_{n=m+1}^{\infty} |c_n|r^n.$$

The right-hand side, which does not depend on z, goes to zero as $m \to \infty$, and hence the series $\sum |c_n||z - p|^n$ converges uniformly on $\overline{\Delta_r(p)}$.

If $K \subset \Delta_R(p)$ is compact, then there exists some $r < R$ such that $K \subset \Delta_r(p)$ (consider an open cover of K by discs $\Delta_r(p)$ for all $r < R$). The result follows.   □

Exercise 2.3.5: Show that the series $\sum_{n=1}^{\infty} \frac{1}{n^2} z^n$ has radius of convergence 1, and show that it converges absolutely on the boundary of the unit disc. Hence it actually converges uniformly on the entire closed unit disc.

Exercise 2.3.6: Show that $\sum_{n=0}^{\infty} n^n z^{n^n}$ has radius of convergence 1, while $\sum_{n=0}^{\infty} n^n z^n$ is not convergent for any $z \neq 0$.

Exercise 2.3.7: Suppose $\sum_{n=0}^{\infty} c_n z^n$ converges at $z = 1$, but not absolutely. Prove that the radius of convergence is 1.

Exercise 2.3.8 (Weierstrass M-test): Let X be a set and suppose that $f_n \colon X \to \mathbb{C}$ is a sequence of functions such that $|f_n(x)| \leq M_n$ for all $x \in X$ and $n \in \mathbb{N}$. If $\sum M_n < \infty$, then $\sum f_n(x)$ converges uniformly absolutely on X ($\sum |f_n(x)|$ converges uniformly on X).

Exercise **2.3.9:** *Suppose $\sum_{n=0}^{\infty} a_n z^n$ and $\sum_{n=0}^{\infty} b_n z^n$ have a radius of convergence at least $r > 0$. Show that the sum series $\sum_{n=0}^{\infty}(a_n + b_n)z^n$ has a radius of convergence at least r and converges to the sum of the two series.*

Exercise **2.3.10:** *Given an $R > 1$, find two power series $\sum_{n=0}^{\infty} a_n z^n$ and $\sum_{n=0}^{\infty} b_n z^n$, such that both have radius of convergence exactly 1, but the sum $\sum_{n=0}^{\infty}(a_n + b_n)z^n$ has a radius of convergence exactly R. Hint: First figure out a series with radius of convergence exactly R.*

Exercise **2.3.11:** *Suppose $\sum_{n=0}^{\infty} a_n z^n$ and $\sum_{n=0}^{\infty} b_n z^n$ have a radius of convergence at least $r > 0$. Let $c_n = \sum_{k=0}^{n} a_{n-k} b_k$. Show that the series $\sum_{n=0}^{\infty} c_n z^n$ has a radius of convergence at least r and converges to the product of the two series. Hint: The key is to look at a point where both series converge absolutely, then use the absolute convergence.*

Exercise **2.3.12:** *If $\sum_{n=0}^{\infty} a_n z^n$ and $\sum_{n=0}^{\infty} b_n z^n$ converge and are equal in a disc $\Delta_r(0)$, then $a_n = b_n$ for all n. Hint: First suppose $b_n = 0$ for all n and find the first n where $a_n \neq 0$.*

Remark 2.3.5. The last few exercises say we can add and multiply power series. Addition of power series is straightforward, and we will need it momentarily. Multiplication will be far easier later, after we will show power series are holomorphic and vice versa, but you should still try the exercise now, it is good practice.

2.4*i*   Analytic functions

2.4.1*i*   Definition

Functions that possess a convergent power series are called analytic. From the beginnings of calculus until the 19th century, when mathematicians considered a "function," they really meant "analytic function" (or something like it) in modern language. We talk about both complex-analytic and real-analytic functions depending on whether the variables are real or complex, and they may depend on one or several variables. We are interested in complex-analytic functions of one variable. As there is little chance of confusion, we say just "analytic" instead of "complex-analytic."

Definition 2.4.1. Let $U \subset \mathbb{C}$ be open. A function $f \colon U \to \mathbb{C}$ is *analytic* if for every $p \in U$, there exists an $r > 0$ and a power series $\sum c_n(z - p)^n$ converging to f on $\Delta_r(p) \subset U$.

Exercise **2.4.1** (Easy)*: Prove that polynomials $P(z)$ are analytic.*

Exercise **2.4.2:** *Prove that $1/z$ is analytic in $\mathbb{C} \setminus \{0\}$ by explicitly writing down a power series at any $p \in \mathbb{C} \setminus \{0\}$ using the geometric series.*

Exercise **2.4.3:** *Suppose $f(z) = \sum_{n=0}^{\infty} a_n z^n$ converges and defines a function on $\mathbb{D}$ such that $f(z) = f(-z)$ for all $z \in \mathbb{D}$ (f is "even"). Prove that there exists a function defined by a power series $g(z) = \sum_{n=0}^{\infty} b_n z^n$ converging in $\mathbb{D}$ such that $f(z) = g(z^2)$.*

2.4.2i   Analytic functions are holomorphic

Eventually, we will see that analytic functions and holomorphic functions are one and the same.* We start by proving that analytic functions are holomorphic, that is, they are complex differentiable.

Proposition 2.4.2. *Let $f : \Delta_R(p) \to \mathbb{C}$ be defined by*

$$f(z) = \sum_{n=0}^{\infty} c_n(z - p)^n, \qquad \text{converging in } \Delta_R(p).$$

Then f is complex differentiable at every $z \in \Delta_R(p)$, and

$$f'(z) = \sum_{n=1}^{\infty} n c_n(z - p)^{n-1}, \qquad \text{converging in } \Delta_R(p).$$

Proof. Without loss of generality, let $p = 0$. Differentiate z^n at some z_0 by considering the difference quotient

$$\frac{z^n - z_0^n}{z - z_0} = \sum_{k=0}^{n-1} z^k z_0^{n-1-k},$$

which goes to $n z_0^{n-1}$ as $z \to z_0$, as the right-hand side is defined even if $z = z_0$. Apply the formula to f term by term. For $z_0, z \in \Delta_R(0)$,

$$\frac{f(z) - f(z_0)}{z - z_0} = \sum_{n=1}^{\infty} c_n \frac{z^n - z_0^n}{z - z_0} = \sum_{n=1}^{\infty} c_n \sum_{k=0}^{n-1} z^k z_0^{n-1-k}.$$

We must show that the expression on the right gives a continuous function of z at z_0. It is continuous at z_0 provided that the series in the expression converges uniformly for z in a neighborhood of z_0.

The setup will be just like in Figure 2.5. Let r and s be such that $0 < s < r < R$ and suppose that z_0 and z are in $\Delta_s(0)$.

$$\left| c_n \sum_{k=0}^{n-1} z^k z_0^{n-1-k} \right| \leq \sum_{k=0}^{n-1} |c_n| s^{n-1} = n |c_n| s^{n-1} = n |c_n| r^{n-1} \left(\frac{s}{r} \right)^{n-1}.$$

The expression $|c_n| r^{n-1}$ is bounded by some $M > 0$ for all n, because the series for f converges in $\Delta_R(0)$ and $r < R$. As

$$\sqrt[n]{n |c_n| s^{n-1}} = \sqrt[n]{n |c_n| r^{n-1} \left(\frac{s}{r} \right)^{n-1}} \leq \sqrt[n]{n M \left(\frac{s}{r} \right)^{n-1}} \underset{\text{as } n \to \infty}{\longrightarrow} \frac{s}{r} < 1,$$

---

*For this reason, some authors define "analytic" to mean complex differentiable, which is no problem eventually, but right now it would be.

the root test shows that $\sum n|c_n|s^{n-1}$ converges. So the series for the difference quotient,

$$\sum_{n=1}^{\infty} c_n \sum_{k=0}^{n-1} z^k z_0^{n-1-k},$$

converges uniformly in z on $\Delta_s(p)$, and we can swap the limit $z \to z_0$ with the series limit:

$$\lim_{z \to z_0} \frac{f(z) - f(z_0)}{z - z_0} = \sum_{n=1}^{\infty} c_n \sum_{k=0}^{n-1} z_0^k z_0^{n-1-k} = \sum_{n=1}^{\infty} n c_n z_0^{n-1}.$$

As s and r were arbitrary, the convergence happens in all of $\Delta_R(0)$. $\qquad\square$

So the derivative of a power series is again given by a power series. By induction, it follows that a power series is infinitely complex differentiable.

Corollary 2.4.3. *Let $f : \Delta_R(p) \to \mathbb{C}$ be defined by*

$$f(z) = \sum_{n=0}^{\infty} c_n(z - p)^n, \qquad \text{converging in } \Delta_R(p).$$

Then f is infinitely complex differentiable in $\Delta_R(p)$, and the kth derivative is given by

$$f^{(k)}(z) = \sum_{n=k}^{\infty} n(n-1)\cdots(n-k+1)c_n(z-p)^{n-k}, \qquad \text{converging in } \Delta_R(p).$$

Furthermore,

$$c_n = \frac{f^{(n)}(p)}{n!}.$$

Exercise 2.4.4: *Fill in the details of the proof of the corollary.*

A consequence of this corollary that should be emphasized is that if f is given by the convergent power series in $\Delta_R(p)$ as above, then the power series is unique. We have seen this conclusion in an exercise before; here it follows from the formula for the coefficients c_n. In fact, the coefficients depend only on the values of f in an arbitrarily small neighborhood of p.

If we apply the corollary to analytic functions at every point, we find that they are infinitely differentiable:

Corollary 2.4.4. *An analytic function is infinitely complex differentiable and each derivative is analytic.*

Remark 2.4.5. A subtle issue is that while we proved that analytic functions are complex differentiable because they have a power series representation, we did not yet prove that a convergent power series defines an analytic function. What is left is to prove that a power series convergent in $\Delta_R(p)$ can be expanded about a different point in $\Delta_R(p)$. That will follow once we prove that holomorphic functions are analytic.

Exercise 2.4.5: *Suppose $f: \Delta_R(0) \to \mathbb{C}$ is given by a convergent power series $f(z) = \sum_{n=0}^{\infty} c_n z^n$. Suppose that for some $\epsilon > 0$, $f(x) = 0$ for all $x \in (-\epsilon, \epsilon)$ (an interval on the real line). Using the corollary, prove that $c_n = 0$ for all n and hence f is identically zero.*

Exercise 2.4.6: *Suppose $f: \Delta_R(p) \to \mathbb{C}$ is given by a convergent power series $f(z) = \sum_{n=0}^{\infty} c_n (z - p)^n$. Antidifferentiate: Show that there exists a power series converging in $\Delta_R(p)$ whose complex derivative is $f(z)$.*

Exercise 2.4.7: *Suppose $f: \Delta_R(0) \to \mathbb{C}$ is given by a convergent power series $f(z) = \sum_{n=0}^{\infty} c_n z^n$ and $R > 1$. Show that there is an $M > 0$ such that $|f^{(n)}(0)| \leq n!M$ for all n.*

2.4.3i　　The exponential

We met the complex exponential e^z before (§ 1.2.1), and we proved that it is holomorphic and its own derivative (Exercise 2.1.4). We can now see this fact from a different vantage point.* We claim we could have defined the exponential using a power series.

Proposition 2.4.6. *The power series*

$$f(z) = \sum_{n=0}^{\infty} \frac{1}{n!} z^n,$$

is the unique convergent power series at the origin such that $f(0) = 1$ and $f' = f$. Moreover, the series converges on $\mathbb{C}$ and $f(z) = e^z$.

Proof. We now know that power series are holomorphic and we know how to differentiate them: We do it term-by-term, that is, "formally." Suppose we possess a convergent power series at the origin

$$f(z) = \sum_{n=0}^{\infty} c_n z^n$$

such that $f(0) = 1$ and $f' = f$. Obviously, $f(0) = 1$ implies that $c_0 = 1$. The trick is to figure out the rest of the series. So,

$$f'(z) = \sum_{n=1}^{\infty} n c_n z^{n-1} = \sum_{n=0}^{\infty} (n + 1) c_{n+1} z^n.$$

As the coefficients of the power series at zero are unique, we get $c_n = (n + 1) c_{n+1}$. By induction, $c_n = \frac{1}{n!}$. It is not hard to check directly (exercise) that the series converges in all of $\mathbb{C}$.

---

*That vantage point being the same as that dark place in your past that is the undergraduate differential equations class when you covered power series methods for solving ODEs.

Let us prove that f is the exponential. Both functions are holomorphic, the exponential is never zero, and both are equal to their derivatives. So,

$$\frac{d}{dz}\left[\frac{f(z)}{\exp(z)}\right] = \frac{f'(z)\exp(z) - f(z)\exp'(z)}{\left(\exp(z)\right)^2} = \frac{f(z)\exp(z) - f(z)\exp(z)}{\left(\exp(z)\right)^2} = 0.$$

Hence, $f(z) = C\exp(z)$ for some constant (Proposition 2.2.1). As $f(0) = \exp(0) = 1$, we conclude $C = 1$. $\qquad\square$

Exercise **2.4.8:** *Prove that the series for the exponential converges by computing the radius of convergence directly (e.g., show that the series converges for every $z \in \mathbb{C}$).*

Exercise **2.4.9:** *Compute the series for $\sin z$ and $\cos z$, then show that these satisfy $f''(z) = -f(z)$.*

Exercise **2.4.10:** *Show that there exists a holomorphic $f : \mathbb{C} \to \mathbb{C}$ that solves $f'(z) + z f(z) = 0$ and such that $f(0) = 1$. Hint: Solve formally as a power series, then see if you can guess the answer in "closed form," that is, in terms of the exponential. Hint #2: What is $f'(0)$?*

Exercise **2.4.11:** *Given $a, b \in \mathbb{C}$, show that there exists a holomorphic $f : \mathbb{C} \to \mathbb{C}$ such that $f''(z) = z f(z)$, and $f(0) = a$ and $f'(0) = b$. Hint: Define formally and show convergence. Note: These are the* Airy *functions, and they have some interesting behavior; on the real line they oscillate like sine and cosine for negative z and behave like an exponential for positive z.*

2.4.4*i*   The identity theorem

One of the main properties of analytic functions is that once you know them in a neighborhood you know them everywhere. In fact, a much more general statement is true; you only need to know an analytic function on a set with a limit point.

Theorem 2.4.7 (Identity). *Suppose $U \subset \mathbb{C}$ is a domain, and $f : U \to \mathbb{C}$ is analytic. If $Z_f = \{z \in U : f(z) = 0\}$ has a limit point in U, then f is identically zero. In other words, all points of Z_f are isolated unless $f \equiv 0$.*

Definition 2.4.8. The points in the set Z_f are called the *zeros* of f.

The "In other words" bit is one consequence of this theorem that we will use often. More concretely, if $f(p) = 0$, but f is not identically zero, then there is a disc $\Delta_r(p)$ such that $f(z) \neq 0$ for all $z \in \Delta_r(p) \setminus \{0\}$.

Another common application of the theorem is the following weaker statement: "If the function is zero on a nonempty open subset, then $f \equiv 0$." Think of the implications: If $f : U \to \mathbb{C}$ is analytic, and we know f in a tiny disc $\Delta_r(p)$ for an arbitrarily small r, then we know f on all of U.

Proof. Suppose f is not identically zero. The set Z_f is closed (in U, of course) as f is continuous. We must show that points of Z_f are isolated. Without loss of generality, suppose $0 \in U$ and $0 \in Z_f$, and suppose 0 is not in the interior of Z_f. Near 0, write

$$f(z) = \sum_{n=0}^{\infty} c_n z^n.$$

As $f(0) = 0$, $c_0 = 0$. Let k be the smallest k such that $c_k \neq 0$; this k exists as otherwise f would be identically zero near 0 and we assumed 0 is not in the interior of Z_f. Then

$$f(z) = z^k \sum_{n=k}^{\infty} c_n z^{n-k} = z^k g(z).$$

The series $g(z)$ is a convergent power series and $g(0) = c_k \neq 0$. A power series is continuous, and hence $g(z) \neq 0$ in a whole neighborhood of 0. As z^k is only zero at 0, we find that 0 is an isolated point of Z_f.

So the only points of Z_f that are not isolated are those that are in the interior of Z_f. Let $Z'_f \subset Z_f$ be the set of nonisolated points of Z_f. This set must be closed as Z_f is closed, and no sequence of points in Z'_f can have an isolated point of Z_f as a limit. We have proved above that nonisolated points must be interior points of Z_f (and hence of Z'_f). So Z'_f is both open and closed. As U is connected and that $Z'_f \neq U$, we conclude $Z'_f = \emptyset$. Thus all points of Z_f are isolated. $\qquad\square$

A useful idea from the proof that is worthwhile emphasizing is that if we have a power series $f(z)$ at a, and f has a zero at a (and not identically zero), we can factor out some power of $z - a$,

$$f(z) = (z - a)^k g(z),$$

where $g(z)$ is a power series at a such that $g(a) \neq 0$.

Exercise 2.4.12: *Suppose $U \subset \mathbb{C}$ is a domain, $f : U \to \mathbb{C}$ and $g : U \to \mathbb{C}$ are analytic, and the set $\{z \in U : f(z) = g(z)\}$ has a limit point in U. Prove that $f \equiv g$.*

Exercise 2.4.13:
   a) *Suppose $U \subset \mathbb{C}$ is a domain, $f : U \to \mathbb{C}$ and $g : U \to \mathbb{C}$ are analytic, and $f(z)g(z) = 0$ for all $z \in U$. Prove that either f or g is identically zero. (In other words, the ring of holomorphic functions on U is an integral domain.)*
   b) *Find an open but disconnected U and holomorphic f and g, such that still $fg = 0$, but neither f nor g is identically zero.*

Exercise 2.4.14: *Suppose $U \subset \mathbb{C}$ is a domain and $f : U \to \mathbb{C}$ is analytic and not constant. Show that if $K \subset U$ is compact, then $Z_f \cap K$ is finite.*

Exercise 2.4.15: *Suppose $U \subset \mathbb{C}$ is a domain, $f : U \to \mathbb{C}$ and $g : U \to \mathbb{C}$ are analytic, and $p \in U$. Suppose that $f^{(k)}(p) = g^{(k)}(p)$ for $k = 0, 1, 2, \ldots$. Prove that $f \equiv g$.*

Exercise 2.4.16: *Suppose $U \subset \mathbb{C}$ is a domain and $f : U \to \mathbb{C}$ is analytic. Prove that if $f'(z) = 0$ for all z in a neighborhood of some $z_0 \in U$, then f is constant.*

Exercise 2.4.17: *Suppose $U \subset \mathbb{C}$ is a domain, $a, b, c \in \mathbb{C}$, and $z_0 \in U$. Show that an analytic solution f on U to the linear equation $f'(z) = a f(z) + b$ given $f(z_0) = c$ is unique. Hint: Show that the power series at z_0 is uniquely determined. Just show uniqueness, no need to show existence.*

One of the downsides of analytic functions is that there are no compactly supported analytic functions on $\mathbb{C}$. The *support* of a function $f : U \to \mathbb{C}$ is the closure (in U) of the set $\{z \in U : f(z) \neq 0\}$, that is, the support is $\overline{U \setminus Z_f} \cap U$.

Exercise 2.4.18: *Let $U \subset \mathbb{C}$ be a domain and $f : U \to \mathbb{C}$ be analytic and not identically zero. Show that the support of f is U. In particular, the support cannot be compact.*

3*i*   Line Integrals and Rudimentary Cauchy Theorems

The Brain: Pinky, are you pondering what I am pondering?

Pinky: Uh, I think so, Brain, but we'll never get a monkey to use dental floss.

3.1*i*   Line integrals

3.1.1*i*   Paths

Definition 3.1.1. A *piecewise-C^1 path* or a *path* for short is a continuous complex-valued piecewise continuously differentiable function $\gamma \colon [a,b] \to \mathbb{C}$ such that $\gamma'(t)$ and all its one-sided limits are never 0.* A path γ is *closed* if $\gamma(a) = \gamma(b)$. A path γ is *simple closed* if $\gamma(a) = \gamma(b)$ and $\gamma|_{(a,b]}$ is injective.

Our paths will essentially all be piecewise-C^1, so we may forget to mention it sometimes. By "piecewise-C^1," we mean that there exist numbers $t_0 = a < t_1 < \cdots < t_k = b$ for some k such that $\gamma|_{[t_{\ell-1},t_\ell]}$ is continuously differentiable (C^1) up to the endpoints for every ℓ and its derivative is never zero. In other words, γ is continuously differentiable inside all the subintervals $(t_{\ell-1}, t_\ell)$, γ' is never zero, the one-sided limits

$$\lim_{t \uparrow t_\ell} \gamma'(t) \qquad \lim_{t \downarrow t_\ell} \gamma'(t)$$

exist for all ℓ (except, of course, only one exists at a or b), and these limits are nonzero. Another way to say it is that $\gamma'(t)$ extends to a nonzero continuous function on each closed interval $[t_{\ell-1}, t_\ell]$. Allowing these "corners" makes working with paths easier as we can define them easily piecewise, and finitely many such corners make no difference for integrals. Do note also that we are saying that paths are continuous. We will allow discontinuities just a little later with "chains."

When we say that γ is a path in $U \subset \mathbb{C}$, we mean that $\gamma \colon [a,b] \to U$ is a path. Another common abuse of notation we will freely and shamelessly commit is that if

---

*Some authors do not require the derivative to be nonzero.

we refer to γ as if it were a set, we mean the image $\gamma([a,b])$. Nowhere[*] will we really use or need that γ' is never zero, but leaving that off would allow some paths that one would generally not wish to call piecewise-C^1, as you will see in the exercises.

Example 3.1.2: The path $\gamma\colon [0,4] \to \mathbb{C}$, given by

$$\gamma(t) = \begin{cases} t & \text{if } t \in [0,1], \\ 1 + i(t-1) & \text{if } t \in (1,2], \\ 3 - t + i & \text{if } t \in (2,3], \\ i(4-t) & \text{if } t \in (3,4], \end{cases}$$

is a piecewise-C^1 simple closed path traversing the sides of the unit square. See Figure 3.1. You should check that the conditions are satisfied. For example, on $t \in (0,1)$, $\gamma'(t) = 1$, and so $\lim_{t\uparrow 1} \gamma'(t) = 1$. Similarly $\lim_{t\downarrow 1} \gamma'(t) = i$, and so on.

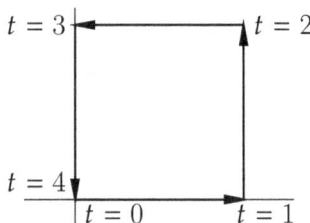

Figure 3.1: The path γ traversing the unit square.

3.1.2*i*   The line integral

Definition 3.1.3. Given a piecewise-C^1 path $\gamma\colon [a,b] \to \mathbb{C}$ and a continuous function f on γ, we define the *line integral*[†]

$$\int_\gamma f(z)\,dz \stackrel{\text{def}}{=} \int_a^b f\big(\gamma(t)\big)\gamma'(t)\,dt.$$

The right-hand side makes sense. The integrand is undefined at finitely many points (where $\gamma'(t)$ does not exist), but it is piecewise continuous, which is enough to be Riemann integrable: On each closed interval $[t_{\ell-1}, t_\ell]$, the integrand extends to a continuous function. Note that the definition makes sense even if $\gamma'(t)$ is zero somewhere, and we will, from time to time, use it in that setting.

Let us compute the most important example of a line integral.

---

[*]That's not strictly true, we will need it in an optional section.
[†]This integral is also called a *path integral*, a *curve integral*, or a *contour integral*.

Example 3.1.4: Let $\gamma\colon [0,2\pi] \to \mathbb{C}$ given by $\gamma(t) = re^{it}$ be the circle of radius r oriented counterclockwise, that is, $\partial\Delta_r(0)$. See Figure 3.2. For $n \in \mathbb{Z}$, we claim that

$$\int_\gamma z^n\, dz = \begin{cases} 2\pi i & \text{if } n = -1, \\ 0 & \text{otherwise.} \end{cases}$$

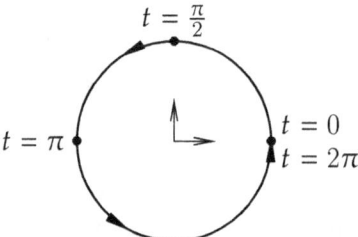

Figure 3.2: The path γ traversing the circle.

First, $\gamma'(t) = ire^{it}$. To prove the claim, we compute

$$\int_\gamma z^n\, dz = \int_0^{2\pi} r^n e^{int} ire^{it}\, dt = ir^{n+1} \int_0^{2\pi} e^{i(n+1)t}\, dt.$$

The result follows as the integral on the right-hand side is zero (exercise below), unless $n + 1 = 0$, in which case, the integral is 2π and $r^{n+1} = 1$. Note in particular that the value of the integral does not depend on r.

Exercise 3.1.1 (Easy): Prove that if $k \neq 0$, then $\int_0^{2\pi} e^{ikt}\, dt = 0$.

Exercise 3.1.2: Let γ be as in Example 3.1.4 and $f(z) = \sum_{n=-d}^{d} c_n z^n$. Compute $\int_\gamma f(z)\, dz$.

Exercise 3.1.3: Compute $\int_\gamma \bar{z}^n\, dz$ for all $n \in \mathbb{Z}$ and γ as in Example 3.1.4.

Our definition of the line integral is equivalent to the definition you have seen in multivariable calculus. Actually, it is a special case of it. Let $\gamma(t) = x(t) + i\, y(t)$, for $t \in [a, b]$, be a path, and let $dz = dx + i\, dy$. Then

$$\int_\gamma f(z)\, dz = \int_\gamma f(z)(dx + i\, dy)$$

$$= \int_\gamma f(z)\, dx + if(z)\, dy = \int_a^b \underbrace{\left(f(\gamma(t))x'(t) + if(\gamma(t))y'(t) \right)}_{f(\gamma(t))\gamma'(t)}\, dt.$$

In the second line you should recognize the definition of a line integral $\int_\gamma P\,dx + Q\,dy$ from calculus. An arbitrary line integral in the plane can be obtained if we also include $d\bar{z}$. See the exercise below.

Exercise 3.1.4: *Let $dz = dx + i\,dy$ and $d\bar{z} = dx - i\,dy$. Show that for every (real- or complex-valued) continuous P and Q, there exist continuous (complex-valued) F and G such that*

$$\int_\gamma P\,dx + Q\,dy = \int_\gamma F\,dz + G\,d\bar{z}.$$

Then show that every line integral can be computed even if you only know how to compute integrals of the form $\int_\gamma f(z)\,dz$.

For injective paths, the value of the integral does not depend on how we parametrize the image of the path, except that it does depend on orientation (which direction we go). It is easy to see why when $\gamma\colon [a,b] \to \mathbb{C}$ is continuously differentiable and we have a continuously differentiable $h\colon [c,d] \to [a,b]$ such that $h' > 0$ (increasing), $h(c) = a$, and $h(d) = b$. Then $\gamma \circ h$ is a new C^1 path that is a different parametrization of γ. Change of variable $t = h(s)$ from calculus says

$$\int_\gamma f(z)\,dz = \int_a^b f\big(\gamma(t)\big)\gamma'(t)\,dt$$

$$= \int_c^d f\big(\gamma(h(s))\big)\gamma'(h(s))h'(s)\,ds = \int_{\gamma\circ h} f(z)\,dz.$$

If, however, $h' < 0$ (decreasing), $h(c) = b$, and $h(d) = a$, then the sign flips in the change of variables:

$$\int_\gamma f(z)\,dz = -\int_{\gamma\circ h} f(z)\,dz.$$

The general version of this result, a version of which we state below, is a bit more difficult to prove. As this result, while morally important, is actually more of a technicality so that we only have to refer to the image rather than the actual path parametrization, we will leave its proof as an exercise.

Proposition 3.1.5 (Reparametrization). *Suppose $\gamma\colon [a,b] \to \mathbb{C}$ and $\alpha\colon [c,d] \to \mathbb{C}$ are piecewise-C^1 paths such that $\gamma\big([a,b]\big) = \alpha\big([c,d]\big)$. Suppose either*

(i) *γ and α are injective, or*

(ii) *$\gamma|_{(a,b]}$ and $\alpha|_{(c,d]}$ are injective and $\gamma(a) = \alpha(c) = \gamma(b) = \alpha(d)$ (simple closed paths).*

Then there exists a strictly monotone continuous $h\colon [c,d] \to [a,b]$ such that $\gamma\big(h(t)\big) = \alpha(t)$ for all $t \in [c,d]$. Furthermore:

(i) If h is increasing, then for every f continuous on the path,

$$\int_\gamma f(z)\, dz = \int_\alpha f(z)\, dz.$$

(ii) If h is decreasing, then for every f continuous on the path,

$$\int_\gamma f(z)\, dz = -\int_\alpha f(z)\, dz.$$

Exercise 3.1.5: *Prove the first case of Proposition 3.1.5, suppose γ and α are injective.*
  a) *Prove the existence of h, its monotonicity, and continuity. Hint: First prove γ^{-1} is a continuous function on $\gamma([a,b])$ using that closed subsets of $[a,b]$ are compact.*
  b) *Prove the proposition for $f \equiv 1$. Hint: the fundamental theorem of calculus.*
  c) *Prove the proposition for continuous f. Hint: Cut the path into small pieces where you can approximate f by a constant and apply the last part.*

Exercise 3.1.6: *Prove the second case of Proposition 3.1.5, that is, for simple closed paths assuming only that $\gamma|_{(a,b]}$ and $\alpha|_{(c,d]}$ are injective and $\gamma(a) = \alpha(c) = \gamma(b) = \alpha(d)$.*

Exercise 3.1.7: *A piecewise-C^1 path $\gamma\colon [a,b] \to \mathbb{C}$ can be reparametrized to a C^1 "path" if the derivative is allowed to vanish. That is, there exists a strictly increasing continuous $h\colon [a,b] \to [a,b]$ such that $\gamma \circ h$ is C^1. Hint: Make the derivative zero at the "corners."*

Exercise 3.1.8 (Tricky): *Consider infinitely many nested circles all touching at one point and let that point be the origin: Suppose r_n is the radius of the nth circle and the nth circle is given by $r_n e^{i(-1)^n \theta} - r_n$, for $0 \le \theta \le 2\pi$ (they are traversed in alternating directions). If $\sum r_n < \infty$, then you can find a continuous C^1 function $\gamma\colon [0,1] \to \mathbb{C}$ that traverses all the circles. If $\sum r_n = \infty$, then you can find a continuous $\gamma\colon [0,1] \to \mathbb{C}$ that is C^1 on $(0,1]$, but necessarily $\lim_{t\to 0} \gamma'(t)$ does not exist.*

Remark 3.1.6. The last two exercises show why we must morally require that $\gamma'(t)$ never vanishes (including the one-sided limits) for piecewise-C^1 paths. We think of the path as the set $\gamma([a,b])$, not the parametrization, and where the path is C^1, we want it to not have "corners." In practical terms, we don't often use this requirement, but it does make some more geometric arguments quite a bit simpler.

Due to the reparametrization result above, we often write down the "boundary" of a certain open set (as long as that boundary is piecewise-C^1 of course) and consider any parametrization going counterclockwise when integrating over it, without explicitly giving the parametrization. For instance, given a disc $\Delta_r(p)$, we parametrize the boundary $\partial\Delta_r(p)$ by $\gamma\colon [0,2\pi] \to \mathbb{C}$ given by $\gamma(t) = p + re^{it}$, and then we write

$$\int_{\partial\Delta_r(p)} f(z)\, dz = \int_\gamma f(z)\, dz.$$

This equality can be simply taken as a definition of integration over $\partial\Delta_r(p)$.

Finally, there is the "u-substitution" from calculus.

Proposition 3.1.7. *Let $U, V \subset \mathbb{C}$ be open, $\gamma: [a, b] \to V$ piecewise-C^1 path, $g: V \to U$ holomorphic (assume g' is continuous*), and $f: U \to \mathbb{C}$ continuous. Then $g \circ \gamma$ is a piecewise-C^1 path in U (possibly with vanishing derivative[†], however, if g' is zero on γ) and*

$$\int_\gamma f(g(z))g'(z)\,dz = \int_{g \circ \gamma} f(w)\,dw.$$

Proof. That $g \circ \gamma$ is a piecewise-C^1 path (except with perhaps vanishing derivative) is obvious. To prove the equality, apply the chain rule, $(g \circ \gamma)'(t) = g'(\gamma(t))\gamma'(t)$:

$$\int_\gamma f(g(z))g'(z)\,dz = \int_a^b f\Big(g(\gamma(t))\Big)g'(\gamma(t))\gamma'(t)\,dt$$

$$= \int_a^b f((g \circ \gamma)(t))(g \circ \gamma)'(t)\,dt = \int_{g \circ \gamma} f(w)\,dw. \qquad \square$$

Exercise 3.1.9: Let $f: \partial \mathbb{D} \to \mathbb{C}$ be a continuous function. Prove that

$$\int_{\partial \mathbb{D}} f(z)\,dz = \int_{\partial \mathbb{D}} \frac{f\left(\frac{1}{z}\right)}{z^2}\,dz = \int_{\partial \mathbb{D}} f(\bar{z})\bar{z}^2\,dz.$$

3.1.3i Arclength integral

We can also integrate with respect to arclength, the ds from calculus. We will write ds as $|dz|$. For an f continuous on a piecewise-C^1 path $\gamma: [a, b] \to \mathbb{C}$, we define

$$\int_\gamma f(z)\,|dz| \stackrel{\text{def}}{=} \int_a^b f(\gamma(t))|\gamma'(t)|\,dt.$$

Proposition 3.1.8 (Triangle inequality for line integrals). *Suppose $\gamma: [a, b] \to \mathbb{C}$ is a piecewise-C^1 path and f is a continuous function on γ. Then*

$$\left| \int_\gamma f(z)\,dz \right| \leq \int_\gamma |f(z)|\,|dz|.$$

In particular, if $|f(z)| \leq M$ on γ and $\ell = \int_\gamma ds = \int_\gamma |dz|$ is the length of γ, then

$$\left| \int_\gamma f(z)\,dz \right| \leq M\ell.$$

---

*We will soon see that g' is always continuous.

[†]There does exist a reparametrization with nonvanishing derivative, if we really really wanted one.

Proof. We estimate using Proposition 1.1.4,

$$\left| \int_{\gamma} f(z)\,dz \right| = \left| \int_a^b f(\gamma(t))\gamma'(t)\,dt \right| \le \underbrace{\int_a^b |f(\gamma(t))|\,|\gamma'(t)|\,dt}_{\int_{\gamma} |f(z)|\,|dz|} \le M \int_a^b |\gamma'(t)|\,dt. \quad \square$$

Arclength is not preserved under uniform convergence of the paths. In other words, just because the images of two paths are very close to each other, it does not mean that the integrals over them will be the same. Same caveat holds for the dz integral as it does for the $|dz|$ integral. So one has to be careful when saying that two paths are close to each other. You would need that the derivatives are close as well since γ' appears under the integral.

> *Exercise 3.1.10:*
> a) *Find a sequence of piecewise-C^1 paths $\gamma_n \colon [0,1] \to \mathbb{C}$ that uniformly converge to a constant function (one could say a path of length zero), but such that $\int_{\gamma_n} |dz| \ge n$.*
> b) *Suppose a sequence of C^1 paths $\gamma_n \colon [0,1] \to \mathbb{C}$ converges uniformly to a C^1 path $\gamma \colon [0,1] \to \mathbb{C}$ such that γ'_n converges uniformly to γ', then $\int_{\gamma_n} |dz| \to \int_{\gamma} |dz|$.*

3.1.4*i*   Chains

It is useful to combine paths; to have a certain "arithmetic" of paths. The resulting objects are called *chains*, and they are just formal combinations of paths. For example, if γ and α are piecewise-C^1 paths, then the chain $\gamma + \alpha$ is an object over which we can integrate functions that are continuous on both paths:

$$\int_{\gamma+\alpha} f(z)\,dz \overset{\text{def}}{=} \int_{\gamma} f(z)\,dz + \int_{\alpha} f(z)\,dz.$$

Definition 3.1.9. A *chain* in $U \subset \mathbb{C}$ is an expression

$$\Gamma = a_1\gamma_1 + \cdots + a_n\gamma_n,$$

where $a_1, \ldots, a_n \in \mathbb{Z}$ and $\gamma_1, \ldots, \gamma_n$ are piecewise-C^1 paths in U. We integrate over Γ as

$$\int_{\Gamma} f(z)\,dz = \int_{a_1\gamma_1+\cdots+a_n\gamma_n} f(z)\,dz \overset{\text{def}}{=} a_1 \int_{\gamma_1} f(z)\,dz + \cdots + a_n \int_{\gamma_n} f(z)\,dz.$$

Two chains Γ_1 and Γ_2 in U are *equivalent* (we will write $\Gamma_1 = \Gamma_2$) if

$$\int_{\Gamma_1} f(z)\,dz = \int_{\Gamma_2} f(z)\,dz$$

for all continuous $f \colon U \to \mathbb{C}$. We define the *zero chain* 0 by defining $\int_0 f(z)\,dz = 0$ for all continuous $f \colon U \to \mathbb{C}$.

The chain arithmetic is done in the obvious way as formal sums of paths: If $\Gamma_1 = 2\gamma_1 + \gamma_2$ and $\Gamma_2 = 3\gamma_2 + \gamma_3$, then $\Gamma_1 + \Gamma_2 = 2\gamma_1 + 4\gamma_2 + \gamma_3$. Similarly for scalar multiplication: $3\Gamma_1 = 6\gamma_1 + 3\gamma_2$. We write $-\Gamma$ for $(-1)\Gamma$. A chain Γ is equivalent to the zero chain if

$$\int_\Gamma f(z)\,dz = 0$$

for all continuous f, and the chains Γ_1 and Γ_2 are equivalent if $\Gamma_1 - \Gamma_2 = 0$. Chains in this book are always composed of piecewise-C^1 paths, although that is not the most general definition used in the literature.

Remark 3.1.10. For the equivalence, the set where the continuous f is defined is not a big deal. We could take f to be continuous on U, $\mathbb{C}$, or just the images of Γ_1 and Γ_2. By Tietze's extension theorem (a theorem in any metric space), every continuous function on a closed subset of $\mathbb{C}$ (such as the images of Γ_1 and Γ_2) extends to a continuous function on $\mathbb{C}$. The way we defined things, we do not need Tietze.

Remark 3.1.11. It is important that the definition of equivalence is for all continuous functions. We will show later that if U is say the disc, then for any closed Γ in U, $\int_\Gamma f(z)\,dz = 0$ for all holomorphic f. Clearly that should not imply that Γ is equivalent to the zero chain. See also Exercise 3.1.13.

Exercise 3.1.11: *Let $\gamma_1\colon [a,b] \to \mathbb{C}$ and $\gamma_2\colon [b,c] \to \mathbb{C}$ be two piecewise-C^1 paths and $\gamma_1(b) = \gamma_2(b)$. Prove that the function $\gamma\colon [a,c] \to \mathbb{C}$ defined by $\gamma(t) = \gamma_1(t)$ if $t \in [a,b]$ and $\gamma(t) = \gamma_2(t)$ if $t \in [b,c]$ is a piecewise-C^1 path, and for all f continuous on the image of γ, we have*

$$\int_\gamma f(z)\,dz = \int_{\gamma_1+\gamma_2} f(z)\,dz.$$

Exercise 3.1.12: *Let $\partial\mathbb{D}$ denote the counterclockwise path around the unit disc. Show that for every integer n, the chain $n\partial\mathbb{D}$ is equivalent to the path $\gamma\colon [0,2\pi] \to \mathbb{C}$ given by $\gamma(t) = e^{int}$, the path that goes n times around the unit disc counterclockwise.*

Exercise 3.1.13:
  a) *Suppose $U \subset \mathbb{C}$ is open and $\gamma\colon [a,b] \to U$ is a piecewise-C^1 path and $\gamma|_{[a,b)}$ is injective, but possibly a closed path, so possibly $\gamma(a) = \gamma(b)$. Show that as chains, γ is not equivalent to the zero chain. Note: Closed γ is trickier.*
  b) *Find a piecewise-C^1 path $\gamma\colon [a,b] \to \mathbb{C}$ that is equivalent to the zero chain. Note that our definition of "path" prevents γ from being constant.*

Using Exercise 3.1.11, any chain that is put together from connecting paths can be converted and integrated as a single path, and so in the sequel we may do this procedure implicitly.

Definition 3.1.12. Given two points $z, w \in \mathbb{C}$, the *segment* $[z, w]$ is the path $\gamma \colon [0, 1] \to \mathbb{C}$ given by $\gamma(t) = (1 - t)z + tw$. For the purposes of chain arithmetic, $-[z, w] = [w, z]$. A path is *polygonal* if it can be written as (is equivalent to) a chain $[z_1, z_2] + [z_2, z_3] + \cdots + [z_{k-1}, z_k]$ for some complex numbers $z_1, \ldots, z_k$.

As the following exercise shows, we can, if we want to, get by with just polygonal paths for most practical purposes. The paths that come up in applications are often constructed out of segments and arcs anyhow.

> *Exercise 3.1.14* (Tricky)*: Suppose $U \subset \mathbb{C}$ is open, $\gamma \colon [a, b] \to U$ is a piecewise-C^1 path, and $f \colon U \to \mathbb{C}$ is continuous. Then for every $\epsilon > 0$, there exists a polygonal path (or chain) α in U with the same beginning and end point, such that*
>
> $$\left| \int_\alpha f(z)\, dz - \int_\gamma f(z)\, dz \right| < \epsilon.$$
>
> *Hint: Consider the Riemann sum. Also f is uniformly continuous on some smaller U.*

3.2i Starter versions of Cauchy

3.2.1i Primitives, cycles, and Cauchy for derivatives

Definition 3.2.1. Let $U \subset \mathbb{C}$ be open and $f \colon U \to \mathbb{C}$ a function. A holomorphic $F \colon U \to \mathbb{C}$ with $f = F'$ is called a (holomorphic)* *primitive* (or an *antiderivative*) of f.

Primitives do not always exist, but if they do, then they are unique up to a constant.

Proposition 3.2.2. *Suppose $U \subset \mathbb{C}$ is a domain, and $F \colon U \to \mathbb{C}$ and $G \colon U \to \mathbb{C}$ are holomorphic functions such that $F' = G'$. Then there is a constant C such that $F(z) = G(z) + C$.*

> *Exercise 3.2.1: Prove the proposition. Make sure you use that U is a domain (connected).*

We have antiderivatives. We have integrals. We are in need of a fundamental theorem.

Theorem 3.2.3 (Fundamental theorem of calculus for line integrals). *Suppose $U \subset \mathbb{C}$ is open and $f \colon U \to \mathbb{C}$ is continuous with a primitive $F \colon U \to \mathbb{C}$ (so $F' = f$). Let $\gamma \colon [a, b] \to U$ be a piecewise-C^1 path. Then*

$$\int_\gamma f(z)\, dz = F\big(\gamma(b)\big) - F\big(\gamma(a)\big).$$

---

*We will usually say just "primitive" as it is generally clear that it must be a holomorphic primitive, and besides, that is the only way that we will use the word anyway.

Proof. We compute:

$$\int_\gamma F'(z)\,dz = \int_a^b F'\big(\gamma(t)\big)\gamma'(t)\,dt = \int_a^b \frac{d}{dt}\Big(F\big(\gamma(t)\big)\Big)\,dt = F\big(\gamma(b)\big) - F\big(\gamma(a)\big).$$

The computation uses the chain rule (Proposition 2.2.3) and the fundamental theorem of calculus, where the standard (real) fundamental theorem of calculus is applied to the real and imaginary parts of the expression. □

Remark 3.2.4. The hypothesis that $f = F'$ is continuous is extraneous. We will soon prove that a derivative of a holomorphic function is holomorphic. As that is not yet proved, we need F' to be at least continuous so that the integral makes sense.*

Definition 3.2.5. A chain Γ is called a *cycle* if it is equivalent to $a_1\gamma_1 + \cdots + a_n\gamma_n$, where $\gamma_1,\ldots,\gamma_n$ are closed piecewise-C^1 paths and $a_1,\ldots,a_n \in \mathbb{Z}$.

Recall that a path $\gamma\colon [a,b] \to \mathbb{C}$ is closed if $\gamma(a) = \gamma(b)$. Note that we are not saying that Γ *is* a sum of closed paths, we are saying it *is equivalent to* a sum of closed paths. The square path in Example 3.1.2 is a cycle, and could be written more conveniently as a chain composed of four straight line segments $[0,1] + [1,1+i] + [1+i,i] + [i,0]$. The fundamental theorem has the following immediate corollary.

Corollary 3.2.6 (Cauchy's theorem for derivatives). *Suppose $U \subset \mathbb{C}$ is open and $f\colon U \to \mathbb{C}$ is continuous with a primitive $F\colon U \to \mathbb{C}$. Let Γ be a cycle in U. Then*

$$\int_\Gamma f(z)\,dz = 0.$$

We will prove several versions of Cauchy's theorem, although this one is somewhat different from the others. Usually there will be a restriction on the U or perhaps the path or cycle Γ, while the function is usually just any holomorphic function. A version of Cauchy's theorem can be taken as an "independence of path" result saying that we can define a function at z by a line integral from some fixed point to z. The result will be that such a function is a primitive. So the other versions of Cauchy's theorem will generally either restrict which Γ can be taken or restrict to only those U where every holomorphic function has a primitive.

The next corollary will be entirely subsumed into the more general version of Cauchy we will prove later, but right now it is rather appealing.

Corollary 3.2.7 (Cauchy's theorem for polynomials). *Suppose $P(z)$ is a polynomial and Γ is a cycle (in $\mathbb{C}$). Then*

$$\int_\Gamma P(z)\,dz = 0.$$

---

*A real derivative may only be integrable by a so-called gauge or Henstock–Kurzweil integral—Riemann or even Lebesgue are not enough—so integrability is not an idle concern. If the reader is willing to hunt ants with a sledgehammer, then the statement and proof of the proposition is perfectly fine at this stage if one uses the gauge integral even without any hypothesis on f.

Exercise **3.2.2:** *Prove Cauchy's theorem for polynomials.*

Exercise **3.2.3:** *Suppose f is given by a power series at p that converges in $\Delta_R(p)$. Let Γ be a cycle in $\Delta_R(p)$. Prove that*

$$\int_\Gamma f(z)\,dz = 0.$$

Exercise **3.2.4:** *Suppose $U \subset \mathbb{C}$ is open, $f \colon U \to \mathbb{C}$ holomorphic, and Γ is a cycle in U. Given a $p \in U$, find a holomorphic $g \colon U \to \mathbb{C}$ with $g(p) = 0$ and $g'(p) = 0$ such that $\int_\Gamma g(z)\,dz = \int_\Gamma f(z)\,dz$.*

Exercise **3.2.5:** *Let $n \neq -1$ be an integer and Γ a cycle in $\mathbb{C} \setminus \{0\}$. Compute*

$$\int_\Gamma z^n\,dz.$$

Exercise **3.2.6:** *Using Example 3.1.4, argue that $1/z$ does not have a primitive in $\mathbb{C} \setminus \{0\}$.*

3.2.2i Cauchy–Goursat, the "Cauchy for triangles"

Definition 3.2.8. A set X is *convex* if the segment $[a, b] \subset X$ for all $a, b \in X$. Let $a, b, c \in \mathbb{C}$ be distinct points in $\mathbb{C}$ that do not lie on a straight line. A *triangle* T with vertices a, b, c is the convex hull of $\{a, b, c\}$, that is, the set of all points

$$t_1 a + t_2 b + t_3 c,$$

where $t_1, t_2, t_3 \in [0, 1]$ and $t_1 + t_2 + t_3 = 1$.

Another way to define the convex hull is the intersection of all convex sets containing $\{a, b, c\}$. In particular, T is the smallest convex set containing the vertices. Do note that we have defined a triangle as the solid triangle, including the inside.

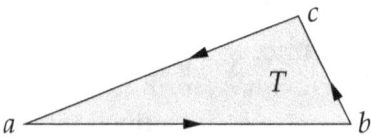

Figure 3.3: Positively oriented triangle.

Order the vertices so that the boundary ∂T has positive orientation; if we travel from a to b to c the inside of the triangle is on the left. More precisely, if we translate so that a is the origin and rotate so that b is on the positive real line, then c has positive imaginary part. See Figure 3.3. Define the boundary cycle of T as

$$\partial T = [a, b] + [b, c] + [c, a].$$

Theorem 3.2.9 (Cauchy–Goursat*). *Suppose $U \subset \mathbb{C}$ is open, $f : U \to \mathbb{C}$ is holomorphic, and $T \subset U$ is a triangle. Then*

$$\int_{\partial T} f(z)\, dz = 0.$$

It is important that $T \subset U$ means that the inside of the triangle is in U, not just the boundary. Otherwise the theorem would not be true.

Proof. We proceed by contrapositive. Suppose f is continuous, and suppose there is a triangle $T \subset U$ over whose boundary the integral is not zero. Write

$$\left| \int_{\partial T} f(z)\, dz \right| = c > 0.$$

We will find a point where f is not complex differentiable.

Cut T into four subtriangles T_1, T_2, T_3, T_4 by cutting each side of T in half. See Figure 3.4.

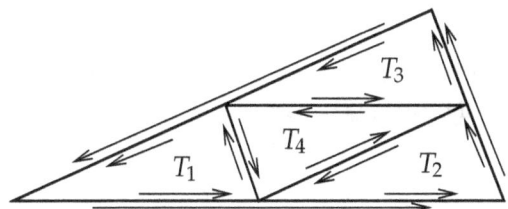

Figure 3.4: Cutting a triangle into four triangles of half the size.

Each T_j is positively oriented. The sides of the inner T_4 triangle have orientation opposite to the orientation of the inner sides of T_1, T_2, and T_3, and so the line integral over these sides cancels. Therefore,

$$c = \left| \int_{\partial T} f(z)\, dz \right| = \left| \int_{\partial T_1} f(z)\, dz + \int_{\partial T_2} f(z)\, dz + \int_{\partial T_3} f(z)\, dz + \int_{\partial T_4} f(z)\, dz \right|.$$

One of the four integrals, say that over ∂T_j, must have modulus at least $c/4$. Label that triangle $T^1 = T_j$:

$$\left| \int_{\partial T^1} f(z)\, dz \right| \geq \frac{c}{4}.$$

Now cut T^1 into subtriangles T^1_1, T^1_2, T^1_3, T^1_4 as above and repeat the procedure. There is one of these four whose integral is at least $\frac{c}{4^2}$, label it T^2. Rinse and repeat. All in all, for the kth triangle T^k,

$$\left| \int_{\partial T^k} f(z)\, dz \right| \geq \frac{c}{4^k}.$$

---

*What makes this theorem the Goursat theorem rather than just another statement of Cauchy's theorem is that in the proof, we are only assuming that the complex derivative exists and not that it is continuous, which is what Cauchy assumed.

Furthermore, $T^k \subset T^{k-1} \subset \cdots \subset T$. In each step, the subtriangles are all similar triangles (same angles), but with sides exactly halved. In particular, the largest possible distance for two points in the triangle gets halved each iteration. In more formal language:*

$$\text{diam}(T^k) = \frac{1}{2}\text{diam}(T^{k-1}) = \frac{1}{2^k}\text{diam}(T).$$

As T is compact and each T^k is closed, the nested intersection of the T^k must be nonempty. As the diameter goes to zero, it must be a single point:

$$\{z_0\} = \bigcap_{k=1}^{\infty} T^k.$$

For some $\alpha \in \mathbb{C}$, let g be defined by

$$f(z) = f(z_0) + \alpha(z - z_0) + g(z).$$

Were f complex differentiable at z_0, there would be an α so that $\frac{g(z)}{z-z_0}$ would go to zero as $z \to z_0$. Let us prove that it does not go to zero for any α. Fix α and thus g. Cauchy's theorem for polynomials says

$$\int_{\partial T^k} f(z)\,dz = \int_{\partial T^k} \left(f(z_0) + \alpha(z - z_0) + g(z)\right) dz = \int_{\partial T^k} g(z)\,dz.$$

And so

$$\frac{c}{4^k} \le \left|\int_{\partial T^k} f(z)\,dz\right| = \left|\int_{\partial T^k} g(z)\,dz\right| \le \int_{\partial T^k} |g(z)|\,|dz|.$$

Let ℓ be the sum of the lengths of the sides of T. The sum of the lengths of the sides of T^k is $\frac{\ell}{2^k}$. By the mean value theorem for integrals[†], there is a $z_k \in \partial T^k$ such that

$$|g(z_k)| = \frac{2^k}{\ell}\int_{\partial T^k} |g(z)|\,|dz|.$$

We have $z_k \ne z_0$ as $g(z_0) = 0$ and $|g(z_k)| \ge \frac{c}{\ell 2^k} > 0$. Let $d = \text{diam}(T)$. Then $|z_k - z_0| \le \frac{d}{2^k}$ and

$$\left|\frac{g(z_k)}{z_k - z_0}\right| \ge \frac{2^k|g(z_k)|}{d} = \frac{4^k}{d\ell}\int_{\partial T^k} |g(z)|\,|dz| \ge \frac{4^k}{d\ell}\frac{c}{4^k} = \frac{c}{d\ell}.$$

Because $z_k \to z_0$, we find that $\frac{g(z)}{z-z_0}$ does not go to zero as $z \to z_0$. So f is not complex differentiable at z_0. $\qquad\square$

---

*Here $\text{diam}(T) = \sup\{|p - q| : p, q \in T\}$ means the maximum distance between two points in T.

[†]If $\varphi\colon [a, b] \to \mathbb{R}$ is continuous, then there is an $x \in [a, b]$ such that $\varphi(x) = \frac{1}{b-a}\int_a^b \varphi(t)\,dt$. To apply it here, parametrize the entire triangle with unit speed.

Exercise 3.2.7: *Suppose $T \subset \mathbb{C}$ is a triangle and $f : T \to \mathbb{C}$ a continuous function whose restriction to the interior of T is holomorphic. Prove that $\int_{\partial T} f(z)\, dz = 0$.*

Exercise 3.2.8: *A closed rectangle $R \subset \mathbb{C}$ is a set $\{ z \in \mathbb{C} : a \le \operatorname{Re} z \le b, c \le \operatorname{Im} z \le d \}$ for real numbers $a < b$, $c < d$. The boundary is again oriented counterclockwise. Prove Cauchy–Goursat for rectangles (replace T in the theorem with R).*

Exercise 3.2.9: *Let R be a rectangle with vertices $-1 - i$, $1 - i$, $1 + i$, and $-1 + i$ and notice that 0 is in the interior. Compute $\int_{\partial R} \frac{1}{z}\, dz$, notice that it is nonzero, and argue why it does not violate the Cauchy–Goursat theorem for rectangles (see the previous exercise). Hint: We do not yet have the complex logarithm, so you can't use that, but notice that for instance:* $\frac{1}{t-i} = \frac{t}{t^2+1} + i \frac{1}{t^2+1}$.

A triangle is one type of a convex set, but as convex sets come up often, let us give some basic properties of convex sets as exercises. These may be good to do in order and possibly use earlier ones in solving the later ones.

Exercise 3.2.10: *Prove:*
    a) An arbitrary intersection of convex sets is convex.
    b) The interior of a convex set is convex.
    c) The closure of a convex set is convex.

Exercise 3.2.11: *Let $X \subset \mathbb{C}$ be a convex set and $\xi \in \partial X$, then prove that there exists a nonzero w such that for all $z \in X$, we have*

$$\operatorname{Re} z\bar{w} \ge \operatorname{Re} \xi\bar{w}.$$

In other words, X is in the closed half-plane bounded by a straight line containing ξ and orthogonal to w. Notice that $\operatorname{Re} z\bar{w}$ is the standard dot product from vector calculus in $\mathbb{R}^2$.

Exercise 3.2.12: *Let $X \subset \mathbb{C}$ be a closed convex set. Prove that X is an intersection of closed half-planes (see previous exercise).*

Exercise 3.2.13: *Union of convex sets is normally not convex, but if $\{X_n\}$ is a sequence of convex sets such that $X_n \subset X_{n+1}$, then prove that the union $\bigcup_n X_n$ is convex.*

3.2.3i Cauchy for star-like sets

Definition 3.2.10. A set $U \subset \mathbb{C}$ is called *star-like* (or more precisely *star-like with respect to z_0*) if there exists a point $z_0 \in U$ such that the segment $[z_0, z] \subset U$ for every $z \in U$. See Figure 3.5.

A convex set is star-like, but not vice versa.

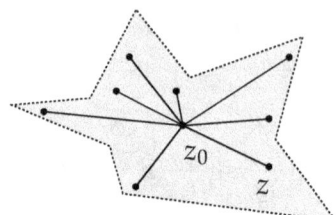

Figure 3.5: A domain that is star-like with respect to z_0.

Exercise 3.2.14: *Prove that if $U \subset \mathbb{C}$ is star-like with respect to z_0 and $[a, b] \subset U$, then the triangle T with vertices z_0, a, and b is entirely contained in U.*

Exercise 3.2.15: *Suppose $U \subset \mathbb{C}$ is open and star-like. Prove that U is connected.*

Exercise 3.2.16:
  a) *Prove that if $U_1, \ldots, U_n \subset \mathbb{C}$ are convex and $U_1 \cap \cdots \cap U_n \neq \emptyset$, then the union $U_1 \cup \cdots \cup U_n$ is star-like.*
  b) *Find an example of convex $U_1, U_2, U_3 \subset \mathbb{C}$ where $U_1 \cap U_2 \neq \emptyset$, $U_1 \cap U_3 \neq \emptyset$, and $U_2 \cap U_3 \neq \emptyset$, but such that $U_1 \cup U_2 \cup U_3$ is not star-like.*

Proposition 3.2.11. *Suppose $U \subset \mathbb{C}$ is open and star-like, $f \colon U \to \mathbb{C}$ is continuous, and*

$$\int_{\partial T} f(z)\, dz = 0$$

for every triangle $T \subset U$. Then f has a primitive, that is, there exists a holomorphic $F \colon U \to \mathbb{C}$ such that $F' = f$.

Proof. Suppose U is star-like with respect to $z_0 \in U$. For $z \in U$, define

$$F(z) = \int_{[z_0, z]} f(\zeta)\, d\zeta.$$

Consider a small disc $\Delta_r(z) \subset U$. If $|h| < r$, then $z + h \in \Delta_r(z)$. The line between z and $z + h$ is in U, and as U is star-like with respect to z_0, the entire triangle (or the line segment if the points are collinear) with vertices z_0, z, and $z + h$ is in U, see Figure 3.6 (and Exercise 3.2.14).

The hypothesis says (trivially true if the points are collinear)

$$\int_{[z_0, z] + [z, z+h] - [z_0, z+h]} f(\zeta)\, d\zeta = 0.$$

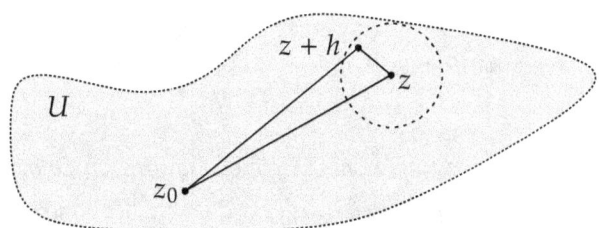

Figure 3.6: Star-like domain and the triangle with vertices z_0, z, and $z + h$.

So

$$\frac{F(z + h) - F(z)}{h} = \frac{1}{h} \int_{[z_0, z+h] - [z_0, z]} f(\zeta) \, d\zeta$$

$$= \frac{1}{h} \int_{[z, z+h]} f(\zeta) \, d\zeta = \frac{1}{h} \int_0^1 f(z + th) h \, dt = \int_0^1 f(z + th) \, dt.$$

In other words,

$$\left| \frac{F(z + h) - F(z)}{h} - f(z) \right| = \left| \int_0^1 f(z + th) \, dt - \int_0^1 f(z) \, dt \right|$$

$$\leq \int_0^1 \left| f(z + th) - f(z) \right| \, dt.$$

By continuity of f at z,

$$\lim_{h \to 0} \frac{F(z + h) - F(z)}{h} = f(z). \qquad \square$$

Cauchy–Goursat (Theorem 3.2.9) says that the integral around triangles is always zero if f is holomorphic. Thus we get the following immediate corollary.

Corollary 3.2.12. *Suppose $U \subset \mathbb{C}$ is open and star-like and $f : U \to \mathbb{C}$ is holomorphic. Then f has a primitive, that is, there exists a holomorphic $F : U \to \mathbb{C}$ such that $F' = f$.*

We also get another corollary, which we call a theorem as it is one of the fundamental results.

Theorem 3.2.13 (Cauchy's theorem for star-like domains). *Suppose $U \subset \mathbb{C}$ is open and star-like, $f : U \to \mathbb{C}$ is holomorphic, and Γ is a cycle in U. Then*

$$\int_\Gamma f(z) \, dz = 0.$$

Proof. Corollary 3.2.12 implies that there is a primitive $F : U \to \mathbb{C}$. Cauchy's theorem for derivatives (Corollary 3.2.6) then implies that the integral is zero. $\qquad \square$

Exercise 3.2.17: *Suppose $f: \mathbb{C} \setminus \{0\} \to \mathbb{C}$ is holomorphic and $\gamma: [a,b] \to \mathbb{C} \setminus \{0\}$ is a closed piecewise-C^1 path such that $\operatorname{Re} \gamma(t) < |\gamma(t)|$ for all $t \in [a,b]$. Show that $\int_\gamma f(z)\, dz = 0$.*

Exercise 3.2.18: *Let γ be the upper semicircle of the unit circle oriented from 1 to -1. Suppose $f: \mathbb{C} \to \mathbb{C}$ is holomorphic and $\int_0^1 f(x^2)\, dx = \pi$. Compute $\int_\gamma f(z^2)\, dz$.*

Exercise 3.2.19: *Suppose $U_1, U_2 \subset \mathbb{C}$ are star-like domains such that $U_1 \cap U_2$ is nonempty and connected. Prove Cauchy's theorem for $U = U_1 \cup U_2$, that is, if $f: U \to \mathbb{C}$ is holomorphic and Γ is a cycle in U, then $\int_\Gamma f(z)\, dz = 0$.*

Exercise 3.2.20: *Suppose $U \subset \mathbb{C}$ is open and star-like and $f: U \to \mathbb{C}$ is antiholomorphic, that is, it is the conjugate of a holomorphic function. Let $d\bar{z} = dx - i\, dy$ as before. Suppose Γ is a cycle in U. Prove that $\int_\Gamma f(z)\, d\bar{z} = 0$.*

Remark 3.2.14. A complex-valued function can be thought of as a vector-field on $\mathbb{R}^2$. Corollary 3.2.12 is in fact a special case of a theorem you have seen in vector calculus, a version of the *Poincaré lemma*: *In a star-like domain $U \subset \mathbb{R}^2$, if a C^1 vector field $(u,v): U \to \mathbb{R}^2$ satisfies $\frac{\partial u}{\partial y} = \frac{\partial v}{\partial x}$ (the vector field is* irrotational*), then there exists a real-valued $f: \mathbb{R}^2 \to \mathbb{R}$ such that $\nabla f = (u,v)$ (the vector field is* conservative*, a gradient).* More concisely, *an irrotational vector field in a star-like domain is conservative.* See the "conservative vector fields" section of your favorite calculus textbook. You can gain a lot of intuition on the current material on holomorphic functions by reviewing the vector calculus analogues.

Exercise 3.2.21: *Use the result on irrotational vector fields from Remark 3.2.14 to prove Corollary 3.2.12. Assume you know that holomorphic functions are C^1.*

3.2.4i   Cauchy's formula in a disc

Perhaps the most fundamental theorem in complex analysis in one variable, and the root cause of all the amazing properties of holomorphic functions is the Cauchy integral formula. Let us state it for a disc, and leave more general statements for later.

Theorem 3.2.15 (Cauchy integral formula in a disc). *Suppose $U \subset \mathbb{C}$ is open, $f: U \to \mathbb{C}$ is holomorphic, and $\overline{\Delta_r(p)} \subset U$. Then for all $z \in \Delta_r(p)$,*

$$f(z) = \frac{1}{2\pi i} \int_{\partial \Delta_r(p)} \frac{f(\zeta)}{\zeta - z}\, d\zeta.$$

What should be surprising about this theorem is that the values of a holomorphic function inside the disc (a large set) are determined by their values on the circle (a relatively small set).

Proof. Fix $z \in \Delta_r(p)$ and write γ for the boundary of $\Delta_r(p)$ oriented counterclockwise. Let $\Delta_s(z)$ be a small disc with $\overline{\Delta_s(z)} \subset \Delta_r(p)$, and write α for the boundary of $\Delta_s(z)$.

We connect α to γ via two straight lines as in Figure 3.7. The two resulting regions between α and γ give closed paths c_1 and c_2 with the counterclockwise orientations marked in the figure.

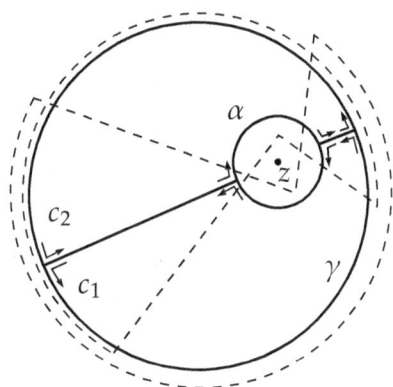

Figure 3.7: Connecting γ and α.

As chains, $c_1 + c_2 = \gamma - \alpha$. Each c_j lies in a star-like domain (some possibilities marked by dashed lines in the figure), where $\frac{f(\zeta)}{\zeta - z}$ is holomorphic as a function of ζ (since z is outside each of these domains). By Cauchy's theorem for star-like sets,

$$\int_\gamma \frac{f(\zeta)}{\zeta - z} \, d\zeta - \int_\alpha \frac{f(\zeta)}{\zeta - z} \, d\zeta = \int_{c_1} \frac{f(\zeta)}{\zeta - z} \, d\zeta + \int_{c_2} \frac{f(\zeta)}{\zeta - z} \, d\zeta = 0.$$

So

$$\frac{1}{2\pi i} \int_\gamma \frac{f(\zeta)}{\zeta - z} \, d\zeta = \frac{1}{2\pi i} \int_\alpha \frac{f(\zeta)}{\zeta - z} \, d\zeta.$$

Let $\alpha(t) = z + se^{it}$ be the parametrization. Then

$$\frac{1}{2\pi i} \int_\alpha \frac{f(\zeta)}{\zeta - z} \, d\zeta = \frac{1}{2\pi i} \int_0^{2\pi} \frac{f(z + se^{it})}{z + se^{it} - z} sie^{it} \, dt = \frac{1}{2\pi} \int_0^{2\pi} f(z + se^{it}) \, dt.$$

As the integral over γ (which does not depend on s) is equal to the integral over α for all $s > 0$ small enough, we can take the limit as $s \to 0$. By continuity of f at z,

$$\lim_{s \downarrow 0} \frac{1}{2\pi i} \int_\alpha \frac{f(\zeta)}{\zeta - z} \, d\zeta = \lim_{s \downarrow 0} \frac{1}{2\pi} \int_0^{2\pi} f(z + se^{it}) \, dt = f(z). \qquad \square$$

Exercise 3.2.22: *Make the construction of c_1 and c_2 and the two star-like domains in the proof explicit. That is, exactly describe the "cut" that makes c_1 and c_2, and describe two star-like domains (you don't have to do the two pictured).*

Exercise 3.2.23: *Show why the theorem should be surprising. Given $a, b \in \mathbb{C}$ and $z \in \mathbb{D}$, construct a continuous $f : \mathbb{C} \to \mathbb{C}$ such that $\frac{1}{2\pi i} \int_{\partial \mathbb{D}} \frac{f(\zeta)}{\zeta - z} \, d\zeta = a$ and $f(z) = b$.*

Exercise 3.2.24: *Suppose f is holomorphic in an open neighborhood of $\overline{\Delta_r(p)}$. Show that f at p is the average of the values on $\partial \Delta_r(p)$. That is, show*

$$f(p) = \frac{1}{2\pi} \int_0^{2\pi} f(p + re^{it}) \, dt.$$

Exercise 3.2.25: *Suppose f is holomorphic in an open neighborhood of $\overline{\Delta_r(p)}$. Show that f at p is the average of the values on $\Delta_r(p)$. That is, show*

$$f(p) = \frac{1}{\pi r^2} \int_{\Delta_r(p)} f(z) \, dA,$$

where $dA = dx \, dy = r \, dr \, d\theta$ is the area measure.

Exercise 3.2.26: *Compute*

$$\int_\gamma \frac{\cos(z^2) + z}{z(z - \sqrt{\pi})} \, dz$$

if γ is:

   a) The circle of radius 1 centered at the origin oriented counterclockwise.
   b) The circle of radius 2 centered at the origin oriented counterclockwise. Hint: partial fractions.
   c) The circle of radius 5 centered at $i + 1$ oriented clockwise.

Exercise 3.2.27: *Strengthen the theorem: Show that the conclusion holds if we only assume that $f : \overline{\Delta_r(p)} \to \mathbb{C}$ is continuous and f is holomorphic on $\Delta_r(p)$.*

3.3i Consequences of Cauchy

3.3.1i Holomorphic functions are analytic

Perhaps the most profound consequence of Cauchy's formula is that holomorphic functions are analytic. We have already seen that analytic functions are holomorphic, and now we prove the converse.

Theorem 3.3.1. *Let $U \subset \mathbb{C}$ be open, $f : U \to \mathbb{C}$ be holomorphic, $p \in U$, and $\Delta_R(p) \subset U$. Then there exists a power series $\sum c_n(z - p)^n$ such that for all $z \in \Delta_R(p)$,*

$$f(z) = \sum_{n=0}^{\infty} c_n(z - p)^n.$$

Moreover,

$$c_n = \frac{1}{2\pi i} \int_{\gamma} \frac{f(z)}{(z - p)^{n+1}} \, dz,$$

where γ is any circle of radius r, $0 < r < R$, centered at p oriented counterclockwise.

Proof. First fix an r such that $0 < r < R$. Thus $\overline{\Delta_r(p)} \subset U$, and in particular, $\partial\Delta_r(p) \subset U$. Fix a $z \in \Delta_r(p)$. For $\zeta \in \partial\Delta_r(p)$,

$$\left| \frac{z - p}{\zeta - p} \right| = \frac{|z - p|}{r} < 1.$$

So the geometric series in $\frac{z-p}{\zeta-p}$ converges, that is,

$$\sum_{n=0}^{\infty} \left(\frac{z - p}{\zeta - p} \right)^n = \frac{1}{1 - \frac{z-p}{\zeta-p}} = \frac{\zeta - p}{\zeta - z}.$$

Write $f(z)$ using the Cauchy integral formula:

$$\begin{aligned}
f(z) &= \frac{1}{2\pi i} \int_{\partial\Delta_r(p)} \frac{f(\zeta)}{\zeta - z} \, d\zeta \\
&= \frac{1}{2\pi i} \int_{\partial\Delta_r(p)} \frac{f(\zeta)}{\zeta - p} \frac{\zeta - p}{\zeta - z} \, d\zeta \\
&= \frac{1}{2\pi i} \int_{\partial\Delta_r(p)} \frac{f(\zeta)}{\zeta - p} \sum_{n=0}^{\infty} \left(\frac{z - p}{\zeta - p} \right)^n \, d\zeta \\
&= \sum_{n=0}^{\infty} \underbrace{\left(\frac{1}{2\pi i} \int_{\partial\Delta_r(p)} \frac{f(\zeta)}{(\zeta - p)^{n+1}} \, d\zeta \right)}_{c_n} (z - p)^n.
\end{aligned}$$

In the last equality, we were allowed to interchange the limit on the sum with the integral via uniform convergence (uniform in the $\zeta \in \partial\Delta_r(p)$): z is fixed and if M is the supremum of $\left| \frac{f(\zeta)}{\zeta-p} \right| = \frac{|f(\zeta)|}{r}$ on $\partial\Delta_r(p)$ (a compact set), then

$$\left| \frac{f(\zeta)}{\zeta - p} \left(\frac{z - p}{\zeta - p} \right)^n \right| \leq M \left(\frac{|z - p|}{r} \right)^n, \qquad \text{and} \qquad \frac{|z - p|}{r} < 1.$$

Thus, $\sum \left| \frac{f(\zeta)}{\zeta-p} \left(\frac{z-p}{\zeta-p} \right)^n \right|$ converges uniformly in $\zeta \in \partial \Delta_r(p)$, and so $\sum \frac{f(\zeta)}{\zeta-p} \left(\frac{z-p}{\zeta-p} \right)^n$ converges uniformly absolutely (and hence uniformly).

We found a power series converging to $f(z)$ for all $z \in \Delta_r(p)$. By uniqueness of the power series (see Corollary 2.4.3), the c_n we compute are the same for every $r < R$. Hence, we get the same series for every r and it converges in $\Delta_R(p)$. $\qquad \square$

The key point in the proof is writing the *Cauchy kernel* $\frac{1}{\zeta-z}$ as

$$\frac{1}{\zeta-z} = \frac{1}{\zeta-p}\frac{\zeta-p}{\zeta-z}$$

and then using the geometric series. The proof illustrates a common technique: Given an integral of a function against a kernel, take a feature of the kernel, in this case having a series, and prove that the integral has that same feature. In the proof above, the trick is to figure out how to massage the kernel so that in the geometric series we get terms that are something times $(z-p)^n$.

Not only have we proved that f has a power series, we computed that the radius of convergence is at least R, where R is the maximum R such that $\Delta_R(p) \subset U$. See Figure 3.8. That is a surprisingly powerful result. Nothing like that is true for power series in a real variable, see Exercise 3.3.3. It allows for computation of the radius of convergence (or at least a lower bound for it) just from knowing the domain of definition of a holomorphic function. The radius of convergence then gives us bounds on the derivatives, and so we know quite a bit about the size of the derivatives of a function just from knowing how far away from a point it is still holomorphic.

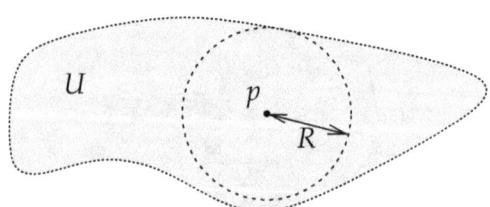

Figure 3.8: Largest disc around p that fits in U is where the series at p for a holomorphic $f: U \to \mathbb{C}$ converges.

Let us state the main conclusion of this subsection once more.

Corollary 3.3.2. *Let $U \subset \mathbb{C}$ be an open set. A function $f: U \to \mathbb{C}$ is holomorphic if and only if f is analytic.*

As a corollary of this corollary, we find that all the results that we proved for analytic functions are true for holomorphic functions. And it goes the other way too. For example, it is easy to show that the composition of holomorphic functions is holomorphic (the chain rule), and so it is true for analytic functions. It is much harder to prove directly that composition of power series is again a power series. Similarly for a product of power series. And we have just proved what we postponed in a remark: A convergent power series defines an analytic function. We only proved before that it defines a holomorphic function.

Exercise 3.3.1: *Consider $f(z) = \frac{\sin(z)}{z}$ defined on $\mathbb{C} \setminus \{0\}$. The theorem gives you that the power series at $z = 1$ converges in a disc of radius 1. Prove that the radius of convergence is actually infinity. Hint: Write $\sin(z)$ as a power series at the origin first.*

Exercise 3.3.2: *Find the radius of convergence of the series at zero of the holomorphic function $f(z) = e^{z^7} \sin\left(\cos\left(\frac{e^z}{z^2-25}\right)\right) e^{\frac{z}{z+4}}$. Hint: Showing that it is at least something is the easier part, showing it can be no larger than what you think it is, that is the harder part.*

Exercise 3.3.3: *Show that for the so-called real-analytic functions, the radius of convergence cannot be read-off from the domain: Prove that the function $f(x) = \frac{1}{1+x^2}$, which is defined on the entire real line, can be expressed (locally) as a real power series $\sum_{n=0}^{\infty} c_n (x - a)^n$ for every $a \in \mathbb{R}$, but this power series always has a finite radius of convergence. Compute this radius of convergence for every a. Hint: Consider the holomorphic function $\frac{1}{1+z^2}$.*

Exercise 3.3.4: *Suppose that $f \colon \mathbb{D} \to \mathbb{C}$ is holomorphic and suppose that f is expanded in a power series around some $p \in \mathbb{D}$.*
   a) Write the best lower estimate of the radius of convergence in terms of $|p|$.
   b) Given p, find a function f whose radius of convergence is precisely given by the formula you found above.

Exercise 3.3.5: *Suppose $f \colon \mathbb{D} \to \mathbb{C}$ is holomorphic such that $f(z) = f(iz)$ for all $z \in \mathbb{D}$. Show that there exists a holomorphic $g \colon \mathbb{D} \to \mathbb{C}$ such that $f(z) = g(z^4)$.*

Exercise 3.3.6: *Suppose $U \subset \mathbb{C}$ is a domain such that $\overline{\mathbb{D}} \subset U$, the function $f \colon U \to \mathbb{C}$ is holomorphic, and*

$$\int_{\partial \mathbb{D}} f(z)\bar{z}^n \, dz = 0$$

for all $n \in \mathbb{N}$. Prove that f is identically zero.

Exercise 3.3.7: *Suppose that $g \colon \mathbb{D} \to \mathbb{C}$ is such that*

$$\lim_{h \to 0} \frac{g(z + h) - g(z)}{\bar{h}}$$

exists for all $z \in \mathbb{D}$ (note that conjugate on the h). Prove that there exists a sequence $\{c_n\}$ such that for all $z \in \mathbb{D}$,

$$g(z) = \sum_{n=0}^{\infty} c_n \bar{z}^n.$$

3.3.2*i*   Derivative is holomorphic and Morera's theorem

Let us restate Corollary 2.4.4 in terms of holomorphic functions, now that we know that holomorphic functions are analytic.

Theorem 3.3.3. *Let $U \subset \mathbb{C}$ be open and $f : U \to \mathbb{C}$ holomorphic. Then f is infinitely complex differentiable. In particular, f' is holomorphic.*

The "in particular" is an important consequence. It is also a somewhat surprising consequence. It says that if f is differentiable in some way, then so is the derivative. Nothing like that is true for the real derivative: Any continuous $g : (a, b) \to \mathbb{R}$ is the derivative of a real differentiable function (e.g., $f(x) = \int_c^x g(t)\,dt$ for $c \in (a, b)$), and continuous functions need not be differentiable anywhere. Even worse, the real derivative could even be discontinuous.

> *Exercise* **3.3.8:** *Let $f : \mathbb{R} \to \mathbb{R}$ be defined by $f(0) = 0$ and $f(x) = x^2 \sin(1/x)$ for $x \neq 0$.*
>   *a) Show that f is differentiable everywhere (including at 0), but f' is not continuous.*
>   *b) Modify f so that it is still differentiable everywhere, but f' is not even bounded.*

The derivatives of a holomorphic function can be computed by integration[*] via the Cauchy integral formula as well. Yeah that does sound weird. It is definitely not something that you should expect for any old functions.

Theorem 3.3.4 (Cauchy integral formula for derivatives). *Suppose $U \subset \mathbb{C}$ is open, $f : U \to \mathbb{C}$ is holomorphic, $\overline{\Delta_r(p)} \subset U$. Then for all $z \in \Delta_r(p)$ and all $k \in \mathbb{N}$,*

$$f^{(k)}(z) = \frac{k!}{2\pi i} \int_{\partial \Delta_r(p)} \frac{f(\zeta)}{(\zeta - z)^{k+1}}\,d\zeta.$$

Proof. We know that f is infinitely complex differentiable, and we can compute the derivatives using the Wirtinger operator. For induction, suppose the theorem holds for some k (the standard formula says it is true for $k = 0$). Fix some $z \in \Delta_r(p)$.

$$\begin{aligned}
f^{(k+1)}(z) = \frac{\partial}{\partial z}\left[f^{(k)}(z)\right] &= \frac{\partial}{\partial z}\left[\frac{k!}{2\pi i} \int_{\partial \Delta_r(p)} \frac{f(\zeta)}{(\zeta - z)^{k+1}}\,d\zeta\right] \\
&= \frac{k!}{2\pi i} \int_{\partial \Delta_r(p)} f(\zeta)\frac{\partial}{\partial z}\left[\frac{1}{(\zeta - z)^{k+1}}\right]\,d\zeta \\
&= \frac{(k+1)!}{2\pi i} \int_{\partial \Delta_r(p)} \frac{f(\zeta)}{(\zeta - z)^{k+2}}\,d\zeta.
\end{aligned}$$

Here, we are really passing the partial derivatives in x and y (where $z = x + iy$) underneath the integral, which can be done by the Leibniz integral rule, Theorem B.2.3,

---

[*]Complex analysis allows you to integrate to find the derivative and to differentiate to find an integral. Now tell that to your calculus students.

for instance. Actually it requires the simple generalization in Exercise B.2.7. We have used that the x and y partial derivatives of $\frac{f(\zeta)}{(\zeta-z)^{k+1}}$ are continuous functions of $(z, \zeta) \in \Delta_r(p) \times \partial\Delta_r(p)$. $\qquad\square$

We could have also used the difference quotient instead of the Wirtinger operator. That approach requires slightly more care—you have to show uniform convergence of a certain limit of functions—but we would not have needed the result that holomorphic functions are analytic. In fact, it could give an independent proof that holomorphic functions are infinitely complex differentiable. We leave it as an exercise.

Exercise 3.3.9: Compute

$$\int_{\partial\mathbb{D}} \frac{z^2 e^z}{(2z-1)^3}\, dz.$$

Exercise 3.3.10: Give a different proof of the Cauchy formula for derivatives by using the difference quotient. First show the formula for f', then again using the difference quotient and the fact that the kernel (the function inside) is holomorphic in z, show the formula for f'', etc. For this procedure to work it is not necessary to assume that f' is holomorphic, it will follow from your work.

Exercise 3.3.11: Suppose $f(z, t)$ is a continuous function of $(z, t) \in U \times (a, b)$, where $U \subset \mathbb{C}$ is open, and for every fixed $t \in (a, b)$, the function $z \mapsto f(z, t)$ is holomorphic. Prove that $\frac{\partial f}{\partial z}$ is a continuous function of $U \times (a, b)$. As a consequence, show that $\frac{\partial f}{\partial x}$ and $\frac{\partial f}{\partial y}$ are continuous (where $z = x + iy$).

Exercise 3.3.12: The previous exercise may seem trivial, but the key is that we can prove (using Cauchy's formula) that the partials are continuous as a function of $U \times (a, b)$ by using continuity of f on $U \times (a, b)$. No such result is true for nonholomorphic functions. Prove that the function defined by $f(x, t) = t \sin(x/t)$ for $t \neq 0$ and $f(x, 0) = 0$ is continuous as a function of $\mathbb{R}^2$, and for each fixed t, the function $x \mapsto f(x, t)$ is differentiable (infinitely differentiable in fact), but $\frac{\partial f}{\partial x}$ is not continuous as a function of both x and t.

The fact that f' is holomorphic, surprisingly, gives us a certain converse to Cauchy. Morera's theorem is quite a useful tool for showing holomorphicity as it is often easier to integrate a continuous function than to compute a derivative.

Theorem 3.3.5 (Morera). *Let $U \subset \mathbb{C}$ be open and $f: U \to \mathbb{C}$ continuous. Suppose that*

$$\int_{\partial T} f(z)\, dz = 0$$

for every triangle such that $T \subset U$. Then f is holomorphic.

Proof. As holomorphicity is a local property, we can assume that U is a disc. Proposition 3.2.11 then says that f has a primitive F in the disc U, and $f = F'$ is thus holomorphic as complex derivatives are holomorphic. $\qquad\square$

We remark that in the proof, the reduction to a disc (or some other simpler set) is important. It is not true that every function satisfying the hypotheses of Morera's theorem has a primitive in U for a general U. For example, $1/z$ is holomorphic in $U = \mathbb{C} \setminus \{0\}$ and satisfies the hypotheses of Morera's theorem; however, it does not have a primitive in $\mathbb{C} \setminus \{0\}$. We will see much more of its (nonexistent) primitive, the logarithm, shortly.

Exercise 3.3.13: *Show that if $f : \mathbb{C} \to \mathbb{C}$ is continuous and holomorphic on $\mathbb{C} \setminus \mathbb{R}$, then f is holomorphic everywhere.*

Exercise 3.3.14: *Let $U \subset \mathbb{C}$ be open and $f : U \to \mathbb{C}$ continuous. Write $d\bar{z} = dx - i\,dy$. Suppose that for every triangle such that $T \subset U$, we have*

$$\int_{\partial T} f(z)\,d\bar{z} = 0.$$

Prove that f is antiholomorphic, that is, the conjugate of f is holomorphic.

Exercise 3.3.15: *Let $U \subset \mathbb{C}$ be a domain and $f : U \to \mathbb{C}$ continuous. Suppose that*

$$\int_{\partial T} \operatorname{Re} f(z)\,dz = 0 \qquad and \qquad \int_{\partial T} \operatorname{Im} f(z)\,dz = 0$$

for every triangle such that $T \subset U$. Prove that f is constant.

Exercise 3.3.16: *Prove Morera for rectangles. That is, suppose that $U \subset \mathbb{C}$ is open, $f : U \to \mathbb{C}$ is continuous, and*

$$\int_{\partial R} f(z)\,dz = 0$$

for every rectangle $R \subset U$ of the form $a < \operatorname{Re} z < b$, $c < \operatorname{Im} z < d$. Prove that f is holomorphic. Hint: You may need to prove an analogue of Proposition 3.2.11 for rectangles, which is trickier with rectangles.

3.3.3i   The maximum modulus principle

A simple and yet surprisingly useful consequence of Cauchy's formula is the so-called *maximum modulus principle* (sometimes just *maximum principle*), which has several different versions. We prove one statement and leave other versions as exercises. The main idea is that the modulus of a holomorphic function never achieves a maximum. In other words, $|f(z)|$ is bounded by the supremum of its values near the boundary of the domain or near ∞. The basic idea of the proof is that Cauchy's integral formula tells us that $f(z)$ is an average of the values of f in a circle around z, and the average can't be bigger than the numbers we're averaging.

Theorem 3.3.6 (Maximum modulus principle). *Suppose $U \subset \mathbb{C}$ is a domain and $f: U \to \mathbb{C}$ is holomorphic. If $|f(z)|$ achieves a local maximum on U, then f is constant.*

Proof. Suppose $|f(z)|$ achieves a local maximum at $p \in U$. Without loss of generality, suppose $p = 0$. Also assume that $f(0)$ is nonnegative and real—otherwise multiply by some $e^{i\theta}$. We write* that as $f(0) \geq 0$.

Take a closed disc $\overline{\Delta_r(0)} \subset U$. As 0 is a local maximum, suppose that r is small enough so that $|f(z)| \leq |f(0)| = f(0)$ whenever $|z| \leq r$. Cauchy's formula says

$$f(0) = |f(0)| = \left| \frac{1}{2\pi i} \int_{\partial \Delta_r(0)} \frac{f(z)}{z} \, dz \right| = \left| \frac{1}{2\pi i} \int_0^{2\pi} \frac{f(re^{it})}{re^{it}} rie^{it} \, dt \right|$$

$$\leq \frac{1}{2\pi} \int_0^{2\pi} |f(re^{it})| \, dt \leq \frac{1}{2\pi} \int_0^{2\pi} f(0) \, dt = f(0).$$

Hence, all the inequalities above are equalities. Because $f(0) - |f(re^{it})| \geq 0$,

$$\int_0^{2\pi} \left(f(0) - |f(re^{it})| \right) dt = 0$$

means $|f(re^{it})| = f(0)$ for all t. By applying Cauchy's formula again, we find

$$\frac{1}{2\pi} \int_0^{2\pi} |f(re^{it})| \, dt = f(0) = \frac{1}{2\pi} \int_0^{2\pi} f(re^{it}) \, dt$$

or

$$0 = \operatorname{Re} \int_0^{2\pi} \left(|f(re^{it})| - f(re^{it}) \right) dt = \int_0^{2\pi} \left(|f(re^{it})| - \operatorname{Re} f(re^{it}) \right) dt.$$

The inequality $|w| - \operatorname{Re} w \geq 0$ holds for all $w \in \mathbb{C}$, so $|f(re^{it})| - \operatorname{Re} f(re^{it}) \geq 0$ for all t and hence, as the integral is zero, $|f(re^{it})| = \operatorname{Re} f(re^{it})$ for all t. Thus $\operatorname{Im} f(re^{it}) = 0$ and $f(re^{it}) = |f(re^{it})| = f(0)$ for all t. This reasoning works for every small enough r, and consequently the set where $f(z) = f(0)$ contains a small disc. As holomorphic functions are analytic, the identity theorem (Theorem 2.4.7) implies that f is constant in U. $\qquad\square$

We will find much use of the following version: If f is holomorphic on a bounded U and continuous on $\overline{U}$, then $|f|$ achieves its maximum on the boundary ∂U.

Corollary 3.3.7 (Maximum modulus principle, part deux). *Suppose $U \subset \mathbb{C}$ is nonempty, open, and bounded (so $\overline{U}$ is compact). If $f: \overline{U} \to \mathbb{C}$ is continuous and the restriction $f|_U$ is holomorphic, then $|f(z)|$ achieves a maximum on ∂U. In other words,*

$$\sup_{z \in U} |f(z)| \leq \sup_{z \in \partial U} |f(z)|.$$

---

*Perhaps you're thinking to yourself: Of course we write that ξ is nonnegative by writing $\xi \geq 0$. But we mean that "$\xi \geq 0$" is a shortcut to "$\xi \in \mathbb{R}$ and $\xi \geq 0$."

Exercise 3.3.17: Prove Corollary 3.3.7.

Exercise 3.3.18 (Minimum modulus principle): *Suppose $U \subset \mathbb{C}$ is a domain and $f : U \to \mathbb{C}$ is holomorphic.*
   a) *Prove that if $|f(z)|$ has a local minimum at $p \in U$ and $f(p) \neq 0$, then f is constant.*
   b) *Show by example that the hypothesis $f(p) \neq 0$ cannot be removed.*

Exercise 3.3.19 (Maximum modulus principle, part trois): *Suppose $U \subset \mathbb{C}$ is a domain, $f : U \to \mathbb{C}$ is holomorphic, and $M > 0$ is such that $\limsup_{z \to p} |f(z)| \leq M$ for all $p \in \partial U$. If U is unbounded, then also $\limsup_{z \to \infty} |f(z)| \leq M$. Prove that $|f(z)| \leq M$ for all $z \in U$. Note: For $g : U \to \mathbb{R}$, by definition, $\limsup_{z \to p} g(z) = \inf_{r > 0} \sup\{g(z) : z \in U \cap \Delta_r(p)\}$ and $\limsup_{z \to \infty} g(z) = \inf_{R > 0} \sup\{g(z) : z \in U \text{ and } |z| > R\}$.*

Exercise 3.3.20: Suppose $U \subset \mathbb{C}$ is a bounded domain, $f : \overline{U} \to \mathbb{C}$ is continuous and the restriction $f|_U$ is holomorphic, and there is a constant M such that $|f(z)| = M$ for all $z \in \partial U$. Prove that f is either constant or $f(z) = 0$ for some $z \in U$.

Exercise 3.3.21: Suppose $U \subset \mathbb{C}$ is open and $f : U \to \mathbb{C}$ is holomorphic. Let $M > 0$ be fixed and define $X = \{z \in U : |f(z)| < M\}$. Prove that X is open and the closure of X (in U, so $\overline{X} \cap U$) is the set $\{z \in U : |f(z)| \leq M\}$.

Exercise 3.3.22: Let $P(z)$ be a nonconstant polynomial. Show that for every $c > 0$, each component of the set $\{z \in \mathbb{C} : |P(z)| < c\}$ contains at least one zero (root) of P. Hint: Do the two previous exercises first.

Exercise 3.3.23: Let $f : \Delta_R(p) \to \mathbb{C}$ be holomorphic and nonconstant. Prove that $M(r) = \sup\{|f(z)| : z \in \partial \Delta_r(p)\}$ is a strictly increasing function of $r \in [0, R)$.

3.3.4i Cauchy estimates, Liouville, and the fundamental theorem of algebra

It may seem we are cramming quite a bit into one subsection, but we have the tools to make three fundamental results just pop out with little work. The triangle inequality on the integral formula for the coefficients of the power series obtains an estimate on their size. These estimates immediately give Liouville's theorem on entire holomorphic functions, which at once gives the fundamental theorem of algebra. Some analysts like to make fun of algebraists at this stage, saying that the standard proof of their fundamental theorem uses analysis. One can go even further. It is not a theorem of algebra at all! It is a theorem in complex analysis.*

---

*There! It's ours now and you can't have it back.

For a set K, denote the *supremum norm* or *uniform norm*:

$$\|f\|_K \overset{\text{def}}{=} \sup_{z \in K} |f(z)|.$$

Theorem 3.3.8 (Cauchy estimates). *Let $U \subset \mathbb{C}$ be open, $f: U \to \mathbb{C}$ be holomorphic, and $\overline{\Delta_r(p)} \subset U$ be a closed disc. Expand $f(z) = \sum c_n (z - p)^n$. Then for all n,*

$$|c_n| = \left| \frac{f^{(n)}(p)}{n!} \right| \leq \frac{\|f\|_{\partial \Delta_r(p)}}{r^n}.$$

In other words, the sequence $\{|c_n| r^n\}$ is bounded by $\|f\|_{\partial \Delta_r(p)}$. Compare to Proposition 2.3.3.

Proof. The proof is a brute force estimation:

$$|c_n| = \left| \frac{1}{2\pi i} \int_{\partial \Delta_r(p)} \frac{f(\zeta)}{(\zeta - p)^{n+1}} \, d\zeta \right| \leq \frac{1}{2\pi} \int_{\partial \Delta_r(p)} \frac{\|f\|_{\partial \Delta_r(p)}}{r^{n+1}} |d\zeta| = \frac{\|f\|_{\partial \Delta_r(p)}}{r^n}. \qquad \square$$

A better estimate is not possible given only the information $M = \|f\|_{\partial \Delta_r(p)}$. Cauchy estimates say that $|c_n| \leq \frac{M}{r^n}$. But if $f(z) = \frac{M}{r^n}(z - p)^n$, then $\|f\|_{\partial \Delta_r(p)} = M$ and $|c_n| = \frac{M}{r^n}$. It is an exercise that up to multiplication by $e^{i\theta}$, this example is the only one.

Exercise 3.3.24 (Easy)*: Suppose $f: \mathbb{D} \to \mathbb{C}$ is holomorphic and for each $M > 0$, there exists an $n \in \mathbb{N}$ such that*

$$\left| \frac{f^{(n)}(0)}{n!} \right| \geq M.$$

Prove that f is unbounded.

Exercise 3.3.25 (Easy)*: Suppose $f: \mathbb{D} \to \mathbb{D}$ is holomorphic.*
  a) Prove that $|f^{(n)}(0)| \leq n!$ for all n.
  b) For every n, find an example $f: \mathbb{D} \to \mathbb{D}$ such that $|f^{(n)}(0)| = n!$.

Exercise 3.3.26: Let $\mathbb{H} = \{z \in \mathbb{C} : \operatorname{Im} z > 0\}$ be the upper half-plane and $f: \mathbb{H} \to \mathbb{D}$ holomorphic. Prove

$$\lim_{\substack{t \to \infty \\ t \in \mathbb{R}, t > 0}} f'(it) = 0.$$

Exercise 3.3.27: Suppose $U \subset \mathbb{C}$ is a domain, $\overline{\Delta_r(0)} \subset U$, and $f: U \to \mathbb{C}$ is holomorphic such that $\|f\|_{\partial \Delta_r(0)} = M$. Cauchy estimates say that for every n, $|f^{(n)}(0)| \leq \frac{n!M}{r^n}$. Prove that if for some n, $|f^{(n)}(0)| = \frac{n!M}{r^n}$, then $f(z) = cz^n$ for some $c \in \mathbb{C}$. Hint: The inequalities are equalities in the proof (there are really two inequalities in the proof).

Definition 3.3.9. A holomorphic function $f : \mathbb{C} \to \mathbb{C}$ is called an *entire holomorphic function* or perhaps just *entire* for short.

Polynomials are one type of entire functions and we saw that nonconstant polynomials are unbounded. While in general the behavior of entire functions such as $\exp z$ as we approach infinity is wilder than that of the polynomials, they are unbounded.

Theorem 3.3.10 (Liouville*)**.** *If f is entire and bounded, then f is constant.*

Proof. Let f be entire and suppose $|f(z)| \leq M$ for all $z \in \mathbb{C}$. Expand f at the origin:

$$f(z) = \sum_{n=0}^{\infty} c_n z^n,$$

which converges for all z. As f is holomorphic on a disc of arbitrary radius, the Cauchy estimates say

$$|c_n| \leq \frac{\|f\|_{\partial \Delta_r(p)}}{r^n} \leq \frac{M}{r^n} \qquad \text{for all } r > 0.$$

Letting $r \to \infty$ shows that $c_n = 0$ for $n \geq 1$. In other words, $f(z) = c_0$ for all z.   □

Exercise 3.3.28: *Suppose f is entire and $|f(z)| \leq e^{\operatorname{Re} z}$ for all $z \in \mathbb{C}$. Show that $f(z) = ce^z$ for some constant $c \in \mathbb{C}$.*

Exercise 3.3.29: *Suppose f is entire, $n \in \mathbb{N}$, $M > 0$, and $|f(z)| \leq M(1 + |z|)^n$ for all $z \in \mathbb{C}$. Show that f is a polynomial of degree at most n.*

Exercise 3.3.30: *Suppose f is entire and $\operatorname{Im} f(z) > 0$ for all $z \in \mathbb{C}$. Prove f is constant.*

Exercise 3.3.31: *Suppose $f : \mathbb{C} \to \mathbb{C}$ is holomorphic and misses a segment, that is, there exists a segment $[a, b]$ such that $f(\mathbb{C}) \subset \mathbb{C} \setminus [a, b]$. Show that f is constant. Hint: See the map from Exercise 2.2.17.*

Exercise 3.3.32: *While there doesn't exist a nonconstant holomorphic function $f : \mathbb{C} \to \mathbb{D}$, there do exist surjective holomorphic functions $f : \mathbb{D} \to \mathbb{C}$. Find one.*

Theorem 3.3.11 (Fundamental theorem of algebra)**.** *If $P(z)$ is a nonconstant polynomial, then P has a root.*

Proof. If $P(z)$ does not have a root, then $R(z) = \frac{1}{P(z)}$ is an entire holomorphic function. Suppose $P(z)$ is nonconstant. In Exercise 1.3.9 you proved that $\lim_{z \to \infty} P(z) = \infty$ and so $\lim_{z \to \infty} R(z) = 0$. In other words, $R(z)$ is bounded. Liouville says that $R(z)$ and therefore $P(z)$ must be constant, a contradiction.   □

---

*Liouville proved a different (though similar) theorem. This particular one was proved by Cauchy (what a showoff). But calling it Cauchy's theorem would be unhelpful.

Exercise 3.3.33: *Prove that a polynomial $P(z)$ of degree d can be written as*

$$P(z) = a \prod_{n=1}^{d}(z - z_n) = a(z - z_1) \cdots (z - z_d),$$

for some $a \in \mathbb{C}$ and $z_1, \ldots, z_d \in \mathbb{C}$. Hint: Prove that $P(z_0) = 0$ implies $P(z) = Q(z)(z-z_0)$ for some polynomial Q of degree $d - 1$.

Exercise 3.3.34 (Easy)*:* *Prove the one-dimensional version of the* Jacobian conjecture: *Suppose that $P(z)$ is a polynomial and $P'(z)$ is nonzero for all z, then P is an automorphism of $\mathbb{C}$, that is, $P(z) = az + b$ and $a \neq 0$.*

Exercise 3.3.35 (Easy)*:* *Let $P \colon \mathbb{C} \to \mathbb{C}$ be a nonconstant polynomial. Show that P is onto.*

Exercise 3.3.36: *Suppose f is entire and is never zero. For $M > 0$, let $X_M = \{z \in \mathbb{C} : |f(z)| = M\}$ (note that X_M is closed).*
   a) Show that X_M is nonempty for all $M > 0$.
   b) Show that for every M, the set X_M has no bounded topological components.
Hint: See the exercises for the maximum modulus principle.

3.4i \ The Cauchy transform and convergence

3.4.1i \ Holomorphic functions via integrals

It is common to define functions using integrals, for instance, the Cauchy integral itself (usually called the Cauchy transform).

Lemma 3.4.1. *Suppose $U \subset \mathbb{C}$ is open, and $\psi \colon U \times [0, 1] \to \mathbb{C}$ is a continuous function such that for each fixed $t \in [0, 1]$, the function $z \mapsto \psi(z, t)$ is holomorphic. Then*

$$h(z) = \int_0^1 \psi(z, t) \, dt$$

is a holomorphic function on U.

This kind of lemma has two common proofs, and as they are both useful in other places, let us do both of them.

Proof A. One proof is to use Morera's theorem (Theorem 3.3.5) together with Fubini Theorem B.2.2), and Cauchy–Goursat (Theorem 3.2.9). Let $T \subset U$ be a triangle. Then

$$\int_{\partial T} h(z) \, dz = \int_{\partial T} \int_0^1 \psi(z, t) \, dt \, dz = \int_0^1 \int_{\partial T} \psi(z, t) \, dz \, dt = \int_0^1 0 \, dt = 0.$$

We used Fubini's theorem to swap the integrals: The integral over ∂T is really a sum of integrals over an interval and the integrand is continuous (as a function of both variables), so Fubini applies. Morera's theorem now says that $h(z)$ is holomorphic. □

Proof B. The second proof* is to apply Wirtinger derivatives and differentiate under the integral:

$$\frac{\partial}{\partial \bar{z}}\big[h(z)\big] = \frac{\partial}{\partial \bar{z}} \int_0^1 \psi(z,t)\,dt = \int_0^1 \frac{\partial}{\partial \bar{z}}\big[\psi(z,t)\big]\,dt = \int_0^1 0\,dt = 0.$$

As once before, we are really passing the partial derivatives in x and y under the integral via the Leibniz integral rule, Theorem B.2.3 (or again really Exercise B.2.7). Leibniz rule applies because Exercise 3.3.11 says that the partial derivatives are continuous as functions of both variables. Leibniz also implies that h is continuously (real) differentiable, and thus the Cauchy–Riemann equations (Proposition 2.2.6) now say that $h(z)$ is holomorphic. □

In either case, the idea is to swap some limits (something that must always be justified), and the two techniques above are two kinds of swaps that come up often (Fubini, and differentiating under the integral). By writing each path in a chain as an integral of one real variable, we obtain the following corollary.

Corollary 3.4.2. *Suppose $U \subset \mathbb{C}$ is open, Γ is a chain, and $\psi \colon U \times \Gamma \to \mathbb{C}$ is a continuous function such that for each fixed $w \in \Gamma$, the function $z \mapsto \psi(z,w)$ is holomorphic. Then*

$$h(z) = \int_\Gamma \psi(z,w)\,dw$$

is a holomorphic function on U.

For a continuous $f \colon \partial \Delta_r(p) \to \mathbb{C}$, define the *Cauchy transform* $Cf \colon \Delta_r(p) \to \mathbb{C}$ by

$$Cf(z) \overset{\text{def}}{=} \frac{1}{2\pi i} \int_{\partial \Delta_r(p)} \frac{f(\zeta)}{\zeta - z}\,d\zeta.$$

Corollary 3.4.3. *For a continuous $f \colon \partial \Delta_r(p) \to \mathbb{C}$, the Cauchy transform $Cf \colon \Delta_r(p) \to \mathbb{C}$ is holomorphic.*

The corollary gives a converse to Cauchy's formula. If $f \colon \overline{\Delta_r(p)} \to \mathbb{C}$ is continuous and

$$f(z) = \frac{1}{2\pi i} \int_{\partial \Delta_r(p)} \frac{f(\zeta)}{\zeta - z}\,d\zeta \qquad \text{for all } z \in \Delta_r(p),$$

then $f|_{\Delta_r(p)}$ is holomorphic.

It is not necessarily true that Cf tends to f as we approach the boundary of the disc unless f came from a holomorphic function to begin with. That is, Cf might (or might not) have limits at the boundary of the disc, and they need not be equal to f.

---

*For no good rational reason, this proof is the one I have seen more often, possibly because complex analysts are often PDE people and they rather differentiate than integrate.

Exercise 3.4.1: *Explicitly compute the Cauchy transform Cf on $\mathbb{D}$ for $f \colon \partial\mathbb{D} \to \mathbb{C}$ given by $f(z) = \bar{z}$. Then note that Cf does not tend to f as z goes to $\partial\mathbb{D}$. Hint: $\bar{\zeta} = \frac{1}{\zeta}$ for $\zeta \in \partial\mathbb{D}$ and $\frac{1/\zeta}{\zeta - z} = \frac{1}{z(\zeta - z)} - \frac{1}{z\zeta}$.*

Exercise 3.4.2: *Suppose $g \colon \partial\Delta_r(p) \to \mathbb{C}$ and there exists a continuous $f \colon \overline{\Delta_r(p)} \to \mathbb{C}$ that is holomorphic in $\Delta_r(p)$, where $g = f|_{\partial\Delta_r(p)}$. Prove that Cg extends to a continuous function on $\overline{\Delta_r(p)}$ such that $Cg(z) = g(z)$ for $z \in \partial\Delta_r(p)$. In other words, if g is the boundary value of a holomorphic function, then Cg does indeed tend to g as z tends to the boundary $\partial\Delta_r(p)$. Hint: See Exercise 3.2.27.*

Exercise 3.4.3 (Easy): *Suppose $g \colon U \times [a, b] \to \mathbb{C}$ is continuous, for each fixed $t \in [a, b]$, $z \mapsto g(z, t)$ is holomorphic, and $|g(z, t)| \leq M$ for all $(z, t) \in U \times [a, b]$. Prove that*

$$f(z) = \int_a^b g(z, t)\, dt$$

is a holomorphic function on U such that $|f(z)| \leq M(b - a)$ for all $z \in U$.

Exercise 3.4.4: *Suppose $g \colon [-1, 1] \to \mathbb{C}$ is continuous and define*

$$f(z) = \int_{-1}^{1} \frac{g(t)}{t - z}\, dt.$$

Show that f is holomorphic in $\mathbb{C} \setminus [-1, 1]$ and $\lim_{z \to \infty} f(z) = 0$.

3.4.2*i*   Convergence of sequences of holomorphic functions

When dealing with a class of functions, any analyst worth their salt* will ask about the right topology for this class of functions. Another consequence of Cauchy's formula is that the right topology for holomorphic functions is the same as that for continuous functions: uniform convergence on compact subsets. In fact, that's the convergence that we used for power series.

Definition 3.4.4. A sequence of functions $f_n \colon U \to \mathbb{C}$ converges *uniformly on compact subsets* to $f \colon U \to \mathbb{C}$ if $f_n|_K$ converges uniformly to $f|_K$ for every compact $K \subset U$.

What do we mean by "the right topology" for a class of functions? Well, we mean the most natural topology that preserves the class (limits in that topology are still in that class). Results from introductory analysis (see Theorem B.1.7 and Corollary B.1.10) say that a uniform limit of a continuous functions is continuous. By concentrating on a compact neighborhood such as $\overline{\Delta_r(p)}$, we can see that uniform

---

*Apparently the salt thing comes from Roman times, soldiers were paid partly with salt to preserve their meats. So if you didn't ask, you will have to eat only vegetables.

convergence on compact subsets is enough. In other words, uniform convergence on compact subsets is the right topology for continuous functions.

It is rather surprising that this topology is the right one for holomorphic functions. For real differentiable functions it is not the right one: $|x|^{1+1/n}$ is C^1 on $\mathbb{R}$ and converges uniformly on compact subsets to $|x|$, which is not differentiable.

Theorem 3.4.5. *Suppose $U \subset \mathbb{C}$ is open and $f_n : U \to \mathbb{C}$ is a sequence of holomorphic functions converging uniformly on compact subsets to $f : U \to \mathbb{C}$. Then f is holomorphic. Moreover, for every ℓ, the ℓth derivative $f_n^{(\ell)}$ converges uniformly on compact subsets to $f^{(\ell)}$.*

Proof. Let $p \in U$ be fixed. Take a closed disc $\overline{\Delta_r(p)} \subset U$. For any $z \in \Delta_r(p)$,

$$f_n(z) = \frac{1}{2\pi i} \int_{\partial \Delta_r(p)} \frac{f_n(\zeta)}{\zeta - z} \, d\zeta.$$

The set $\partial \Delta_r(p)$ is compact. Let $\delta > 0$ be the distance of z to $\partial \Delta_r(p)$. For $\zeta \in \partial \Delta_r(p)$,

$$\left| \frac{f_n(\zeta)}{\zeta - z} - \frac{f(\zeta)}{\zeta - z} \right| = \frac{|f_n(\zeta) - f(\zeta)|}{|\zeta - z|} \leq \frac{1}{\delta} |f_n(\zeta) - f(\zeta)|.$$

So as $f_n \to f$ uniformly on $\partial \Delta_r(p)$, $\zeta \mapsto \frac{f_n(\zeta)}{\zeta - z}$ converges to $\zeta \mapsto \frac{f(\zeta)}{\zeta - z}$ uniformly on $\partial \Delta_r(p)$. We can, therefore, take the limit as $n \to \infty$ underneath the integral to obtain

$$f(z) = \frac{1}{2\pi i} \int_{\partial \Delta_r(p)} \frac{f(\zeta)}{\zeta - z} \, d\zeta.$$

This formula holds for all $z \in \Delta_r(p)$. The function $f|_{\partial \Delta_r(p)}$ is continuous by uniform convergence, and f on $\Delta_r(p)$ is equal to the Cauchy transform $C[f|_{\partial \Delta_r(p)}]$, which is holomorphic. In other words, f is holomorphic.

Let's attack the "Moreover." Suppose $K \subset U$ is compact. If $U \neq \mathbb{C}$, then the distance of K and ∂U is positive, say $d > 0$. If $U = \mathbb{C}$, take an arbitrary $d > 0$. Consider

$$K' = \bigcup_{z \in K} \overline{\Delta_{d/2}(z)}.$$

Clearly $K \subset K' \subset U$. See Figure 3.9.

The set K' is also compact: It is clearly bounded, let us show it is closed. Suppose that p is not in K'. By compactness of K, there is a $q \in K$ such that $|p - q|$ is the distance of p to K. As $p \notin K'$, $|p - q| > d/2$. Every point in $\Delta_{|p-q|-d/2}(p)$ is also further than $d/2$ from K, so the complement of K' is open.

As $\{f_n\}$ converges uniformly on compact subsets, it converges uniformly on K'. Given an $\epsilon > 0$, find an N such that $|f_n(z) - f(z)| < \epsilon$ for all $z \in K'$ and all $n \geq N$. For any $p \in K$, use the Cauchy estimates in a $d/2$ disc on $f_n - f$, and note that $\partial \Delta_{d/2}(p) \subset K'$:

$$\left| f_n^{(\ell)}(p) - f^{(\ell)}(p) \right| \leq \frac{\ell! \, \|f_n - f\|_{\partial \Delta_{d/2}(p)}}{(d/2)^\ell} \leq \frac{\ell! 2^\ell}{d^\ell} \epsilon.$$

Thus the sequence $\{f_n^{(\ell)}\}$ converges uniformly to $f^{(\ell)}$ on K. $\qquad\square$

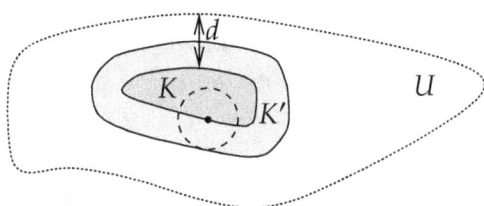

Figure 3.9: Enlarging the set K by half the distance to the boundary. One of the closed discs $\overline{\Delta_{d/2}(p)}$ is marked in dashed line.

The fact that we can write all the derivatives as integrals of the function, and hence obtain Cauchy estimates, allows us to use a far weaker topology than one would think. Integration is a far nicer operation than differentiation, and for holomorphic functions, we can differentiate by integrating.

Exercise 3.4.5: Let $U \subset \mathbb{C}$ be open, $K \subset U$ be compact, $r > 0$, and $K' = \bigcup_{z \in K} \overline{\Delta_r(z)}$ is such that $K' \subset U$. If $f : U \to \mathbb{C}$ is holomorphic, prove that for every nonnegative integer ℓ

$$\| f^{(\ell)} \|_K \leq \frac{\ell!}{r^\ell} \| f \|_{K'}.$$

Exercise 3.4.6: Suppose $U \subset \mathbb{C}$ is a bounded domain, and $f_n : \overline{U} \to \mathbb{C}$ are continuous functions holomorphic on U such that the restrictions $f_n|_{\partial U}$ converge uniformly. Prove that f_n converge uniformly on $\overline{U}$ to a continuous function $f : \overline{U} \to \mathbb{C}$ that is holomorphic in U. Hint: Maximum modulus principle gives that the sequence $\{f_n\}$ is Cauchy at each point (actually uniformly Cauchy). Feel free to use the fact that uniform limit of continuous functions is continuous.

Exercise 3.4.7: Consider $f_n(z) = \frac{\sin(nz)}{n}$. Note that $f_n|_{\mathbb{R}}$ converge uniformly to zero.
   a) Show that for no interval $[a, b] \subset \mathbb{R}$ is there a $\delta > 0$ such that f_n converge uniformly on the rectangle $R = \{ z \in \mathbb{C} : a \leq \operatorname{Re} z \leq b, -\delta \leq \operatorname{Im} z \leq \delta \}$.
   b) In every $[a, b] \subset \mathbb{R}$, find an x such that $\{f'_n(x)\}$ does not converge.

Exercise 3.4.8: Weierstrass approximation theorem says that every continuous function on an interval $[a, b] \subset \mathbb{R}$ is uniformly approximated (on $[a, b]$) by polynomials $P(z)$. Prove why such a theorem cannot hold on a closed curve such as the unit circle $\partial \mathbb{D}$. That is, find a continuous function $f : \partial \mathbb{D} \to \mathbb{C}$ that is not the uniform limit of a sequence of polynomials $P_n(z)$ on $\partial \mathbb{D}$.

Exercise 3.4.9: Prove:
   a) For every holomorphic $f : \mathbb{D} \to \mathbb{C}$ there is a sequence of polynomials $P_n(z)$ that converges uniformly on compact subsets of $\mathbb{D}$ to f.
   b) For every holomorphic $f : \mathbb{H} \to \mathbb{C}$ ($\mathbb{H}$ is the upper half-plane) there is a sequence of polynomials $P_n(z)$ that converges uniformly on compact subsets of $\mathbb{H}$ to f.

Exercise 3.4.10: *Suppose $f : [0, \infty) \to \mathbb{R}$ is a continuous function such that there is some $c > 0$, some $M > 0$, and some $T > 0$ such that $|f(t)| \leq Me^{ct}$ for all $t \geq T$ (f is of exponential order). Let $U = \{z \in \mathbb{C} : \operatorname{Re} z > c\}$. Prove that the Laplace transform exists on U and is holomorphic, that is, prove that*

$$F(z) = \int_0^\infty f(t)e^{-zt}\, dt = \lim_{r \to \infty} \int_0^r f(t)e^{-zt}\, dt$$

converges uniformly on compact subsets to a holomorphic function on U. Note: To use our setup, consider every sequence $\{r_n\}$ of real numbers converging to $+\infty$.

3.5i Schwarz's lemma and automorphisms of the disc

3.5.1i Schwarz's lemma

The following statement may seem technical and specialized, but surprisingly it is incredibly powerful.* Any disc can be translated and rescaled to the unit disc $\mathbb{D}$, and any bounded function can be rescaled to be valued in $\mathbb{D}$.

Lemma 3.5.1 (Schwarz's lemma). *Suppose $f : \mathbb{D} \to \mathbb{D}$ is holomorphic and $f(0) = 0$, then*

(i) $|f(z)| \leq |z|$ *for all $z \in \mathbb{D}$, and*

(ii) $|f'(0)| \leq 1$.

Furthermore, if $|f(z_0)| = |z_0|$ for some $z_0 \in \mathbb{D} \setminus \{0\}$ or $|f'(0)| = 1$, then there is a $\theta \in \mathbb{R}$ such that $f(z) = e^{i\theta}z$ for all $z \in \mathbb{D}$.

Proof. As $f(0) = 0$, the constant term is zero when f is expanded at 0, and hence

$$f(z) = \sum_{n=1}^\infty c_n z^n = z \sum_{n=1}^\infty c_n z^{n-1} = zg(z),$$

where $g(z)$ is a holomorphic function of $\mathbb{D}$. Consider $0 < r < 1$. For $z \in \partial\Delta_r(0)$,

$$|g(z)| = \frac{|f(z)|}{|z|} \leq \frac{1}{r}.$$

The maximum modulus principle implies $|g(z)| \leq \frac{1}{r}$ holds for all $z \in \Delta_r(0)$. Fix $z \in \mathbb{D}$ and take the limit as $r \uparrow 1$. You find $|g(z)| \leq 1$, or $|f(z)| \leq |z|$, for all $z \in \mathbb{D}$. Then

$$|f'(0)| = \left| \lim_{z \to 0} \frac{f(z)}{z} \right| = |g(0)| \leq 1.$$

---

*In graduate school, on an exam in complex analysis, I solved all the problems with a combination of Schwarz's lemma and the Riemann mapping theorem (which we will see later). My advisor felt compelled to remind me that there do exist other theorems in complex analysis.

If $|f(z_0)| = |z_0|$ for some $z_0 \in \mathbb{D} \setminus \{0\}$, then g attains a maximum inside $\mathbb{D}$ and hence is constant. It must be that $f(z) = e^{i\theta}z$. As $g(0) = f'(0)$, the same conclusion, for the same reason, holds if $|f'(0)| = 1$. $\qquad\square$

To illustrate the lemma, consider the statement for $f(z) = z^n$ for an integer $n > 1$. The function f takes the disc to the disc and $f(0) = 0$. For $z \in \mathbb{D} \setminus \{0\}$,

$$|z^n| = |z|^n < |z|.$$

As $f'(z) = nz^{n-1}$, we get $|f'(0)| = 0 < 1$, but notice that a bound on the derivative does not hold at other points: By picking the right z and n, we can make $|f'(z)|$ as large as we want. For a bound at other points, see the Schwarz–Pick lemma in Exercise 3.5.10. We can make $|z^n|$ arbitrarily small for a fixed $z \in \mathbb{D}$ by picking a large enough n, though we cannot make it bigger than $|z|$. What is interesting is that Schwarz's lemma says that all holomorphic functions behave this way, not just z^n.

Exercise 3.5.1: *State and prove a version of Schwarz's lemma for a holomorphic function $f \colon \Delta_r(p) \to \Delta_s(q)$ with $f(p) = q$.*

Exercise 3.5.2: *Let $\mathbb{H} = \{z \in \mathbb{C} : \operatorname{Im} z > 0\}$ be the upper half-plane. Prove that if $f \colon \mathbb{H} \to \mathbb{H}$ is holomorphic such that $f(i) = i$, then*

$$\left| \frac{f(z) - i}{\overline{f(z)} - i} \right| \le \left| \frac{z - i}{\overline{z} - i} \right| \qquad \text{and} \qquad |f'(i)| \le 1.$$

Exercise 3.5.3 (Tricky)*: Prove a certain generalization of Schwarz's lemma called the Cartan's uniqueness theorem: Suppose $U \subset \mathbb{C}$ is a bounded domain, $f \colon U \to U$ is holomorphic, $p \in U$, and $f(p) = p$.*
a) Show $|f'(p)| \le 1$.
b) Show that if $f'(p) = 1$, then $f(z) = z$ for all $z \in U$.
c) Find counterexamples to both statements for some unbounded U.
Hint: Normalize to have $p = 0$. Consider the power series expansions of f^ℓ, the ℓth composition of f with itself, $f(f(f(\cdots f(z)\cdots)))$. For a), consider the linear term of f^ℓ when $|f'(p)| > 1$. For b), consider the first term other than the linear term that is nonzero, and compute it for f^ℓ in terms of the one for f. Then apply Cauchy estimates.

3.5.2i Automorphisms of the disc

Let us compute the automorphism* group of the disc using Schwarz's lemma. We start with certain specific automorphisms. For $a \in \mathbb{D}$, define[†]

$$\varphi_a(z) \stackrel{\text{def}}{=} \frac{z - a}{1 - \overline{a}z}.$$

---

*Recall an automorphism of U is a biholomorphism of U to itself.
[†]Careful when reading literature, some authors use $\frac{a-z}{1-\overline{a}z}$ as the definition.

Proposition 3.5.2. *For every $a \in \mathbb{D}$,*

(i) $\varphi_a(a) = 0$, $\varphi_a(0) = -a$, $\varphi_a'(0) = 1 - |a|^2$, $\varphi_a'(a) = \frac{1}{1-|a|^2}$,

(ii) $\varphi_a(\partial\mathbb{D}) = \partial\mathbb{D}$, *and* $\varphi_a(\mathbb{D}) = \mathbb{D}$,

(iii) φ_a *restricted to* $\mathbb{D}$ *is an automorphism of the disc and*

$$\varphi_a^{-1} = \varphi_{-a}.$$

> *Exercise 3.5.4: Prove the proposition. Hint: (i) is a direct computation, for (ii) remember that φ_a is an LFT and what an LFT does to circles, and (iii) is a direct computation.*

See Figure 3.10 for an example of what φ_a does to the disc. Next, we prove that up to a rotation, all automorphisms of $\mathbb{D}$ are φ_a.

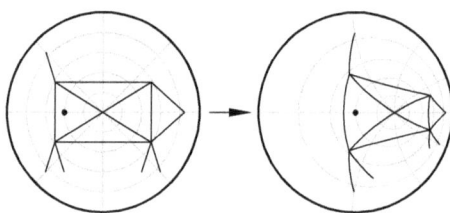

Figure 3.10: What φ_a does to the unit disc when $a = -0.4$. The positions of a and $0 = \varphi_a(a)$ are marked with dots.

Proposition 3.5.3. *If $f \in \mathrm{Aut}(\mathbb{D})$, then there exists an $a \in \mathbb{D}$ and $\theta \in \mathbb{R}$ such that*

$$f(z) = e^{i\theta} \frac{z - a}{1 - \bar{a}z} = e^{i\theta}\varphi_a(z).$$

Proof. Let $b = f(0)$. Consider $g = \varphi_b \circ f$, which is an automorphism, and $g(0) = 0$ as $\varphi_b(b) = 0$. As in the proof of Schwarz's lemma, we find a holomorphic $h(z)$ such that $g(z) = zh(z)$. By Schwarz's lemma, if $z \in \mathbb{D} \setminus \{0\}$, then

$$|h(z)| = \frac{|g(z)|}{|z|} \leq 1.$$

Consequently, h is a map of the disc to the closed disc.

But h can have no zeros: $h(z) = \frac{g(z)}{z}$ cannot be zero for $z \neq 0$ as g is injective and it cannot have a zero at $z = 0$ as $h(0) = \lim_{z \to 0} \frac{g(z)}{z} = g'(0) \neq 0$. As g is a biholomorphism, g^{-1} is continuous. So $g^{-1}(K)$ is compact for any compact $K \subset \mathbb{D}$. In other words, $|g(z)|$ must approach 1 as z approaches the boundary $\partial\mathbb{D}$. Then so must $|h(z)|$. The function $|h(z)|$ must, therefore, attain a minimum inside $\mathbb{D}$, or in other words $\left|\frac{1}{h(z)}\right|$ must attain a maximum inside $\mathbb{D}$. So $h(z)$ is a constant, and $g(z) = \alpha z$ for some constant α. Clearly, $|\alpha| = 1$ or $\alpha = e^{i\theta}$. Applying φ_{-b} to both sides of $e^{i\theta} z = \varphi_b \circ f$ we obtain $f(z) = \varphi_{-b}(ze^{i\theta}) = e^{i\theta}\varphi_{-be^{-i\theta}}(z)$. $\qquad\square$

Exercise 3.5.5: *Justify the claim in the proof. If a continuous $g: \mathbb{D} \to \mathbb{D}$ is such that $g^{-1}(K)$ is compact for every compact $K \subset \mathbb{D}$, then if $\{z_n\}$ is a sequence in $\mathbb{D}$ such that $|z_n| \to 1$, then $|g(z_n)| \to 1$.*

Exercise 3.5.6: *Given two distinct $a, b \in \mathbb{D}$, show that there exists a unique $f \in \text{Aut}(\mathbb{D})$ such that $f(a) = b$ and $f(b) = a$.*

Exercise 3.5.7: *Prove that if $\mathbb{H} = \{z \in \mathbb{C} : \text{Im } z > 0\}$ is the upper half-plane and $f: \mathbb{H} \to \mathbb{H}$ is an automorphism of $\mathbb{H}$, then*

$$f(z) = \frac{az + b}{cz + d}$$

for real numbers a, b, c, d such that $ad - bc > 0$.

Exercise 3.5.8: *Suppose $U \subset \mathbb{C}$ is a domain, $\overline{\mathbb{D}} \subset U$, and $f: U \to \mathbb{C}$ is holomorphic. Suppose $|f(z)| = 1$ whenever $|z| = 1$, that is, $f(\partial\mathbb{D}) \subset \partial\mathbb{D}$. Find a formula for f. Use the following outline:*

   a) *Show that f must have finitely many zeros in $\mathbb{D}$. That is, $f(z) = 0$ for at most finitely many $z \in \mathbb{D}$.*
   b) *Suppose that f has no zeros in $\mathbb{D}$. Prove that f is constant (and what sort of constant).*
   c) *If $f(a) = 0$, then prove that $z \mapsto \frac{f(z)}{\varphi_a(z)}$ is still holomorphic in U (i.e., can be defined at a to be holomorphic in U) and still takes the circle to the circle.*
   d) *Now find a general formula for f.*

Exercise 3.5.9: *Suppose $f: \mathbb{D} \to \mathbb{D}$ is a holomorphic function with zeros at $z_1, \ldots, z_n$, that is, $f(z_\ell) = 0$ for $\ell = 1, \ldots, n$. Prove that*

$$|f(z)| \leq |\varphi_{z_1}(z)\varphi_{z_2}(z)\cdots\varphi_{z_n}(z)|.$$

Exercise 3.5.10: *Prove the* Schwarz–Pick *lemma: If $f: \mathbb{D} \to \mathbb{D}$ is holomorphic, then*

$$\left|\frac{f(z) - f(\zeta)}{1 - \overline{f(\zeta)}f(z)}\right| \leq \left|\frac{z - \zeta}{1 - \overline{\zeta}z}\right| \qquad \text{and} \qquad \frac{|f'(z)|}{1 - |f(z)|^2} \leq \frac{1}{1 - |z|^2}$$

for all $z, \zeta \in \mathbb{D}$. If equality holds in one of the inequalities for some $z \neq \zeta$, then f is an automorphism of $\mathbb{D}$. Conversely, if f is an automorphism of $\mathbb{D}$, then equality holds in both inequalities for all $z, \zeta \in \mathbb{D}$.

In particular, the Schwarz–Pick lemma (Exercise 3.5.10) gives a bound on the derivative at all points. If $f : \mathbb{D} \to \mathbb{D}$ is holomorphic, nonconstant, and $f(a) = b$, then

$$|f'(a)| \leq \frac{1 - |b|^2}{1 - |a|^2}.$$

If equality holds, then $f(z) = \varphi_{-b}\left(e^{i\theta}\varphi_a(z)\right)$ for some $\theta \in \mathbb{R}$.

4i The Logarithm and Cauchy

Never doubt the courage of the French. They were the ones who discovered that snails are edible.

—Doug Larson

4.1i The logarithm and the winding number

4.1.1i The logarithm

Let us ponder over the primitives of z^n for $n \in \mathbb{Z}$.* When $n \geq 0$, then z^n is defined in the entire plane, and a primitive is simply $\frac{z^{n+1}}{n+1}$. If $n < -1$, then z^n is defined in the punctured plane $\mathbb{C} \setminus \{0\}$, and again a primitive is $\frac{z^{n+1}}{n+1}$. What about $z^{-1} = 1/z$? It has a primitive, but never one defined in the entire punctured plane.

We demonstrated that in any star-like domain, a holomorphic function has a primitive. Consider the so-called *slit plane*

$$U = \mathbb{C} \setminus (-\infty, 0] = \mathbb{C} \setminus \{z \in \mathbb{C} : \operatorname{Re} z \leq 0, \operatorname{Im} z = 0\}.$$

It is a star-like domain and so there exists a primitive for $1/z$ in U. If we require that this primitive is 0 at $z = 1$, we get a function

$$\operatorname{Log} \colon U \to \mathbb{C},$$

called the *principal branch* of the logarithm. We saw another gadget before called the "principal branch," the principal branch of the argument, Arg. Let us show that

$$\operatorname{Log} z = \log|z| + i \operatorname{Arg} z,$$

where $\log|z|$ is just the standard real logarithm of $|z|$. Set $L(z) = \log|z| + i \operatorname{Arg} z$, and let us show that $L = \operatorname{Log}$. We have $L(1) = 0 = \operatorname{Log}(1)$, so far so good. Observe

$$e^{L(z)} = e^{\log|z|} e^{i \operatorname{Arg} z} = |z| e^{i \operatorname{Arg} z} = z.$$

---

*It appears, doesn't it, that elementary complex analysis is the study of z^n.

So L is the inverse of the exponential, at least for $z \in U$. In particular, L is holomorphic by the inverse function theorem. Take the derivative of both sides of $z = e^{L(z)}$,

$$1 = L'(z)e^{L(z)} = L'(z)z.$$

Et voilà!* We have $L'(z) = 1/z$, so $L = \text{Log}$.

If we use a different branch of the argument we get another antiderivative of $1/z$. We make the definition

$$\log z \overset{\text{def}}{=} \log|z| + i \arg z.$$

This definition is totally bonkers at first glance. First, the log on the left is a different log than the log on the right. On the right, it is the standard real log, that is, $\log \colon (0, \infty) \to \mathbb{R}$, where $\log 1 = 0$. But the log on the left is not even a function, it has infinitely many values for every z, since the arg on the right-hand side has infinitely many values. The value of $\log(-1)$ is πi, but also $-\pi i$, $3\pi i$, or $(\pi + 2\pi k)i$ for any $k \in \mathbb{Z}$. So log is a function just as much as arg is a function. See Figure 4.1. The double duty of "log" is almost never a problem and it is generally clear which log one is talking about based on what sort of things are being plugged into it.

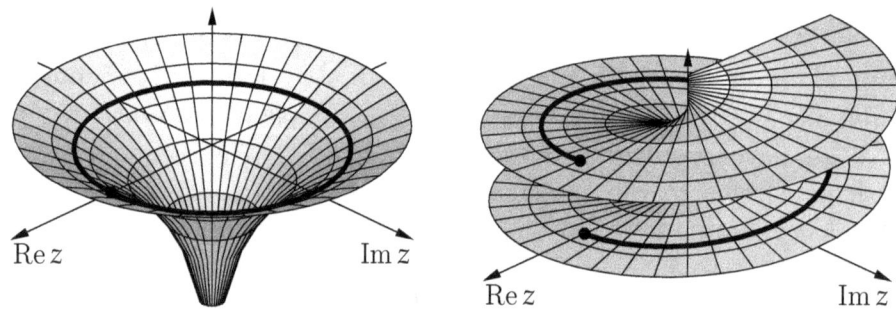

Figure 4.1: "Graphs" of the real part (left) and imaginary part (right) of the complex logarithm $\log z = \log|z| + i \arg z$. The imaginary part is an infinite spiral, only two turns are pictured. A path on the graph around the unit circle is marked.

While log is not really a function—it is a multivalued function[†]—it is the definition that we want. The principal branch, useful when one wants to get some actual numbers, is often not what we need; it is not as useful as one would think. And beware that computers like to give back the principal branch even when it doesn't make any sense.

So how do we use log? Well it comes up in line integrals, which are used to count and classify zeros (roots) and/or singularities of functions, or vice versa—zeros and singularities are used to compute line integrals. Let us compute the integral of $1/z$

---

*Cauchy was French, n'est-ce pas?

[†]Cauchy: Quel Malheur! Je déteste le logarithme! Je veux devenir plombier.

around the unit circle $\partial \mathbb{D}$, oriented counterclockwise as usual (parametrized by e^{it}). Suppose we start and end the integration at $z = 1$:

$$\int_{\partial \mathbb{D}} \frac{1}{z} \, dz = \log 1 - \log 1 = 2\pi i.$$

That makes no sense, no? Well, it should really only be done with quotation marks:

$$\int_{\partial \mathbb{D}} \frac{1}{z} \, dz \text{ "=" } \log 1 - \log 1 \text{ "=" } 2\pi i.$$

That's a lot better, no?* The equalities are only true morally. Interpreted correctly, it is exactly what is happening. You really do subtract one of the values of $\log 1$ from another value of $\log 1$. To figure out which from which, start with say $\log 1 = 0$, and follow the function along the circle slowly and notice that $i \arg z$ grows from 0 to $2\pi i$. So the $\log 1$ at the end is $2\pi i$. See the path marked on Figure 4.1, the jump in the imaginary part between the beginning and the end is precisely that $2\pi i$.

To make working with log easier, we usually talk about a *branch of the logarithm*. So $L : U \to \mathbb{C}$ is a branch of the logarithm if L is holomorphic, $L'(z) = 1/z$, and $L(z)$ is equal to some value of $\log z$ for every $z \in U$. It is not possible to define a branch of the logarithm in every U, but for example we can do it in every star-like U where $0 \notin U$. In general, one can define a branch of the logarithm in every simply connected domain, that is, a domain without holes, that does not contain zero. More on that later. Similarly, we define branches of $\log(z - p)$, a primitive of $\frac{1}{z-p}$, in which case the domain should not contain p.

We may also talk about branches more loosely, and talk about following them along a path. We don't define a single branch—we define a branch in some small open set, follow its values for a while, then switch to another branch that happens to agree with the first branch at the point where we switch. See Figure 4.2. Really, we did precisely that in the "computation" above: We followed log from 1 along the circle until we ended up at 1 again, and the branches that we followed ended up $2\pi i$ off. We will see this procedure used more formally in just a moment.

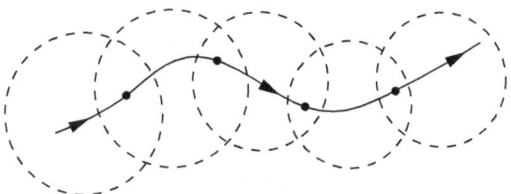

Figure 4.2: Following a branch. The branches are defined in the discs (they do not have to be discs). Points where the branches are supposed to equal are marked.

---

*Non! Je veux aussi devenir plombier maintenant!

Exercise **4.1.1:** *Suppose $a \in \mathbb{C} \setminus \{0\}$, and $R_a = \{\lambda a \in \mathbb{C} : \lambda \geq 0\}$ is the ray from the origin through a. Prove that there exists a branch of the $\log$ in $\mathbb{C} \setminus R_a$.*

Exercise **4.1.2:** *For $n \in \mathbb{N}$, let $\gamma : [0, 2\pi] \to \mathbb{C}$ be $\gamma(t) = e^{int}$, the unit circle traversed n times counterclockwise. Compute $\int_\gamma \frac{1}{z}\, dz$. Argue by splitting up the integral into pieces and using branches of the $\log$.*

Exercise **4.1.3:** *Suppose $U \subset \mathbb{C}$ is open with $\partial \mathbb{D} \subset U$, $f : U \to \mathbb{C}$ is holomorphic, such that $f(z)$ is never negative real or zero. Compute $\int_{\partial \mathbb{D}} \frac{f'(z)}{f(z)}\, dz$.*

Exercise **4.1.4:** *Suppose $\gamma : [a, b] \to \mathbb{C} \setminus \{0\}$ is a piecewise-C^1 path such that $\gamma(a) = \gamma(b) = -1$, but $\gamma(t)$ is never negative real for any $t \in (a, b)$. Using the principal branch of the $\log$, prove that*

$$\int_\gamma \frac{1}{z}\, dz = -2\pi i, 0,\ or\ 2\pi i.$$

Explicitly find three γs that achieve each of these three possibilities.

4.1.2*i*   Winding numbers

OK. Let's get more rigorous.

Definition 4.1.1. Let Γ be a cycle, and $p \notin \Gamma$. Then

$$n(\Gamma; p) \overset{\text{def}}{=} \frac{1}{2\pi i} \int_\Gamma \frac{1}{z - p}\, dz$$

is called the *winding number* of Γ around p, or the *index* of Γ with respect to p.

Intuitively, the winding number is the number of times that Γ winds around p, counterclockwise. This intuition is confirmed by integrating $1/z$ for the path e^{it} for $t \in [0, 2\pi]$ to get a winding number 1 around $p = 0$, as it goes once counterclockwise around zero. If we do the integral with e^{2it}, we go around zero twice counterclockwise, and the winding number really is 2. Similarly, if we use e^{-it}, then we go around zero once in the clockwise direction, and the winding number is -1.

The first thing to observe is that the winding number is an integer.

Proposition 4.1.2. *Suppose Γ is a cycle and $p \notin \Gamma$. Then $n(\Gamma; p)$ is an integer.*

The proof is to take a closed path γ and to follow a branch of $\log$ around γ, and see by how much it changes. See Figure 4.2, where we go all the way around a loop. Since we follow the argument and we go some number of times around p along γ, the argument changes by some multiple of 2π.

Proof. A cycle is (equivalent to) a linear combination (over the integers) of closed paths, so we only need to consider closed piecewise-C^1 paths. Let $\gamma : [0, 1] \to \mathbb{C}$ be the path.

The path γ as a set is compact. It can be covered by finitely many discs $D_1, \ldots, D_n$, none of which contain p, and such that there is a partition $0 = t_0 < t_1 < t_2 < \cdots < t_n = 1$ where $\gamma([t_{j-1}, t_j]) \subset D_j$. Each D_j is star-like and does not contain p, so in each one a branch of $\log(z - p)$ exists, call it L_j. Pick L_1 arbitrarily, then pick $L_2, \ldots, L_n$ in sequence accordingly so that $L_j(\gamma(t_j)) = L_{j+1}(\gamma(t_j))$. Call $z_0 = \gamma(0) = \gamma(1)$. So

$$n(\gamma; p) = \frac{1}{2\pi i} \int_\gamma \frac{1}{z - p} \, dz = \frac{1}{2\pi i} \int_0^1 \frac{\gamma'(t)}{\gamma(t) - p} \, dt = \frac{1}{2\pi i} \sum_{j=1}^n \int_{t_{j-1}}^{t_j} \frac{\gamma'(t)}{\gamma(t) - p} \, dt$$

$$= \frac{1}{2\pi i} \sum_{j=1}^n L_j(\gamma(t_j)) - L_j(\gamma(t_{j-1})) = \frac{1}{2\pi i} (L_n(z_0) - L_1(z_0)).$$

As L_n and L_1 are both branches of $\log(z - p)$, their difference is $2\pi k i$ for some $k \in \mathbb{Z}$, as each is $\log|z_0 - p| + i \arg(z_0 - p)$ for some value of arg. $\qquad \square$

Exercise 4.1.5: Fill in the details in the existence of the partition. That is, once you cover γ by finitely many discs that do not contain p show that the partition $t_0, \ldots, t_n$ exists. Hint: Some of the discs may "repeat," but make sure that you do not get "stuck" before reaching 1.

The second thing to observe is that $n(\Gamma; z)$ is constant as long as we do not cross Γ.

Proposition 4.1.3. *Given a cycle Γ, the function $z \mapsto n(\Gamma; z)$ is constant on the topological components of $\mathbb{C} \setminus \Gamma$. Furthermore, $n(\Gamma; z) = 0$ for z on the unbounded component of $\mathbb{C} \setminus \Gamma$.*

As Γ is compact, there must be a unique unbounded component of the complement $\mathbb{C} \setminus \Gamma$, and possibly several bounded components. See Figure 4.3 for example.

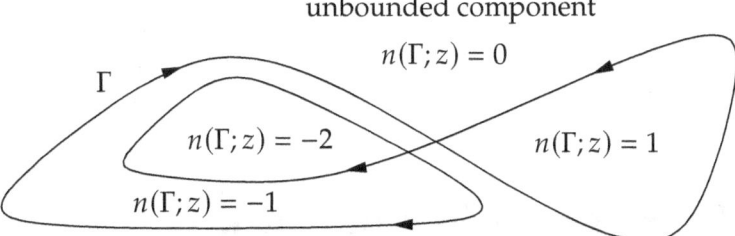

Figure 4.3: Components of $\mathbb{C} \setminus \Gamma$ with the winding number around points in those components marked.

Proof. Let us show that the function

$$p \mapsto n(\Gamma; p) = \frac{1}{2\pi i} \int_\Gamma \frac{1}{z - p} \, dz$$

is continuous on $\mathbb{C} \setminus \Gamma$. Fix $p_0 \in \mathbb{C} \setminus \Gamma$, and let $d = d(p_0, \Gamma)$ be the distance from p_0 to Γ, namely $d = \inf\{|z - p_0| : z \in \Gamma\}$. As Γ is compact, $d > 0$. For any $p \in \Delta_{d/2}(p_0)$, we have $|z - p| \geq d/2$ for every $z \in \Gamma$. Let ℓ be the length of Γ, that is, $\ell = \int_\Gamma |dz|$. Then

$$\left| n(\Gamma; p_0) - n(\Gamma; p) \right| = \left| \frac{1}{2\pi i} \int_\Gamma \frac{p_0 - p}{(z - p_0)(z - p)} \, dz \right| \leq \frac{1}{2\pi} \int_\Gamma \frac{|p_0 - p|}{|z - p_0||z - p|} \, |dz|$$

$$\leq \frac{\ell}{\pi d^2} |p_0 - p|.$$

So, $n(\Gamma; p)$ is a continuous function of p. As it is continuous and integer-valued, it is constant on every connected component of $\mathbb{C} \setminus \Gamma$ (the set where it is defined).

For any $p \in \mathbb{C} \setminus \Gamma$,

$$|n(\Gamma; p)| \leq \frac{1}{2\pi} \int_\Gamma \frac{1}{|z - p|} \, |dz| \leq \frac{1}{2\pi} \frac{\ell}{d(p, \Gamma)}.$$

On the unbounded component—as Γ is compact—there are p with $d(p, \Gamma)$ arbitrarily large, so $n(\Gamma; p)$ is arbitrarily small on this component. As it is constant, it is zero.   □

Exercise 4.1.6: Show that $n\big(\partial \Delta_r(p); z\big) = 0$ if $z \notin \overline{\Delta_r(p)}$ and $n\big(\partial \Delta_r(p); z\big) = 1$ if $z \in \Delta_r(p)$.

Exercise 4.1.7: Let $n \in \mathbb{Z}$. The path $\gamma : [0, 2\pi] \to \mathbb{C}$, where $\gamma(t) = p + re^{int}$, goes n times counterclockwise around $\partial \Delta_r(p)$. Prove that $n(\gamma; z) = n$ if $z \in \Delta_r(p)$.

Exercise 4.1.8: Suppose $0 < r_1 < r_2 < \infty$. Let $\Gamma = \partial \Delta_{r_2}(p) - \partial \Delta_{r_1}(p)$ (that is, the outside circle goes counterclockwise, the inside circle goes clockwise). Prove that if $z \in \mathbb{C}$ is such that $|z - p| < r_1$, then $n(\Gamma; z) = 0$. If $r_1 < |z - p| < r_2$, then $n(\Gamma; z) = 1$. If $r_2 < |z - p|$, then $n(\Gamma; z) = 0$.

Exercise 4.1.9: Suppose $\gamma : [a, b] \to \mathbb{C}$ is a closed C^1 path such that $\gamma(a) = \gamma(b)$ is some real negative number. Suppose that $\gamma(t)$ is real and negative for only k distinct t (that includes $t = a$ and $t = b$, so $k \geq 2$), and whenever $\gamma(t)$ is real and negative, then $\operatorname{Im} \gamma'(t) < 0$. Prove that $n(\gamma; 0) = k - 1$. Hint: Use the principal branch.

4.2i Homology versions of Cauchy

Definition 4.2.1. Let $U \subset \mathbb{C}$ be open and Γ a cycle in U such that $n(\Gamma; p) = 0$ for all $p \in \mathbb{C} \setminus U$. We then say Γ is *homologous to zero in U*.

What homologous to zero means is that Γ does not wind around any point in the complement of U. Do note that "homologous to zero" does not mean "equivalent to zero." For instance, if $U = \mathbb{C}$, then every Γ is homologous to zero trivially. Also note the dependence on U. The unit circle is homologous to zero in $U = \mathbb{C}$, but it is not homologous to zero in $U = \mathbb{C} \setminus \{0\}$.

Theorem 4.2.2 (Cauchy integral formula (homology version)). *Suppose $U \subset \mathbb{C}$ is open, $f : U \to \mathbb{C}$ is holomorphic, and Γ is a cycle in U homologous to zero in U. Then for $z \in U \setminus \Gamma$,*

$$n(\Gamma; z) f(z) = \frac{1}{2\pi i} \int_\Gamma \frac{f(\zeta)}{\zeta - z} \, d\zeta.$$

In the proof*, rather strangely, we will take the difference of the two sides of the equation and extend it to an entire function (even though U may be small) and then we will use Liouville's theorem (Theorem 3.3.10) to show that this difference is zero.

Proof. Define $g : U \times U \to \mathbb{C}$ by

$$g(\zeta, z) = \begin{cases} \frac{f(\zeta) - f(z)}{\zeta - z} & \text{if } \zeta \neq z, \\ f'(\zeta) & \text{if } \zeta = z. \end{cases}$$

Exercise 4.2.1: *Prove that $g(\zeta, z)$ is continuous in $U \times U$, and that the function $z \mapsto g(\zeta, z)$ is holomorphic for every fixed $\zeta \in U$. Hint: The only nontrivial piece of this proof is showing that $z \mapsto g(\zeta, z)$ is holomorphic at $z = \zeta$.*

Let

$$h(z) = \begin{cases} \int_\Gamma g(\zeta, z) \, d\zeta & \text{if } z \in U, \\ \int_\Gamma \frac{f(\zeta)}{\zeta - z} \, d\zeta & \text{if } z \notin \Gamma \text{ and } n(\Gamma; z) = 0. \end{cases} \tag{4.1}$$

As $n(\Gamma; z) = 0$ for all $z \in \mathbb{C} \setminus U$ (Γ is homologous to zero) the function $h(z)$ is defined for every $z \in \mathbb{C}$. Unfortunately, at some points we have two definitions. To show that $h(z)$ is well-defined, we must show that if $z \in U \setminus \Gamma$ and $n(\Gamma; z) = 0$, then the two definitions agree. Consider such a z (in particular, $z \notin \Gamma$). Then,

$$\int_\Gamma g(\zeta, z) \, d\zeta = \int_\Gamma \frac{f(\zeta) - f(z)}{\zeta - z} \, d\zeta = \int_\Gamma \frac{f(\zeta)}{\zeta - z} \, d\zeta - f(z)(2\pi i) n(\Gamma; z) = \int_\Gamma \frac{f(\zeta)}{\zeta - z} \, d\zeta.$$

So $h : \mathbb{C} \to \mathbb{C}$ is well-defined.

Next we show that h is holomorphic. Holomorphicity is a local property, so we only need to prove it in a neighborhood of every point. The set where $n(\Gamma; z) = 0$ is open as it is a union of some topological components of $\mathbb{C} \setminus \Gamma$. So each point in $\mathbb{C}$ has a neighborhood where h is defined entirely by one or the other expression in (4.1). Given any point in $\mathbb{C}$, take a neighborhood where one of the expressions defines h and apply Corollary 3.4.2.

The unbounded component of $\mathbb{C} \setminus \Gamma$ is contained in the set where $n(\Gamma; z) = 0$, so on this component, h is defined by the second expression. Consider a z in this component. Suppose $|f(\zeta)| \leq M$ for $\zeta \in \Gamma$, let ℓ be the length of Γ, and let $d(z, \Gamma)$ be the distance of z to Γ. Then

$$|h(z)| = \left| \int_\Gamma \frac{f(\zeta)}{\zeta - z} \, d\zeta \right| \leq \int_\Gamma \left| \frac{f(\zeta)}{\zeta - z} \right| \, |d\zeta| \leq \frac{M\ell}{d(z, \Gamma)}.$$

---

*This elegant proof is relatively recent, from 1971, due to John Dixon.

As $z \to \infty$, so does $d(z, \Gamma) \to \infty$, and so $h(z) \to 0$. In particular, h is an entire bounded function and Liouville says that h is constant. Moreover, that constant must be zero. So suppose $z \in U \setminus \Gamma$. Then

$$0 = h(z) = \int_\Gamma \frac{f(\zeta) - f(z)}{\zeta - z} \, d\zeta = \int_\Gamma \frac{f(\zeta)}{\zeta - z} \, d\zeta - f(z)(2\pi i) n(\Gamma; z). \qquad \square$$

Cauchy's theorem follows immediately using the integral formula. In fact, the two theorems are equivalent, see the exercises.

Theorem 4.2.3 (Cauchy's theorem (homology version)). *Suppose $U \subset \mathbb{C}$ is open, $f : U \to \mathbb{C}$ is holomorphic, and Γ is a cycle in U. If Γ is homologous to zero in U, then*

$$\int_\Gamma f(z) \, dz = 0.$$

Proof. Fix $z \in U \setminus \Gamma$. Apply the Cauchy integral formula for the function $\zeta \mapsto (\zeta - z) f(\zeta)$ at $\zeta = z$:

$$0 = n(\Gamma; z)(z - z) f(z) = \frac{1}{2\pi i} \int_\Gamma \frac{(\zeta - z) f(\zeta)}{\zeta - z} \, d\zeta = \frac{1}{2\pi i} \int_\Gamma f(\zeta) \, d\zeta. \qquad \square$$

Definition 4.2.4. Two cycles Γ_0 and Γ_1 in $U \subset \mathbb{C}$ are *homologous* in U if $n(\Gamma_0; p) = n(\Gamma_1; p)$ for all $p \in \mathbb{C} \setminus U$.

Equivalently, Γ_0 and Γ_1 are homologous in U if $n(\Gamma_0 - \Gamma_1; p) = 0$ for all $p \in \mathbb{C} \setminus U$, that is, $\Gamma_0 - \Gamma_1$ is homologous to zero in U.

Corollary 4.2.5. *Let $U \subset \mathbb{C}$ be open and $f : U \to \mathbb{C}$ holomorphic, and let Γ_0 and Γ_1 be cycles in U. If Γ_0 and Γ_1 are homologous in U, then*

$$\int_{\Gamma_0} f(z) \, dz = \int_{\Gamma_1} f(z) \, dz.$$

The proof is immediate by applying Cauchy's theorem to $\Gamma_0 - \Gamma_1$.

Exercise 4.2.2 (Easy): *Suppose that Γ is a cycle such that $0 \notin \Gamma$ and $n(\Gamma; 0) = k$. Compute*

$$\int_\Gamma \frac{\cos z}{z} \, dz.$$

Exercise 4.2.3: Let Γ be a cycle in $\mathbb{C} \setminus \{0\}$. Prove that Γ is homologous in $\mathbb{C} \setminus \{0\}$ to $n\partial \mathbb{D}$ for some $n \in \mathbb{Z}$.

Exercise 4.2.4:
    a) Show that being homologous in U is an equivalence relation on cycles.
    b) Prove that the addition of cycles makes the set of equivalence classes into an abelian group, the first homology group of U, usually written $H_1(U)$.
    c) Compute $H_1(\mathbb{C} \setminus \{0\})$ (that is, find what group it is isomorphic to).

Exercise 4.2.5: *Prove that the two theorems (the homology versions of Cauchy's theorem and the Cauchy integral formula) are equivalent logically, that is, one follows from the other. We have already proved that the Cauchy integral formula implies Cauchy's theorem. So prove that Cauchy's theorem implies the Cauchy integral formula.*

Exercise 4.2.6: *Let $U \subset \mathbb{C}$ be open and Γ a cycle in U homologous to zero in U. Suppose that $n(\Gamma; z_1) = k_1$ and $n(\Gamma; z_2) = k_2$ for some two distinct $z_1, z_2 \in U \setminus \Gamma$. Let $f : U \setminus \{z_1, z_2\} \to \mathbb{C}$ be holomorphic. Suppose $0 < \epsilon < |z_1 - z_2|$ is small enough that $\overline{\Delta_\epsilon(z_j)} \subset U$ for $j = 1, 2$, and that $\int_{\partial \Delta_\epsilon(z_1)} f(z) \, dz = A$ and $\int_{\partial \Delta_\epsilon(z_2)} f(z) \, dz = B$. In terms of k_1, k_2, A, and B, compute*

$$\int_\Gamma f(z) \, dz.$$

4.3*i* **Simply connected domains**

Roughly, a simply connected domain[*] is one without any holes. The following is perhaps not the standard definition, but for domains in $\mathbb{C}$ (connected open sets) it is equivalent to the correct one. We will define the term properly once we get to homotopy.[†] We may sometimes say "simply connected in the sense of homology" to emphasize that we are using this particular definition.

Definition 4.3.1. A domain $U \subset \mathbb{C}$ is *simply connected* if every cycle in U is homologous to zero in U.

In other words, U is simply connected if $n(\Gamma; p) = 0$ for every cycle Γ in U and every $p \in \mathbb{C} \setminus U$. So in a simply connected domain, no cycle in U can wind around any point of $\mathbb{C} \setminus U$. Examples of simply connected domains are $\mathbb{C}$, $\mathbb{D}$, or $\mathbb{H}$. An example of a domain that is not simply connected is $\mathbb{C} \setminus \{0\}$. See the exercises below.

Exercise 4.3.1:
    a) *Prove that every star-like domain (e.g., $\mathbb{C}$, $\mathbb{D}$, and $\mathbb{H}$) in $\mathbb{C}$ is simply connected.*
    b) *Prove that $\mathbb{C} \setminus \{0\}$ is not simply connected.*

Exercise 4.3.2: *Prove that if $U \subset \mathbb{C}$ is biholomorphic to $\mathbb{D}$, then U is simply connected.*

Exercise 4.3.3: *Prove that a domain $U \subset \mathbb{C}$ is simply connected if and only if the first homology group $H_1(U)$ is isomorphic to the trivial group $\{0\}$. See Exercise 4.2.4.*

---

[*]There is no agreement among various mathematicians (I've asked a few) if a (path-)disconnected set can be "simply connected." To avoid heated arguments with topologists of various stripes, it's best to just not define the term for disconnected sets. Hence, we only define it for domains.

[†]Homotopy is in an optional section, which is the reason why we make this "wrong" definition.

A special (but common) case of the homology version of Cauchy, Theorem 4.2.3, can be stated as the simply connected case of Cauchy.

Theorem 4.3.2 (Cauchy's theorem (simply connected version)). *Let $U \subset \mathbb{C}$ be a simply connected domain and $f \colon U \to \mathbb{C}$ holomorphic. If Γ is a cycle in U, then*

$$\int_\Gamma f(z)\, dz = 0.$$

The proof follows at once from Theorem 4.2.3, since if U is simply connected, then every Γ in U is homologous to zero in U. In simply connected domains, as Cauchy's theorem holds for all cycles, we have primitives (antiderivatives).

Theorem 4.3.3. *Let $U \subset \mathbb{C}$ be a simply connected domain and let $f \colon U \to \mathbb{C}$ be holomorphic. Then f has a primitive in U.*

Proof. Fix some $p \in U$. As U is path connected, for every $z \in U$, pick some piecewise-C^1 path γ from p to z and define

$$F(z) = \int_\gamma f(\zeta)\, d\zeta.$$

A priori, the function $F(z)$ depends on γ, but if α is another path from p to z, then $\gamma - \alpha$ is a cycle in U and Cauchy's theorem says that

$$\int_\gamma f(\zeta)\, d\zeta - \int_\alpha f(\zeta)\, d\zeta = \int_{\gamma-\alpha} f(\zeta)\, d\zeta = 0.$$

So F is well-defined without specifying the path.

Let us reduce the proof to the proof for star-like domains (Proposition 3.2.11 and Corollary 3.2.12). Let $q \in U$ be a point and consider a disc $\Delta_r(q) \subset U$ (which is star-like with respect to q in particular). We take γ to be the path from p to q. As F does not depend on the path taken, then for $z \in \Delta_r(q)$,

$$F(z) = \int_{\gamma+[q,z]} f(\zeta)\, d\zeta = \int_\gamma f(\zeta)\, d\zeta + \int_{[q,z]} f(\zeta)\, d\zeta.$$

The first term in the sum is a constant, and the second term is precisely the primitive of f from the proof of Proposition 3.2.11, that is, a primitive in $\Delta_r(q)$. See Figure 4.4.   □

Corollary 4.3.4. *Let $U \subset \mathbb{C}$ be a simply connected domain and $f \colon U \to \mathbb{C}$ a nowhere zero holomorphic function. Then there exists a holomorphic $g \colon U \to \mathbb{C}$ such that*

$$e^{g(z)} = f(z).$$

In particular, if $U \subset \mathbb{C} \setminus \{0\}$ is a simply connected domain, then there exists a branch of the logarithm, that is, a holomorphic $L \colon U \to \mathbb{C}$ such that

$$e^{L(z)} = z.$$

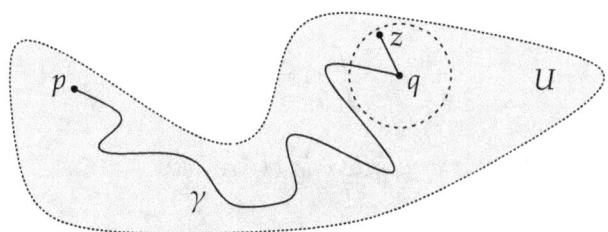

Figure 4.4: Existence of primitive in a simply connected domain with $\Delta_r(q)$ marked.

Proof. The function $\frac{f'(z)}{f(z)}$ is holomorphic on U. Find its primitive $g(z)$. Compute,

$$\frac{d}{dz}\left[\frac{e^{g(z)}}{f(z)}\right] = \frac{e^{g(z)}g'(z)f(z) - e^{g(z)}f'(z)}{\left(f(z)\right)^2} = \frac{e^{g(z)}f'(z) - e^{g(z)}f'(z)}{\left(f(z)\right)^2} = 0.$$

Thus, $\frac{e^{g(z)}}{f(z)}$ is constant (Proposition 2.2.1). It follows that there is a $C \in \mathbb{C}$ such that

$$e^{g(z)+C} = f(z). \qquad \square$$

If we have the logarithm, we can take roots.

Corollary 4.3.5. *Let $U \subset \mathbb{C}$ be a simply connected domain, $f \colon U \to \mathbb{C}$ a nowhere zero holomorphic function, and $k \in \mathbb{N}$. Then there exists a holomorphic $g \colon U \to \mathbb{C}$ such that*

$$\left(g(z)\right)^k = f(z).$$

Proof. Find a $\psi \colon U \to \mathbb{C}$ such that $e^{\psi(z)} = f(z)$. Let $g(z) = e^{\frac{1}{k}\psi(z)}$. Check:

$$\left(g(z)\right)^k = \left(e^{\frac{1}{k}\psi(z)}\right)^k = e^{\psi(z)} = f(z). \qquad \square$$

On the other hand, the existence of primitives, or Cauchy's theorem without restriction on Γ, or existence of logs guarantees simply-connectedness. In particular, we have the following set of equivalent versions of simply-connectedness for domains.

Proposition 4.3.6. *Let $U \subset \mathbb{C}$ be a domain. The following are equivalent:*

(i) *U is simply connected (in the homology sense).*

(ii) *Every holomorphic $f \colon U \to \mathbb{C}$ has a primitive.*

(iii) *For every nowhere zero holomorphic $f \colon U \to \mathbb{C}$, there exists a holomorphic $g \colon U \to \mathbb{C}$ such that $e^{g(z)} = f(z)$.*

(iv) *$\frac{1}{z-p}$ has a primitive in U for every $p \in \mathbb{C} \setminus U$.*

(v) *For every holomorphic $f : U \to \mathbb{C}$ and every cycle Γ in U, we have*

$$\int_\Gamma f(z)\, dz = 0.$$

(vi) *For every $p \in \mathbb{C} \setminus U$ and every cycle Γ in U, we have*

$$\int_\Gamma \frac{1}{z - p}\, dz = 0.$$

Proof. The logic of the proof is the following diagram:

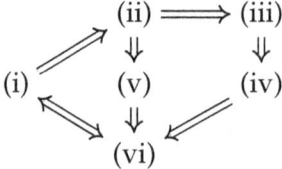

We proved (i) $\Rightarrow$ (ii) above, and (ii) $\Rightarrow$ (iii) is the same as the proof of Corollary 4.3.4. Next, suppose (iii) is true. For $p \in \mathbb{C} \setminus U$, find a g such that $e^{g(z)} = z - p$. Differentiate

$$1 = \frac{d}{dz}\left[z - p\right] = \frac{d}{dz}\left[e^{g(z)}\right] = e^{g(z)}g'(z) = (z - p)g'(z).$$

So (iv) follows. Using Cauchy's theorem for derivatives (Corollary 3.2.6), (iv) implies

$$\int_\Gamma \frac{1}{z - p}\, dz = 0$$

for every $p \in \mathbb{C} \setminus U$, and hence (vi) is true. As

$$n(\Gamma; p) = \frac{1}{2\pi i}\int_\Gamma \frac{1}{z - p}\, dz,$$

(vi) is simply a restatement of (i). Again by Cauchy's theorem for derivatives, (ii) $\Rightarrow$ (v). Finally, (v) $\Rightarrow$ (vi) is immediate. □

The existence of roots, in particular the square root, can also be put on the list, but the proof is a little trickier. For the square root it will follow for example from the proof of the Riemann mapping theorem in § 6.3.1.

There is a simple topological criterion for simply-connectedness of domains in the complex plane (simply-connectedness is, after all, a topological concept). The proposition below ought to be an "if and only if," but the converse is more difficult to prove. It requires finding a cycle around a given compact subset, which is harder than it sounds. We will prove it in § 6.3.3.

Proposition 4.3.7. *Let $U \subset \mathbb{C}$ be a domain. If $\mathbb{C}_\infty \setminus U$ is connected, then U is simply connected.*

Proof. Take $S = \mathbb{C}_\infty \setminus U$ and let Γ be a cycle in U. The function $\varphi(z) = n(\Gamma; z)$ is continuous on $\mathbb{C} \setminus \Gamma$. On the unbounded component of $\mathbb{C} \setminus \Gamma$, the function is zero, and so φ is zero in a neighborhood of ∞. If we set $\varphi(\infty) = 0$, the function is continuous on $\mathbb{C}_\infty \setminus \Gamma$ and hence on S. As S is connected, it is contained in a single component of $\mathbb{C}_\infty \setminus \Gamma$. So φ is constant on S. As $\varphi(\infty) = 0$ and $\infty \in S$, we have $\varphi|_S \equiv 0$. In other words, U is simply connected. $\qquad\square$

It is important to use $\mathbb{C}_\infty$ and not $\mathbb{C}$ in the proposition. If $U = \mathbb{C} \setminus \{0\}$ is the punctured plane, then $\mathbb{C} \setminus U = \{0\}$ is connected, but $\mathbb{C}_\infty \setminus U = \{0, \infty\}$ is not connected.

Exercise 4.3.4: *Suppose $U \subset \mathbb{C}$ is a domain, $\partial\Delta_r(p) \subset U$, but there is a $z \in \Delta_r(p)$ such that $z \notin U$. Prove that U is not simply connected.*

Exercise 4.3.5: *Let $K \subset \mathbb{C}$ be nonempty and compact. Prove that the unbounded component of $\mathbb{C} \setminus K$ is not a simply connected domain.*

Exercise 4.3.6: *Let $U_1, U_2 \subset \mathbb{C}$ be two simply connected domains such that $U_1 \cap U_2$ is nonempty and connected. Prove that $U = U_1 \cup U_2$ is a simply connected domain.*

Exercise 4.3.7: *Let $U_1, U_2 \subset \mathbb{C}$ be two simply connected domains such that $U_1 \cap U_2$ is nonempty and connected. Prove that $U = U_1 \cap U_2$ is a simply connected domain. Note: This result is true in the plane, but it is no longer true in the Riemann sphere.*

Exercise 4.3.8: *Find two nonempty simply connected domains $U_1, U_2 \subset \mathbb{C}$ such that $U_1 \cap U_2$ is nonempty and both*
   1) $U_1 \cup U_2$ is not a simply connected domain.
   2) $U_1 \cap U_2$ is not a simply connected domain (emphasis on domain).

Exercise 4.3.9: *Suppose $U \subset \mathbb{C}$ is a simply connected domain such that $0 \notin U$, the set $(0, \infty) \cap U$ is nonempty and connected, and suppose $r \in \mathbb{R}$. Show that there exists a holomorphic $f \colon U \to \mathbb{C}$ such that $f(x) = x^r$ for all $x > 0$ in U.*

Exercise 4.3.10: *Find a simply connected domain $U \subset \mathbb{C}$ such that $\mathbb{C} \setminus U$ has infinitely many components ($\mathbb{C}_\infty \setminus U$ is still going to have just one component).*

4.4i \ Laurent series

One can also define a series for a holomorphic function around a hole, or a singularity.

Definition 4.4.1. Given $0 \le r_1 < r_2 \le \infty$ and $p \in \mathbb{C}$, define

$$\operatorname{ann}(p; r_1, r_2) \overset{\text{def}}{=} \{z \in \mathbb{C} : r_1 < |z - p| < r_2\}.$$

When $0 < r_1 < r_2 < \infty$ we call this set an *annulus**.

---

*Q: What do you call a banana with a hole? A: A banannulus.

A common case is when $r_1 = 0$, that is, the punctured disc

$$\text{ann}(p;0,r) = \Delta_r(p) \setminus \{p\}.$$

When $r_2 = \infty$ on the other hand, $\text{ann}(p;r,\infty) = \mathbb{C} \setminus \overline{\Delta_r(p)}$ (if $r > 0$). We will, however, avoid the temptation of calling $\text{ann}(p;r,\infty)$ an "annulus."*

Theorem 4.4.2 (Existence of Laurent series). *Suppose that $0 \le r_1 < r_2 \le \infty$ and $f: \text{ann}(p;r_1,r_2) \to \mathbb{C}$ is holomorphic. Then there exist unique numbers $c_n \in \mathbb{C}$ for $n \in \mathbb{Z}$ such that*

$$f(z) = \sum_{n=-\infty}^{\infty} c_n(z-p)^n,$$

converging uniformly absolutely on compact subsets of $\text{ann}(p;r_1,r_2)$. *The numbers c_n are given by*

$$c_n = \frac{1}{2\pi i} \int_\gamma \frac{f(z)}{(z-p)^{n+1}}\, dz,$$

where γ is any circle of radius s, $r_1 < s < r_2$, centered at p oriented counterclockwise.

Recall that convergence of a double series such as

$$\sum_{n=-\infty}^{\infty} a_n$$

means

$$\sum_{n=-\infty}^{\infty} a_n = \lim_{N \to -\infty} \sum_{n=N}^{-1} a_n + \lim_{M \to \infty} \sum_{n=0}^{M} a_n.$$

That is, the limits are taken independently. For the Laurent series, we will generally have absolute convergence, so the limit may be taken in any way and in any order. However, it is still useful to split the Laurent series into two parts like that. Write a Laurent series as

$$\sum_{n=-\infty}^{\infty} c_n(z-p)^n = \sum_{n=0}^{\infty} c_n(z-p)^n + \sum_{n=-\infty}^{-1} c_n(z-p)^n$$

$$= \sum_{n=0}^{\infty} c_n(z-p)^n + \sum_{n=1}^{\infty} c_{-n}\left(\frac{1}{z-p}\right)^n.$$

So the Laurent series behaves like two power series: one in $z-p$ and one in $\frac{1}{z-p}$. You can therefore apply what you know about power series. For example, the first one converges (uniformly absolutely on compact subsets) in $\Delta_{r_2}(p)$ for some r_2, and the second one converges in $\mathbb{C} \setminus \overline{\Delta_{r_1}(p)}$ for some r_1. If $r_1 < r_2$, then the full series converges (uniformly absolutely on compact subsets) in $\text{ann}(p;r_1,r_2)$.

---

*"Holey plane" perhaps? A punctured disc also ought not to be called an "annulus," and calling $\text{ann}(0;0,\infty) = \mathbb{C} \setminus \{0\}$ an "annulus" is right out!

Proof of the theorem. Choose two numbers s_1 and s_2 such that $r_1 < s_1 < s_2 < r_2$. Define the cycle

$$\Gamma = \partial \Delta_{s_2}(p) - \partial \Delta_{s_1}(p).$$

That is, Γ goes around the larger (s_2) circle counterclockwise and around the smaller (s_1) circle clockwise. See Figure 4.5.

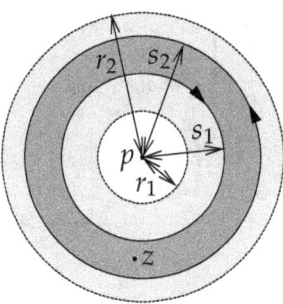

Figure 4.5: The two annuli, the smaller annulus is shaded darker. The two pieces of Γ are noted with the circular arrows.

If $q \in \mathbb{C} \setminus \text{ann}(p; r_1, r_2)$, then $n(\Gamma; q) = 0$: If q is in the "hole" of the annulus $\text{ann}(p; r_1, r_2)$, then $n(\partial \Delta_{s_j}(p); q) = 1$ for both $j = 1, 2$, and if q is outside the annulus altogether, then $n(\partial \Delta_{s_j}(p); q) = 0$ for both $j = 1, 2$ (see Exercise 4.1.6 or Exercise 4.1.8). So Γ is homologous to zero in the annulus $\text{ann}(p; r_1, r_2)$. On the other hand, if z is in the (smaller) annulus $\text{ann}(p; s_1, s_2)$, then for similar reasons, $n(\Gamma; z) = 1$.

Via Cauchy's integral formula (Theorem 4.2.2), for every $z \in \text{ann}(p; s_1, s_2)$,

$$f(z) = \frac{1}{2\pi i} \int_\Gamma \frac{f(\zeta)}{\zeta - z} \, d\zeta = \frac{1}{2\pi i} \int_{\partial \Delta_{s_2}(p)} \frac{f(\zeta)}{\zeta - z} \, d\zeta - \frac{1}{2\pi i} \int_{\partial \Delta_{s_1}(p)} \frac{f(\zeta)}{\zeta - z} \, d\zeta.$$

We expand the two bits separately. First, if $\zeta \in \partial \Delta_{s_2}$, then $\left| \frac{z-p}{\zeta-p} \right| = \frac{|z-p|}{s_2} < 1$ and so we follow the logic of Theorem 3.3.1. The reason that we can swap the integral and the series limit is the same as in that theorem.

$$\frac{1}{2\pi i} \int_{\partial \Delta_{s_2}(p)} \frac{f(\zeta)}{\zeta - z} \, d\zeta = \frac{1}{2\pi i} \int_{\partial \Delta_{s_2}(p)} \frac{f(\zeta)}{\zeta - p} \frac{1}{1 - \frac{z-p}{\zeta-p}} \, d\zeta$$

$$= \frac{1}{2\pi i} \int_{\partial \Delta_{s_2}(p)} \frac{f(\zeta)}{\zeta - p} \sum_{n=0}^{\infty} \left(\frac{z - p}{\zeta - p} \right)^n \, d\zeta$$

$$= \sum_{n=0}^{\infty} \underbrace{\left(\frac{1}{2\pi i} \int_{\partial \Delta_{s_2}(p)} \frac{f(\zeta)}{(\zeta - p)^{n+1}} \, d\zeta \right)}_{c_n} (z - p)^n.$$

Similarly, if $\zeta \in \partial \Delta_{s_1}$, then $\left| \frac{\zeta - p}{z - p} \right| = \frac{s_1}{|z - p|} < 1$, and so

$$-\frac{1}{2\pi i} \int_{\partial \Delta_{s_1}(p)} \frac{f(\zeta)}{\zeta - z} \, d\zeta = \frac{1}{2\pi i} \int_{\partial \Delta_{s_1}(p)} \frac{f(\zeta)}{z - p} \frac{1}{1 - \frac{\zeta - p}{z - p}} \, d\zeta$$

$$= \frac{1}{2\pi i} \int_{\partial \Delta_{s_1}(p)} \frac{f(\zeta)}{z - p} \sum_{m=0}^{\infty} \left(\frac{\zeta - p}{z - p} \right)^m d\zeta$$

$$= \sum_{m=0}^{\infty} \left(\frac{1}{2\pi i} \int_{\partial \Delta_{s_1}(p)} f(\zeta)(\zeta - p)^m \, d\zeta \right) (z - p)^{-m-1}$$

$$= \sum_{n=-\infty}^{-1} \underbrace{\left(\frac{1}{2\pi i} \int_{\partial \Delta_{s_1}(p)} \frac{f(\zeta)}{(\zeta - p)^{n+1}} \, d\zeta \right)}_{c_n} (z - p)^n.$$

Adding these together we have the right thing, except that the formula for c_n is not quite right. Given any s such that $r_1 < s < r_2$, the cycle $\partial \Delta_s(p) - \partial \Delta_{s_1}(p)$ is homologous to zero in $\text{ann}(p; r_1, r_2)$, and $z \mapsto \frac{f(z)}{(z-p)^{n+1}}$ is holomorphic in $\text{ann}(p; r_1, r_2)$. Cauchy's theorem (Theorem 4.2.3) thus says

$$0 = \int_{\partial \Delta_s(p) - \partial \Delta_{s_1}(p)} \frac{f(\zeta)}{(\zeta - p)^{n+1}} \, d\zeta = \int_{\partial \Delta_s(p)} \frac{f(\zeta)}{(\zeta - p)^{n+1}} \, d\zeta - \int_{\partial \Delta_{s_1}(p)} \frac{f(\zeta)}{(\zeta - p)^{n+1}} \, d\zeta.$$

Similarly for s_2, and hence

$$c_n = \frac{1}{2\pi i} \int_{\partial \Delta_s(p)} \frac{f(\zeta)}{(\zeta - p)^{n+1}} \, d\zeta.$$

So we get the same c_n no matter which s we pick.

Next, convergence. For any $\epsilon > 0$, the geometric series used for the first part converges uniformly absolutely when $\left| \frac{z-p}{\zeta-p} \right| = \frac{|z-p|}{s_2} \leq 1 - \epsilon$. In other words, the series converges uniformly absolutely on compact subsets of $\Delta_{s_2}(p)$ (when $|z - p| < s_2$). The geometric series used for the second part converges uniformly absolutely when $\left| \frac{\zeta-p}{z-p} \right| = \frac{s_1}{|z-p|} \leq 1 - \epsilon$. In other words, the series converges uniformly absolutely on compact subsets of $\mathbb{C} \setminus \overline{\Delta_{s_1}(p)}$ (when $|z - p| > s_1$). Hence both parts (and so the entire series) converge uniformly absolutely on compact subsets of $\text{ann}(p; s_1, s_2)$. As s_1 and s_2 were arbitrary such that $r_1 < s_1 < s_2 < r_2$, we get that the series converges uniformly absolutely on compact subsets of $\text{ann}(p; r_1, r_2)$.

Finally, uniqueness of c_n. Suppose $\{d_n\}$ is another sequence such that

$$f(z) = \sum_{n=-\infty}^{\infty} d_n (z - p)^n,$$

converging uniformly absolutely on compact subsets of $\text{ann}(p; r_1, r_2)$. Then

$$c_m = \frac{1}{2\pi i} \int_{\partial \Delta_s(p)} \frac{f(\zeta)}{(\zeta - p)^{m+1}} \, d\zeta = \frac{1}{2\pi i} \int_{\partial \Delta_s(p)} \left(\sum_{n=-\infty}^{\infty} d_n (\zeta - p)^n \right) \frac{1}{(\zeta - p)^{m+1}} \, d\zeta$$

$$= \frac{1}{2\pi i} \sum_{n=-\infty}^{\infty} d_n \int_{\partial \Delta_s(p)} (\zeta - p)^{n-m-1} \, d\zeta$$

$$= d_m,$$

as the only n for which $\int_{\partial \Delta_s(p)} (\zeta - p)^{n-m-1} \, d\zeta$ is nonzero is when $n = m$, that is, when we are integrating $(\zeta - p)^{-1}$, in which case we get $2\pi i$. □

Similarly to the power series, due to the uniqueness of the Laurent series, it does not matter how we obtain it. For example, the function $e^{1/z}$ has the Laurent series

$$e^{1/z} = \sum_{n=0}^{\infty} \frac{1}{n!} \left(\frac{1}{z} \right)^n = \sum_{n=-\infty}^{0} \frac{1}{(-n)!} z^n,$$

which converges uniformly absolutely on compact subsets of $\mathbb{C} \setminus \{0\}$.

The rational function $\frac{1}{1-z}$ that leads to the geometric series can be expanded in a slightly different way if we want its Laurent series expansion in $\text{ann}(0; 1, \infty) = \mathbb{C} \setminus \overline{\mathbb{D}}$:

$$\frac{1}{1-z} = \frac{-1}{z} \frac{1}{1 - \frac{1}{z}} = \frac{-1}{z} \sum_{n=0}^{\infty} \left(\frac{1}{z} \right)^n = \sum_{n=-\infty}^{-1} -z^n.$$

While in general a Laurent series is not a power series, it could very well be when all the c_n for negative n are zero.

Finally, we can differentiate and antidifferentiate formally, in the same way as we did it for power series. The one minor hiccup is that we cannot antidifferentiate the $c_{-1}(z - p)^{-1}$ term. The proof is left as an exercise.

Proposition 4.4.3. *Suppose $p \in \mathbb{C}, 0 \le r_1 < r_2 \le \infty$, and $f : \text{ann}(p; r_1, r_2) \to \mathbb{C}$ is defined by*

$$f(z) = \sum_{n=-\infty}^{\infty} c_n (z - p)^n,$$

converging uniformly on compact subsets of $\text{ann}(p; r_1, r_2)$. Then:

(i) The function f is holomorphic and its derivative is defined by

$$f'(z) = \sum_{n=-\infty}^{\infty} n c_n (z - p)^{n-1},$$

converging uniformly on compact subsets of $\text{ann}(p; r_1, r_2)$.

(ii) If $c_{-1} = 0$, then

$$F(z) = \sum_{n=-\infty, n\neq-1}^{\infty} \frac{c_n}{n+1}(z-p)^{n+1}$$

converges uniformly on compact subsets of $\mathrm{ann}(p; r_1, r_2)$ and $F' = f$.

Exercise 4.4.1: *Prove Proposition 4.4.3. Hint: Remember § 3.4.2.*

Exercise 4.4.2 (Easy): *Suppose $f \colon \Delta_r(p) \to \mathbb{C}$ is holomorphic, and suppose you expand f in a Laurent series in $\mathrm{ann}(p; r_1, r_2)$ for $0 \leq r_1 < r_2 \leq r$. Prove that $c_n = 0$ for all negative n and that c_n for nonnegative n are the coefficients of the power series of f at p.*

Exercise 4.4.3 (Easy): *Suppose f and g are holomorphic functions defined on $\mathrm{ann}(p; r_1, r_2)$. Let a_n be the coefficients in the Laurent series for f and b_n be the coefficients in the Laurent series for g. Suppose that $\alpha, \beta \in \mathbb{C}$. Show that the Laurent series for the function $\alpha f + \beta g$ has coefficients $\alpha a_n + \beta b_n$.*

Exercise 4.4.4 (Easy): *Suppose $\sum_{n=-\infty}^{m} c_n(z-p)^n$ is a Laurent series with only finitely many positive terms. Show that either the series converges nowhere in $\mathbb{C} \setminus \{p\}$, or there exists a number $r \geq 0$ such that the series converges uniformly and absolutely on compact subsets of $\mathrm{ann}(p; r, \infty)$.*

Exercise 4.4.5: *Expand the function $\frac{1}{(z-1)(z-2)}$ using Laurent (or power) series in*
  a) $\mathrm{ann}(0; 0, 1) = \mathbb{D} \setminus \{0\}$,
  b) $\mathrm{ann}(0; 1, 2)$,
  c) $\mathrm{ann}(0; 2, \infty)$.

Exercise 4.4.6: *Suppose $U = \mathrm{ann}(p; r_1, r_2)$ and $r_1 < r < r_2$. Show that every cycle Γ in U is homologous in U to $k\partial\Delta_r(p)$ for some integer k.*

Exercise 4.4.7: *Suppose $f \colon \mathrm{ann}(p; r_1, r_2) \to \mathbb{C}$ is holomorphic, $r_1 < s < r_2$, and*

$$\int_{\partial\Delta_s(p)} f(z)(z-p)^n \, dz = 0$$

for all nonnegative integers n. Prove that f extends through the hole: There exists a holomorphic $g \colon \Delta_{r_2}(p) \to \mathbb{C}$ such that $f = g$ on $\mathrm{ann}(p; r_1, r_2)$.

Exercise 4.4.8: *Suppose f is a holomorphic function defined in a domain that contains the unit circle $\partial\mathbb{D}$, such that*

$$\int_{\partial\mathbb{D}} f(z)\bar{z}^n \, dz = 0$$

for all integers $n \in \mathbb{Z}$. Prove that $f \equiv 0$.

Exercise 4.4.9: *Show that for a Laurent series it is again enough to show convergence somewhere. Suppose $\sum_{n=-\infty}^{\infty} c_n(z-p)^n$ is a Laurent series that converges at two points, z_1 and z_2, where $0 < |z_1 - p| < |z_2 - p| < \infty$. Prove that the series converges uniformly absolutely on compact subsets of $\mathrm{ann}(p; |z_1 - p|, |z_2 - p|)$.*

4.5*i*   Homotopy version of Cauchy ★

4.5.1*i*   Homotopy

Slowly (continuously) deforming one path into another path is called *homotopy*. Let us define it for closed paths, where in this section by "path" we mean only continuous and not necessarily piecewise-C^1.

Definition 4.5.1. Let $U \subset \mathbb{C}$ be open. Two continuous functions $\gamma_0 \colon [a,b] \to U$ and $\gamma_1 \colon [a,b] \to U$ where $\gamma_j(a) = \gamma_j(b)$ (two closed paths) are *homotopic* in U (or relative to U) if there exists a continuous function $H \colon [a,b] \times [0,1] \to U$ such that for all $t \in [a,b]$ and $s \in [0,1]$,

$$H(t,0) = \gamma_0(t), \qquad H(t,1) = \gamma_1(t), \qquad \text{and} \qquad H(a,s) = H(b,s).$$

See Figure 4.6. We also write γ_s, where $\gamma_s(t) = H(t,s)$, for the paths in the homotopy.

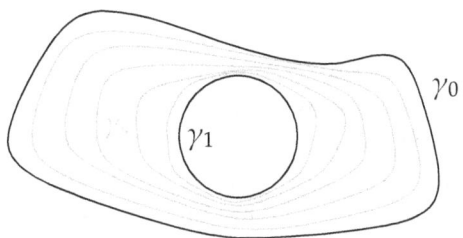

Figure 4.6: Homotopy of two closed paths γ_0 and γ_1 with intermediate paths γ_s marked in gray.

Exercise 4.5.1: *Show that homotopy is an equivalence relation on continuous functions* $\gamma \colon [a,b] \to \mathbb{C}$ *with* $\gamma(a) = \gamma(b)$.

Example 4.5.2: Let $\gamma \colon [a,b] \to \mathbb{D}$ be continuous and $\gamma(a) = \gamma(b)$. Define $H \colon [a,b] \times [0,1] \to \mathbb{D}$ by

$$H(t,s) = (1-s)\gamma(t).$$

This is clearly a homotopy in $\mathbb{D}$, $H(t,0) = \gamma(t)$, and $H(t,1) = 0$, so γ is homotopic to the zero function. So every closed path in $\mathbb{D}$ is homotopic to a constant.

What we want to do is to prove that if γ_0 and γ_1 are piecewise-C^1 paths homotopic in U and $f \colon U \to \mathbb{C}$ is holomorphic, then $\int_{\gamma_0} f(z)\, dz = \int_{\gamma_1} f(z)\, dz$. Consider the intermediate paths $\gamma_s(t) = H(t,s)$. The path γ_s is very close to $\gamma_{s+\epsilon}$ and so it should not be hard to prove that their winding number around various points is the same.

Everything is going swimmingly until we realize that $\int_{\gamma_s} f(z)\, dz$ makes no sense whatsoever. The problem is that γ_s is only continuous and not a piecewise-C^1 path. We can't even define $n(\gamma_s; z)$ using our prior definition. OK, so first we need to define $n(\gamma_s; z)$ in a way that makes sense for any continuous closed path.

Lemma 4.5.3. *Suppose $\gamma \colon [a,b] \to \mathbb{C}$ is continuous and $p \notin \gamma$. Given any $\theta_0 \in \mathbb{R}$ such that $\gamma(a) - p = |\gamma(a) - p| e^{i\theta_0}$, that is, θ_0 is an argument of $\gamma(a) - p$. Then there exists a continuous $\theta \colon [a,b] \to \mathbb{R}$ with $\theta(a) = \theta_0$ such that*

$$\gamma(t) - p = |\gamma(t) - p| e^{i\theta(t)}$$

for all $t \in [a,b]$, that is, $\theta(t)$ is an argument of $\gamma(t) - p$.

Furthermore, if γ is a piecewise-C^1 path, then

$$\frac{1}{2\pi i} \int_\gamma \frac{1}{z-p}\, dz = \frac{\theta(b) - \theta(a)}{2\pi} - i\frac{\log|\gamma(b) - p| - \log|\gamma(a) - p|}{2\pi}.$$

Again, what we'll do is follow log (or arg) around γ and see how much it changes. The proof follows the same logic as in Proposition 4.1.2.

Proof. The image $\gamma([a,b])$ is compact, so it can be covered by finitely many discs $D_1, \ldots, D_n$, none of which contain p, and such that there is a partition $a = t_0 < t_1 < t_2 < \cdots < t_n = b$ such that $\gamma([t_{j-1}, t_j]) \subset D_j$. Each D_j is star-like and does not contain p, so in each one there exists a branch of $\log(z - p)$, call it L_j, such that $L_j(\gamma(t_j)) = L_{j+1}(\gamma(t_j))$. We also ensure that $\operatorname{Im} L_1(\gamma(a)) = \theta_0$. On each $[t_{j-1}, t_j]$ define

$$\theta(t) = \operatorname{Im} L_j(\gamma(t)).$$

On $[t_{j-1}, t_j]$ the function θ is continuous as L_j is continuous on $\gamma([t_{j-1}, t_j])$. The definitions match up at t_{j-1} and t_j with L_{j-1} and L_{j+1} respectively. Thus θ is a continuous function on $[a,b]$. The formula $\gamma(t) - p = |\gamma(t) - p| e^{i\theta(t)}$ follows as L_j is a branch of the log.

The "Furthermore" bit follows as before:

$$\frac{1}{2\pi i} \int_\gamma \frac{1}{z-p}\, dz = \frac{1}{2\pi i} \sum_{j=1}^{n} \int_{t_{j-1}}^{t_j} \frac{\gamma'(t)}{\gamma(t) - p}\, dt$$

$$= \frac{1}{2\pi i} \sum_{j=1}^{n} L_j(\gamma(t_j)) - L_j(\gamma(t_{j-1})) = \frac{1}{2\pi i} \Big(L_n(\gamma(b)) - L_1(\gamma(a))\Big)$$

$$= \frac{\theta(b) - \theta(a)}{2\pi} - i\frac{\log|\gamma(b) - p| - \log|\gamma(a) - p|}{2\pi}. \qquad \square$$

The lemma allows us to define the winding number for continuous closed paths by using the function θ. The "Furthermore" part of the lemma makes sure that the following definition agrees with our previous definition (Definition 4.1.1).

Definition 4.5.4. Let $\gamma \colon [a,b] \to \mathbb{C}$ be continuous, $\gamma(a) = \gamma(b)$, and $p \notin \gamma([a,b])$. Let θ be as in Lemma 4.5.3. Define the *winding number* of γ around p or the *index* of γ with respect to p as

$$n(\gamma;p) \stackrel{\text{def}}{=} \frac{\theta(b) - \theta(a)}{2\pi}.$$

For a closed γ, as two different arguments of a complex number differ by a multiple of 2π, we see that $n(\gamma;p)$ is always an integer. Let us see how the θ, and therefore $n(\gamma;p)$, changes as γ changes (for instance in a homotopy).

Lemma 4.5.5. *Suppose $\gamma \colon [a,b] \to \mathbb{C}$ and $\theta \colon [a,b] \to \mathbb{R}$ are continuous, $p \notin \gamma$, and $\gamma(t)-p = |\gamma(t) - p|e^{i\theta(t)}$ for all t. For every $\epsilon > 0$ there is a $\delta > 0$ such that if $\widetilde{\gamma} \colon [a,b] \to \mathbb{C}$ is continuous, $p \notin \widetilde{\gamma}$, and $|\gamma(t) - \widetilde{\gamma}(t)| < \delta$ for all $t \in [a,b]$, there exists a $\widetilde{\theta} \colon [a,b] \to \mathbb{R}$ such that $|\theta(t) - \widetilde{\theta}(t)| < \epsilon$ for all $t \in [a,b]$ and $\widetilde{\gamma}(t) - p = |\widetilde{\gamma}(t) - p|e^{i\widetilde{\theta}(t)}$.*

Proof. Let t_j, D_j, L_j be the same as in the proof of Lemma 4.5.3, and fix some $\epsilon > 0$. Let $\delta > 0$ be small enough so that if $|z - \zeta| < \delta$ and $z, \zeta \in D_j$, then $|L_j(z) - L_j(\zeta)| < \epsilon$. Such a δ exists as D_j could be picked slightly smaller if needed to make sure that $p \notin \overline{D_j}$ and so that each L_j is uniformly continuous on $\overline{D_j}$ and therefore on D_j.

Next make $\delta > 0$ possibly even smaller so that a δ-neighborhood of each $\gamma([t_{j-1}, t_j])$ is within D_j. Then for $\widetilde{\gamma}$ that is uniformly within δ of γ, we get $\widetilde{\gamma}([t_{j-1}, t_j]) \subset D_j$. The L_j and L_{j-1} agree at one point of $D_{j-1} \cap D_j$ (at $\gamma(t_{j-1})$) and since they are both branches of $\log(z - p)$, they agree on the entire connected set $D_{j-1} \cap D_j$. Thus they also agree at $\widetilde{\gamma}(t_{j-1})$. So

$$\widetilde{\theta}(t) = \operatorname{Im} L_j\big(\widetilde{\gamma}(t)\big)$$

is the function from Lemma 4.5.3, as long as we make $\widetilde{\theta}_0 = \operatorname{Im} L_1\big(\widetilde{\gamma}(a)\big)$. Hence,

$$\big|\theta(t) - \widetilde{\theta}(t)\big| \le \big|L_j\big(\gamma(t)\big) - L_j\big(\widetilde{\gamma}(t)\big)\big| < \epsilon. \qquad \square$$

We can now check how $n(\gamma;p)$ changes, or not, by a homotopy.

Proposition 4.5.6. *Suppose $U \subset \mathbb{C}$ is open and suppose γ_0 and γ_1 are closed paths in U that are homotopic in U. Then*

$$n(\gamma_0;p) = n(\gamma_1;p) \qquad \text{for all } p \in \mathbb{C} \setminus U.$$

Proof. Let $\gamma_s(t) = H(t,s)$ be the maps from the homotopy. Lemma 4.5.5 says that $s \mapsto n(\gamma_s;p)$ is a continuous function. As $s \mapsto n(\gamma_s;p)$ is integer-valued it must be constant. $\qquad \square$

In particular, we've proved that γ_0 and γ_1 are homologous if they are homotopic. The converse is not true. Let us just mention that the path in Figure 4.7 is not homotopic to a constant but it is homologous to the zero chain.

The following corollary is an immediate consequence of Corollary 4.2.5 and Proposition 4.5.6.

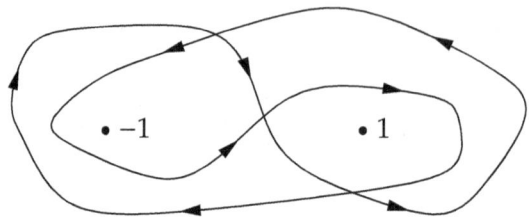

Figure 4.7: A path that is homologous to the zero chain in $\mathbb{C} \setminus \{-1, 1\}$ but not homotopic to a constant in $\mathbb{C} \setminus \{-1, 1\}$.

Corollary 4.5.7. *Suppose $U \subset \mathbb{C}$ is open, $f \colon U \to \mathbb{C}$ is holomorphic, and suppose γ_0 and γ_1 are closed piecewise-C^1 paths in U that are homotopic in U. Then*

$$\int_{\gamma_0} f(z)\, dz = \int_{\gamma_1} f(z)\, dz.$$

A corollary of the corollary is the homotopy version of Cauchy's theorem. While a constant is not technically a path in the way that we defined "path," the integral can easily be defined on it (it is zero), and the integral of any function over it is zero. The following version of Cauchy is then just a special case of the corollary above.

Theorem 4.5.8 (Cauchy's theorem (homotopy version)). *Suppose $U \subset \mathbb{C}$ is open, $f \colon U \to \mathbb{C}$ is holomorphic, and γ is a piecewise-C^1 path in U that is homotopic in U to a constant. Then*

$$\int_{\gamma} f(z)\, dz = 0.$$

Exercise 4.5.2: Let $U \subset \mathbb{C}$ be open and let $\gamma \colon [a, b] \to U$ is continuous and $\gamma(a) = \gamma(b)$. Prove that γ is homotopic in U to a closed piecewise-C^1 path α in U. Hint: Make α a polygonal path.

Exercise 4.5.3: We could take a different approach to solving our issues with the regularity of the homotopy. Let $U \subset \mathbb{C}$ be open and let γ_0 and γ_1 be closed piecewise-C^1 paths in U that are homotopic in U. Show that there exists a homotopy (possibly different one) such that each $\gamma_s(t) = H(t, s)$ is a closed piecewise-C^1 path. Hint: See previous exercise.

Exercise 4.5.4: Let γ be a closed piecewise-C^1 path in $\mathbb{C} \setminus \{0\}$.
a) Show that γ is homotopic in $\mathbb{C} \setminus \{0\}$ to a piecewise-C^1 path whose image is in $\partial \mathbb{D}$. The tricky bit is to make sure that the derivative is never zero.
b) Using part a), show that γ is in fact homotopic to a path $\alpha \colon [0, 2\pi] \to \mathbb{C}$ given by $\alpha(t) = e^{int}$ for some $n \in \mathbb{Z}$.

4.5.2i   The real definition of simply connected

Let us give the real definition of simply connected. For domains in $\mathbb{C}$ it turns out that both definitions we give are equivalent. We will prove one direction of this equivalence in this section, and we will wait with the other direction until we prove the Riemann mapping theorem in section 6.3, because that theorem makes the other direction trivial, see Corollary 6.3.5.

Definition 4.5.9. A domain $U \subset \mathbb{C}$ is *simply connected (in the sense of homotopy)* if every continuous $\gamma \colon [a, b] \to U$ such that $\gamma(a) = \gamma(b)$ is homotopic in U to a constant function.*

Without further ado, here is the simple direction of the equivalence.

Proposition 4.5.10. *If a domain $U \subset \mathbb{C}$ is simply connected in the sense of homotopy, then it is simply connected in the sense of Definition 4.3.1.*

Proof. Let γ be a closed piecewise-C^1 path. We need to show that $n(\gamma; p) = 0$ for all $p \in \mathbb{C} \setminus U$. We know that γ is homotopic to a constant $C \in U$, and it is trivial to see that $n(C; p) = 0$ (the θ is a constant, also notice that $p \neq C$). Thus $n(\gamma; p) = 0$. □

Example 4.5.11: In lieu of a proof of the other direction, simply note that $\mathbb{C}$, $\mathbb{D}$, or the upper half-plane $\mathbb{H}$ are simply connected in the sense of homotopy. For $\mathbb{C}$ and $\mathbb{D}$, we employ the homotopy of example Example 4.5.2. For the half-plane, we modify the homotopy to $H(t, s) = (1 - s)\gamma(t) + si$ to get γ homotopic to the constant i.

A consequence of the proposition is that the simply connected version of Cauchy's theorem holds in the same sense if we define simply-connectedness in terms of homotopy. For completeness, here is the theorem again in the context of this section.

Theorem 4.5.12 (Cauchy's theorem (simply connected version)). *Suppose $U \subset \mathbb{C}$ is a simply connected domain (in the sense of homotopy), $f \colon U \to \mathbb{C}$ is holomorphic, and γ is a piecewise-C^1 path in U. Then*

$$\int_\gamma f(z) \, dz = 0.$$

Exercise 4.5.5: Prove that a star-like domain is simply connected in the sense of homotopy.

Exercise 4.5.6: Let $U, V \subset \mathbb{C}$ be domains such there exists a homeomorphism $f \colon U \to V$, that is, f is bijective, and f and f^{-1} are continuous. Prove that U is simply connected in the sense of homotopy if and only if V is simply connected in the sense of homotopy.

---

*This definition is the correct one (i.e., unlikely to anger a topologist) for any path connected topological space U.

4.6i Cauchy via Green's ⋆

4.6.1i Green's theorem in the complex plane

Cauchy's theorem and Cauchy's integral formula can be obtained via Green's theorem. We review Green's theorem first. Write $dz = dx + i\,dy$ and $d\bar{z} = dx - i\,dy$ as before. Given a piecewise-C^1 path $\gamma\colon [a,b] \to \mathbb{C}$, we define

$$\int_\gamma F(z)\,dz + G(z)\,d\bar{z} \stackrel{\text{def}}{=} \int_a^b \left(F(\gamma(t))\gamma'(t) + G(\gamma(t))\overline{\gamma'(t)} \right) dt,$$

$$\int_\gamma P(z)\,dx + Q(z)\,dy \stackrel{\text{def}}{=} \int_a^b \left(P(\gamma(t))\,\mathrm{Re}\,\gamma'(t) + Q(\gamma(t))\,\mathrm{Im}\,\gamma'(t) \right) dt.$$

Actually, we only need to define one and then get the other via a simple computation, see Exercise 3.1.4.

Let us state a version of Green's theorem without proof. The hypotheses on the domain U and the f are given variously in the literature, so if the reader is working off of a different version of Green's, then the hypotheses of its corollaries in this section must be modified to suit. We're using a version that is the simplest to state in our context. See the next section for a perhaps more common version of the hypotheses.

Theorem 4.6.1 (Green's theorem). *Let Γ be a cycle such that $n(\Gamma;z) = 1$ or 0 for all $z \notin \Gamma$ and let $U = \{z \in \mathbb{C} \setminus \Gamma : n(\Gamma;z) = 1\}$. Suppose P,Q are continuously differentiable functions defined in a neighborhood of $\overline{U}$. Then*

$$\int_\Gamma P(z)\,dx + Q(z)\,dy = \int_U \left(\frac{\partial Q}{\partial x}(z) - \frac{\partial P}{\partial y}(z) \right) dA.$$

Suppose F,G are continuously differentiable functions defined in a neighborhood of $\overline{U}$. In terms of the Wirtinger derivatives, dz, and $d\bar{z}$,

$$\int_\Gamma F(z)\,dz + G(z)\,d\bar{z} = (-2i) \int_U \left(\frac{\partial G}{\partial z}(z) - \frac{\partial F}{\partial \bar{z}}(z) \right) dA.$$

Exercise 4.6.1: *Show that the second form of Green's theorem in terms of the Wirtinger derivatives (the second equation in the theorem) is equivalent to the first form.*

Exercise 4.6.2: *Show that to prove Green's, it would be sufficient to prove*

$$\int_\Gamma F(z)\,dz = 2i \int_U \frac{\partial F}{\partial \bar{z}}(z)\,dA.$$

Cauchy's theorem is an immediate corollary of Green's theorem: In the Wirtinger version of the formula let $G = 0$ and let F be holomorphic.

Corollary 4.6.2. *Let Γ be a cycle such that $n(\Gamma; z) = 1$ or 0 for all $z \notin \Gamma$ and let $U = \{z \in \mathbb{C} \setminus \Gamma : n(\Gamma; z) = 1\}$. Suppose f is a holomorphic function defined in a neighborhood of $\overline{U}$.*

$$\int_\Gamma f(z)\,dz = 0.$$

4.6.2*i*   Generalized Cauchy integral formula

Let us prove a more general version of Cauchy's formula for all functions, not just holomorphic functions. This version is called the *Cauchy–Pompeiu integral formula*.

Theorem 4.6.3 (Cauchy–Pompeiu). *Let Γ be a cycle such that $n(\Gamma; z) = 1$ or 0 for all $z \notin \Gamma$ and let $U = \{z \in \mathbb{C} \setminus \Gamma : n(\Gamma; z) = 1\}$. Suppose f is a continuously differentiable function defined in a neighborhood of $\overline{U}$. Then for $z \in U$:*

$$f(z) = \frac{1}{2\pi i} \int_\Gamma \frac{f(\zeta)}{\zeta - z}\,d\zeta - \frac{1}{\pi} \int_U \frac{\frac{\partial f}{\partial \bar\zeta}(\zeta)}{\zeta - z}\,dA.$$

If f is holomorphic, then the second term is zero, and we obtain the standard Cauchy integral formula. Note that we cheated a little bit in the statement. The integral on the right-hand side is not an integral of a continuous function. There is a singularity, but it turns out that it is still integrable. That is, we write the improper integral

$$\int_U \frac{\frac{\partial f}{\partial \bar\zeta}(\zeta)}{\zeta - z}\,dA = \lim_{r \downarrow 0} \int_{U \setminus \Delta_r(z)} \frac{\frac{\partial f}{\partial \bar\zeta}(\zeta)}{\zeta - z}\,dA.$$

That the integral exists is left as an exercise.

> *Exercise 4.6.3: Observe the singularity in the second term of the Cauchy–Pompeiu formula, and prove that the integral still makes sense (the function is integrable). Hint: Use polar coordinates.*
>
> *Exercise 4.6.4: The reader may be tempted to differentiate in $\bar{z}$ under the second integral in the Cauchy–Pompeiu formula. Why is that not possible? Notice that it would lead to an impossible result.*

Proof. Fix $z \in U$. We wish to apply Green's theorem, but the integrand is not even continuous at z. Let $\Delta_r(z)$ be a small disc such that $\overline{\Delta_r(z)} \subset U$. Green's now applies on $U \setminus \Delta_r(z)$. See Figure 4.8. We compute

$$\int_\Gamma \frac{f(\zeta)}{\zeta - z}\,d\zeta - \int_{\partial \Delta_r(z)} \frac{f(\zeta)}{\zeta - z}\,d\zeta = 2i \int_{U \setminus \Delta_r(z)} \frac{\partial}{\partial \bar\zeta}\left(\frac{f(\zeta)}{\zeta - z}\right)\,dA = 2i \int_{U \setminus \Delta_r(z)} \frac{\frac{\partial f}{\partial \bar\zeta}(\zeta)}{\zeta - z}\,dA.$$

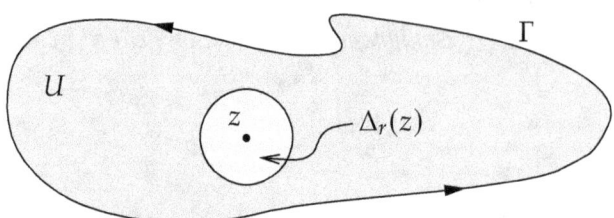

Figure 4.8: Proof of Cauchy–Pompeiu.

The second equality follows because the denominator is holomorphic in ζ. We now let the radius r go to zero. The integral over U is computed as the improper integral

$$\int_U \frac{\frac{\partial f}{\partial \bar{\zeta}}(\zeta)}{\zeta - z} \, dA = \lim_{r \downarrow 0} \int_{U \setminus \Delta_r(z)} \frac{\frac{\partial f}{\partial \bar{\zeta}}(\zeta)}{\zeta - z} \, dA = \frac{1}{2i} \int_\Gamma \frac{f(\zeta)}{\zeta - z} \, d\zeta - \lim_{r \downarrow 0} \frac{1}{2i} \int_{\partial \Delta_r(z)} \frac{f(\zeta)}{\zeta - z} \, d\zeta.$$

By continuity of f,

$$\lim_{r \downarrow 0} \frac{1}{2\pi i} \int_{\partial \Delta_r(z)} \frac{f(\zeta)}{\zeta - z} \, d\zeta = \lim_{r \downarrow 0} \frac{1}{2\pi} \int_0^{2\pi} f(z + re^{i\theta}) \, d\theta = f(z).$$

The theorem follows. $\square$

Exercise 4.6.5: Let $U \subset \mathbb{C}, \Gamma$, and f be as in the theorem, but let $z \notin \overline{U}$. Show that

$$\frac{1}{2\pi i} \int_\Gamma \frac{f(\zeta)}{\zeta - z} \, d\zeta - \frac{1}{\pi} \int_U \frac{\frac{\partial f}{\partial \bar{\zeta}}(\zeta)}{\zeta - z} \, dA = 0.$$

4.7i \ Domains with piecewise-C^1 boundary ⋆

The way Green's theorem, and hence Cauchy's theorem, is often given, and the way it is most often used, is for a domain with piecewise-C^1 boundary. To treat open sets with piecewise-C^1 boundary, we must prove the so-called Jordan curve theorem that is the rather obvious (but surprisingly nontrivial to prove) statement that a simple closed path divides the plane into two components, the interior and the exterior.

Definition 4.7.1. A bounded* open set $U \subset \mathbb{C}$ has *piecewise-C^1 boundary* if ∂U is a disjoint union of finitely many simple closed piecewise-C^1 paths and every $p \in \partial U$ is in the closure of $\mathbb{C} \setminus \overline{U}$.

---

*It is trickier to handle the unbounded case, see the exercises.

Recall that $\gamma\colon [a,b] \to \mathbb{C}$ is simple closed if $\gamma(a) = \gamma(b)$ and $\gamma|_{(a,b]}$ is injective. The condition that every p in the boundary is in the closure of $\mathbb{C} \setminus \overline{U}$ means that at every point, the boundary divides the plane into what's inside U and what's outside U. Let us consider the local question of what a path does, that is, an injective path divides the plane into two pieces.

Lemma 4.7.2. *Let $\gamma\colon [a,b] \to \mathbb{C}$ be an injective piecewise-C^1 path. Then for every $p \in \gamma((a,b))$, there is a connected open neighborhood W of p such that $W \setminus \gamma$ has exactly two components.*

For a piecewise-C^1 path, γ' is never zero, including the one-sided limits of the derivative at the "corners" or the end points. A C^1 path is locally a graph: If $\gamma\colon [-1,1] \to \mathbb{C}$ is a continuously differentiable function, $\gamma(0) = 0$, and $\operatorname{Re}\gamma'(0) \ne 0$, then $z = x + iy = \gamma(t)$, or really the two equations $x = \operatorname{Re}\gamma(t)$ and $y = \operatorname{Im}\gamma(t)$, can be solved for t and y in terms of x near 0 by the implicit function theorem. In other words, γ near 0 is a graph $y = f(x)$ for a C^1 function f. If $\operatorname{Im}\gamma'(0) \ne 0$, the roles of x and y are swapped.

Proof. By the discussion above, a piecewise-C^1 path is a finite union of C^1 paths that are graphs of C^1 functions, either $y = f(x)$ for x in some closed interval, or with x and y reversed. If p is in the interior of such a path (where γ' exists), then without loss of generality it is a graph $y = f(x)$ for perhaps $x \in [-1,1]$ and $f(0) = 0$. For some small $\epsilon > 0$ and $\delta > 0$, we have that $W = (-\epsilon,\epsilon) \times (-\delta,\delta)$ contains no paths making up the original γ and $-\delta < f(x) < \delta$ for $x \in (-\epsilon,\epsilon)$. Clearly, $W \setminus \gamma$ has precisely two components, those points where $f(x) > y$ and those where $f(x) < y$.

So suppose that p is a point where two of those paths making up γ meet and without loss of generality assume p is the origin and $\gamma(0) = p$. As the one sided limits of γ' are never zero, we assume after a rotation assume that $\operatorname{Re}\gamma'(0)$ is not zero for t near 0. That is, the two paths meeting are graphs of the form $y = f_1(x)$ and $y = f_2(x)$. Also assume that the left-hand one-sided limit of γ' at 0 has positive real part and so, without loss of generality, assume that the relevant piece of γ is given by $y = f_1(x)$ for $x \in [-1,0]$. If the right-hand one-sided limit of γ' at 0 has a positive real part, then we may assume that the second piece of γ meeting at p is given by $y = f_2(x)$ for $x \in [0,1]$. The two pieces together are given by a graph $y = f(x)$ where f is f_1 for $x \le 0$ and f_2 for $x \ge 0$ and the argument follows as before.

Suppose that the right-hand one-sided limit of γ' at 0 has negative real part. We assume without loss of generality that the second piece of γ is given by $y = f_2(x)$ for $x \in [-1,0]$. Also without loss of generality, $f_2(x) > f_1(x)$ for some $x < 0$. By injectivity of γ and the intermediate value theorem, $f_2(x) > f_1(x)$ for all $x \in [-1,0)$. Again for some small $\epsilon > 0$ and $\delta > 0$, the set $W = (-\epsilon,\epsilon) \times (-\delta,\delta)$ contains no other of the paths making up γ and $-\delta < f_1(x), f_2(x) < \delta$ for $x \in (-\epsilon,\epsilon)$. It is not hard to see that $W \setminus \gamma$ has precisely two components: One component is the set where $x < 0$ and $f_1(x) < y < f_2(x)$ and the second component is the rest. See Figure 4.9. □

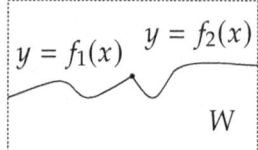 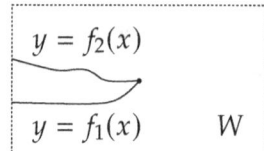

Figure 4.9: A path divides the plane into two pieces, the first case, Re $\gamma'(t) > 0$ for all t, is on the left, and the second case, Re $\gamma'(t) < 0$ for $t > 0$, is on the right.

Exercise 4.7.1: Show that W can be also chosen to be a disc. It is not as trivial as it may seem. Think about γ oscillating wildly and going in and out of the disc.

Exercise 4.7.2: Find an example injective piecewise-C^1 path that is not a graph of a function near some point even after an arbitrary rotation, that is, an example where the second case of the proof is really necessary.

Exercise 4.7.3: Find an example piecewise-C^1 path that is not injective and such that the lemma fails.

Exercise 4.7.4: Find an example of a bounded open set U whose boundary is a union of infinitely many disjoint simple closed curves such that there is some point on $p \in \partial U$ such that for every neighborhood W of p, the set $W \setminus \partial U$ has infinitely many components. (In particular, U is not an open set with piecewise-C^1 boundary.)

The next theorem is usually stated for just continuous paths, but that makes it harder to prove. For piecewise-C^1 paths, the proof, apart from some technicalities, is not hard, and we have actually already proved the local part of it. The main issue with paths that are only continuous is that locally they may not be a graph of a function. A C^1 path is always locally a graph with respect to one of the variables, and a piecewise-C^1 path, as we saw above, is almost a graph.

Theorem 4.7.3 (Jordan curve theorem for piecewise-C^1 paths). *Suppose γ is a simple closed piecewise-C^1 path. Then $\mathbb{C} \setminus \gamma$ has two components, and $n(\gamma; z) = 1$ or $n(\gamma; z) = -1$ for z on the bounded component.*

The bounded component of $\mathbb{C} \setminus \gamma$ is called the *interior* of γ and the unbounded component of $\mathbb{C} \setminus \gamma$ is called the *exterior* of γ.

Proof. A ray is a set that is a straight line starting at some point q going to infinity at some angle. Consider intersections of γ with horizontal lines. For some horizontal line, the intersection with γ with the smallest x coordinate ($z = x + iy$) is at a point where γ' exists and does not point along the horizontal line (otherwise simply move the line a little—there are only finitely many points where γ' does not exist). Near this intersection, the set γ can be written as a graph $x = f(y)$ using the implicit function theorem. Pick a point q to be some point to the right of this graph, and let R be the ray going horizontally left from q. Note that R intersects γ exactly once. See Figure 4.10.

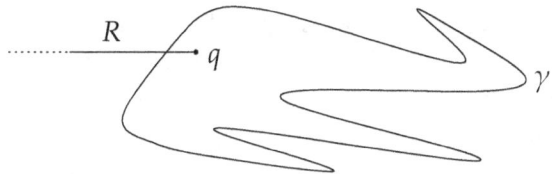

Figure 4.10: A ray that goes from the inside of γ and intersects γ only once.

Without loss of generality, let q be the origin and the ray R be the negative real axis. Then γ intersects the negative real axis at exactly one point. We reparametrize so that this negative real intersection is the beginning and ending point of $\gamma: [0, 1] \to \mathbb{C}$. This point is a point where γ is C^1 and hence $\gamma'(0) = \gamma'(1)$. Also $\gamma'(0) = \gamma'(1)$ does not point along the ray, that is, either $\operatorname{Im} \gamma'(0) = \operatorname{Im} \gamma'(1) > 0$ or $\operatorname{Im} \gamma'(0) = \operatorname{Im} \gamma'(1) < 0$. We then apply Exercise 4.1.9 to find that $n(\gamma; 0) = 1$ or $n(\gamma; 0) = -1$.

In any case, $\mathbb{C} \setminus \gamma$ has the unbounded component and at least one bounded component that contains q, in which the winding number is 1 or -1. We need to show there is no other component. Let U be the component of $\mathbb{C} \setminus \gamma$ that contains q.

For any point $p \in \gamma$, we find a small connected open neighborhood W such that $W \setminus \gamma$ has exactly two components. If p is the point on the ray (the negative real axis), then one of the components of $W \setminus \gamma$ is a subset of U and one of them is a subset of the unbounded component of $\mathbb{C} \setminus \gamma$.

As γ is compact, only finitely many such W are needed to cover it. Suppose W_1 is the neighborhood that contains the negative real point of the boundary. Take a W_2 be one of these neighborhoods such that the intersection $W_1 \cap W_2 \cap \gamma$ is nonempty. As some point of γ is in both W_1 and W_2, then each component of $W_2 \setminus \gamma$ must intersect one of the components of $W_1 \setminus \gamma$. In particular, one of the components of $W_2 \setminus \gamma$ must be a subset of U, and one must be a subset of the unbounded component of $\mathbb{C} \setminus \gamma$. After finitely many steps, as γ is connected, we make this conclusion for all W_j.

In particular, $\gamma \subset \partial U$. But as U is a component of $\mathbb{C} \setminus \gamma$, we get $\partial U \subset \gamma$ and hence $\partial U = \gamma$. If U' is any component of $\mathbb{C} \setminus \gamma$, then it must have nonempty boundary and $\partial U' \subset \gamma$. But the points near γ are only points of U or the unbounded component, so it must be that $U' = U$ or U' is the unbounded component. $\qquad \square$

Using the Jordan curve theorem, we can piece together boundaries.

Proposition 4.7.4. *Suppose $U \subset \mathbb{C}$ is a bounded open set with piecewise-C^1 boundary. Then there exists a cycle Γ that is composed of the paths comprising the boundary ∂U such that $n(\Gamma; z) = 1$ for all $z \in U$ and $n(\Gamma; z) = 0$ for all $z \notin \overline{U}$.*

By a slight abuse of notation we write ∂U for Γ and we use the boundary as a cycle. Because the winding number is 1 around the interior, we say that the boundary is *positively oriented*.

Proof. Without loss of generality, we assume that U is connected by taking one component. Let $\gamma_1, \ldots, \gamma_k$ be the disjoint simple closed piecewise-C^1 paths comprising the boundary of U, suppose that they are all positively oriented, that is, the winding number around their interiors is 1. The set $\mathbb{C} \setminus U$ has one unbounded component and suppose γ_1 is a subset of this unbounded component of $\mathbb{C} \setminus U$. Let U' be the interior of γ_1. As U is connected it must therefore be that $U \subset U'$. In particular, no other γ_j is in the exterior of γ_1 and so $\gamma_2, \ldots, \gamma_k \subset U'$. See Figure 4.11.

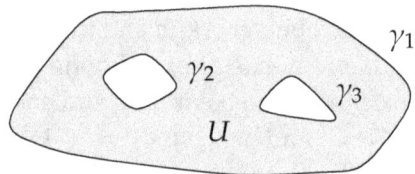

Figure 4.11: The curves comprising the boundary.

Let $\Gamma = \gamma_1 - \gamma_2 - \cdots - \gamma_k$. We need to prove that the set of z such that $n(\Gamma; z) = 1$ is equal to U and $n(\Gamma; z) = 0$ for all z in the complement of $\overline{U}$.

Suppose first that $z \in U$. Then $z \in U'$ so $n(\gamma_1; z) = 1$. If $j \neq 1$, then no point of the interior of γ_j can be in U, as the exterior of γ_j contains γ_1 and therefore points of U. So U is a subset of the intersection of U' and the exterior of γ_j. So z is in the exterior of γ_j and $n(\gamma_j; z) = 0$. Hence, $n(\Gamma; z) = 1$.

On the other hand, if z is in the complement of $\overline{U}$, then either it is in the exterior of γ_1 and hence in the exterior of all the γ_j, or it is in the interior of γ_1 and also in the interior of some γ_j. As U is connected and contains exterior points of all $\gamma_2, \ldots, \gamma_k$, it is not possible for a point to be in the interior of two of these paths. Thus, if $z \in U'$, there is exactly one $j \neq 1$ such that z is in the interior of γ_j. In either case, $n(\Gamma; z) = 0$. $\qquad\qquad\qquad\square$

Green's theorem and the Cauchy integral formula for open sets with piecewise-C^1 boundary follow immediately using ∂U instead of Γ. For example, let us state Green's theorem in the short version as in Exercise 4.6.2 for brevity.

Theorem 4.7.5 (Green's theorem). *Let $U \subset \mathbb{C}$ be a bounded open set with piecewise-C^1 boundary oriented positively. If F is a continuously differentiable function defined in a neighborhood of $\overline{U}$, then*

$$\int_{\partial U} F(z)\, dz = 2i \int_U \frac{\partial F}{\partial \overline{z}}(z)\, dA.$$

The conditions on F can be weakened considerably. A common and easy to prove statement is with the hypothesis that F is continuous on $\overline{U}$ and C^1 only inside U, but it needs to have bounded partial derivatives in order for the right-hand side to be integrable.

> **Exercise 4.7.5:** *If one uses the Lebesgue integral, or a carefully defined improper integral, we can easily reduce the regularity of F needed in Green's theorem above. Using the theorem, obtain the conclusion with the assumption on F being that F is continuous on $\overline{U}$, continuously differentiable on U with bounded partial derivatives on U.*
>
> **Exercise 4.7.6:** *Suppose we drop the requirement that each point in ∂U is in the closure of $\mathbb{C} \setminus \overline{U}$ but require only that ∂U is still composed of disjoint simple closed piecewise-C^1 curves (and still bounded). Find an explicit counterexample to the conclusion of Green's theorem as given above with such a U. Hint: Consider the boundary being two circles, and one of these circles is not in the closure of $\mathbb{C} \setminus \overline{U}$.*
>
> **Exercise 4.7.7:** *Prove that the following definition of piecewise-C^1 boundary is equivalent to the one we gave for bounded open sets: An open $U \subset \mathbb{C}$ has piecewise-C^1 boundary if for every $p \in \partial U$ there exists an open neighborhood W of p and an injective piecewise-C^1 path $\gamma \colon [a,b] \to \mathbb{C}$ such that $W \cap \partial U = \gamma\big((a,b)\big)$, and every $p \in \partial U$ is in the closure of $\mathbb{C} \setminus \overline{U}$.*

5*i*   Counting Zeros and Singularities

If you can count your money, you don't have a billion dollars.

—J. Paul Getty

5.1*i*   Zeros of holomorphic functions

Per the identity theorem, zeros of a holomorphic function are isolated (or the function is identically zero). Let us investigate how a holomorphic function behaves near a zero. We expand upon an idea we used several times before.

Lemma 5.1.1. *Let $U \subset \mathbb{C}$ be open, $f: U \to \mathbb{C}$ be holomorphic, $p \in U$, and f has an isolated zero at p. Then there exists a unique $k \in \mathbb{N}$ and a holomorphic $g: U \to \mathbb{C}$ such that*

$$f(z) = (z - p)^k g(z)$$

and $g(p) \neq 0$. Furthermore, k is the smallest integer such that the kth derivative $f^{(k)}(p) \neq 0$.

Before we prove the lemma, let us give a name to this integer k.

Definition 5.1.2. Suppose f and k are as in the lemma. The k is called the *order* of the zero at p. If the order is 1, we say p is a *simple zero*. We will also call the order, for reasons that will be obvious very soon, the *multiplicity** of the zero at p.

Another way[†] of saying that a zero of f has order k at p is to say that k is the largest integer such that $\frac{f(z)}{(z-p)^k}$ is bounded near p. That these possible definitions are equivalent follows from the lemma. See Exercise 5.1.1 below.

The conclusion of the lemma holds also when $f(p) \neq 0$, in which case $k = 0$ and $g = f$. So one can (if one is really inclined to) say that when $f(p) \neq 0$, then p is a zero of order 0, which sounds somewhat idiotic, but it all fits a general picture, and we will see that negative orders might also make sense in just a little while. However, when we say "f has a zero at p," we never mean that it has a "zero of order zero," we mean an honest zero, $f(p) = 0$.

---

*To be anally retentive: f is of order k at p, and p is a zero of f of multiplicity k. Potayto, potahto.
[†]Which is really the standard way of defining order of a zero for nonholomorphic functions.

If $f^{(k)}(p) = 0$ for all k, then all coefficients of the power series of f at p are zero and f is identically zero. In other words, *every isolated zero of a holomorphic function is of finite order*. In yet other words, *only the zero function has a zero of infinite order*. No such thing is true for real differentiable functions (see the exercises).

Proof of the lemma. On $U \setminus \{p\}$ the function $g(z) = \frac{f(z)}{(z-p)^k}$ is holomorphic for any k, so the trick is only near p. We expand f at p, that is, for z in some disc $\Delta_r(p)$,

$$f(z) = \sum_{n=0}^{\infty} c_n(z-p)^n = \sum_{n=k}^{\infty} c_n(z-p)^n = (z-p)^k \sum_{n=0}^{\infty} c_{n+k}(z-p)^n,$$

where k is the smallest n such that $c_n \neq 0$ (hence the "Furthermore"). Clearly $k > 0$. The series $\sum_{n=0}^{\infty} c_{n+k}(z-p)^n$ is equal to $\frac{f(z)}{(z-p)^k}$ on the punctured disc $\Delta_r(p) \setminus \{p\}$, where the series for f converges. So it can define g near p. Uniqueness follows rather easily: Suppose $(z-p)^{k_1} g_1(z) = (z-p)^{k_2} g_2(z)$, where $g_1(p) \neq 0$ and $g_2(p) \neq 0$. Without loss of generality, $k_1 \leq k_2$. Then $g_1(z) = (z-p)^{k_2-k_1} g_2(z)$. Plug in $z = p$ to see $k_2 = k_1$. $\square$

Next, near a zero of order k, a holomorphic function really acts like the function z^k acts near the origin.

Theorem 5.1.3. *Suppose $U \subset \mathbb{C}$ is open and $f : U \to \mathbb{C}$ is holomorphic. Suppose $p \in U$ is a zero* of f of order $k \in \mathbb{N}$. Then there exists an open neighborhood V of p and a holomorphic $g : V \to \mathbb{C}$ with $g(p) = 0$ and $g'(p) \neq 0$, such that*

$$f(z) = \big(g(z)\big)^k.$$

In more fancy language, g is a local biholomorphic change of variables near p (see Theorem 2.2.8) that makes p into the origin, and makes f into z^k.

Proof. Let $V = \Delta_r(p)$ be a disc such that $f(z) \neq 0$ for any $z \in \Delta_r(p) \setminus \{p\}$. Use the lemma to get a function h holomorphic on $\Delta_r(p)$ such that $h(p) \neq 0$ and $f(z) = (z-p)^k h(z)$. In particular, $h(z) \neq 0$ for any $z \in \Delta_r(p)$. As h is nowhere zero on $\Delta_r(p)$, which is simply connected, there exists a holomorphic function φ on $\Delta_r(p)$ such that $\varphi^k = h$. Let $g(z) = (z-p)\varphi(z)$. As $g'(z) = (z-p)\varphi'(z) + \varphi(z)$, we have $g'(p) = \varphi(p) \neq 0$. $\square$

Exercise 5.1.1: Suppose $U \subset \mathbb{C}$ is open and $f : U \to \mathbb{C}$ holomorphic. Show that f has a zero of order k at $p \in U$ if and only if there is a disc $\Delta_r(p) \subset U$ and some $C_1 > 0$ and $C_2 > 0$ such that for all $z \in \Delta_r(p)$,

$$C_1|z-p|^k \leq |f(z)| \leq C_2|z-p|^k.$$

Exercise 5.1.2 (Easy): Suppose $U \subset \mathbb{C}$ is a domain and f has zeros at $z_1, z_2, \dots, z_n$ of orders $k_1, k_2, \dots, k_n$ and no other zeros. Then there exists a holomorphic $g : U \to \mathbb{C}$ that is never zero such that $f(z) = (z-z_1)^{k_1}(z-z_2)^{k_2} \cdots (z-z_n)^{k_n} g(z)$.

---

*As we said, by this we mean that this is an honest zero of f and $k > 0$.

Exercise 5.1.3 (Easy): *Strengthen the statement of the theorem. Suppose $U \subset \mathbb{C}$ is simply connected and $f : U \to \mathbb{C}$ is holomorphic.*

a) *Suppose f has a zero at p of order k and no other zeros. Show that there exists a holomorphic $g : U \to \mathbb{C}$ such that $g(p) = 0$, $g'(p) \neq 0$, and $f(z) = \big(g(z)\big)^k$.*

b) *Suppose f has a zero at p of order k, and, other than p, it has zeros at $z_1, z_2, \ldots, z_n$ of orders $k_1, k_2, \ldots, k_n$ and no other zeros. Prove that there exists a holomorphic $g : U \to \mathbb{C}$ such that $g(p) = 0$, $g'(p) \neq 0$, and $f(z) = (z - z_1)^{k_1}(z - z_2)^{k_2} \cdots (z - z_n)^{k_n} \big(g(z)\big)^k$.*

Exercise 5.1.4: *The zero set of $\operatorname{Re} f$ looks like $\operatorname{Re} z^k$: Show that if $f : U \to \mathbb{C}$ has a zero of order $k \in \mathbb{N}$ at $p \in U$, then there exist C^1 curves $\gamma_j : (-\epsilon, \epsilon) \to \mathbb{C}$ for $j = 1, \ldots, k$ with $\gamma_j(0) = p$ and $\gamma_j'(0) \neq 0$, such that the curves only intersect one another at p, and such that $\operatorname{Re} f\big(\gamma_j(t)\big) = 0$ for all $t \in (-\epsilon, \epsilon)$.*

Exercise 5.1.5: *Suppose f is holomorphic in an open neighborhood of a point p and suppose $\operatorname{Re} f$ has a critical point at p (derivative is zero). Prove that p is a saddle point of $\operatorname{Re} f$ (neither a local minimum nor a local maximum). Hint: Apply the theorem and show that $k \geq 2$. Note that $\operatorname{Re} z^k$ has a saddle point at the origin.*

Exercise 5.1.6:

a) *For $x \in \mathbb{R}$, let $f(x) = e^{-1/x^2}$ if $x \neq 0$ and $f(0) = 0$. Prove that f is infinitely (real) differentiable, the origin is an isolated zero (the only zero in fact), and $f^{(k)}(0) = 0$ for all $k = 0, 1, 2, \ldots$. That is, the origin is a zero of infinite order.*

b) *Prove that if we define $f(z) = e^{-1/z^2}$ for $z \in \mathbb{C} \setminus \{0\}$, then the function, while holomorphic in $\mathbb{C} \setminus \{0\}$, cannot be made continuous at the origin, no matter how we'd try to define $f(0)$.*

Exercise 5.1.7: *Prove L'Hôpital's rule: If f and g are holomorphic near p, both with an isolated zero at p, then $\lim_{z \to p} \frac{f(z)}{g(z)}$ exists (including possibly ∞) and equals $\lim_{z \to p} \frac{f'(z)}{g'(z)}$.*

5.2i Isolated singularities

5.2.1i Types of singularities and Riemann extension

Definition 5.2.1. Suppose $U \subset \mathbb{C}$ is open and $p \in U$. A holomorphic function $f : U \setminus \{p\} \to \mathbb{C}$ is said to have an *isolated singularity* at p. An isolated singularity is *removable* if there exists a holomorphic $F : U \to \mathbb{C}$ such that $f(z) = F(z)$ for all $z \in U \setminus \{p\}$. An isolated singularity p is a *pole* if

$$\lim_{z \to p} f(z) = \infty.$$

An isolated singularity that is neither removable nor a pole is an *essential singularity*.

In other words, f has an isolated singularity at p if it is defined and holomorphic in a punctured neighborhood $\Delta_r(p) \setminus \{p\}$. It is removable if the function extends across, a pole if f goes to infinity (e.g., $1/z$), and essential otherwise (e.g., $e^{1/z}$). A holomorphic function must blow up in some way if a singularity is not removable. That is a rather surprising property of holomorphic functions with a rather surprisingly simple proof.

Theorem 5.2.2 (Riemann extension theorem). *Suppose $U \subset \mathbb{C}$ is open, $p \in U$, and $f: U \setminus \{p\} \to \mathbb{C}$ is holomorphic. If f is bounded (near p suffices), then p is a removable singularity of f.*

Proof. Define $g(z) = (z - p)^2 f(z)$ for $z \neq p$ and $g(p) = 0$. The function g is clearly holomorphic for $z \neq p$. Consider the difference quotient $\frac{g(z)-g(p)}{z-p} = (z - p)f(z)$. Supposing f is bounded,

$$\lim_{z \to p} \frac{g(z) - g(p)}{z - p} = \lim_{z \to p}(z - p)f(z) = 0.$$

So g is also complex differentiable at p and so holomorphic on U. The order k of the zero of g at p is at least 2 (as $g(p) = 0$ and $g'(p) = 0$). Write $g(z) = (z - p)^k h(z)$, where h is holomorphic on U. Then $f(z) = (z - p)^{k-2} h(z)$, or in other words, p is a removable singularity. $\square$

Exercise 5.2.1: Suppose that $S \subset \mathbb{C}$ is a closed discrete set (each point is isolated), $f: \mathbb{C} \setminus S \to \mathbb{C}$ is holomorphic, and $f(\mathbb{C} \setminus S) \subset \mathbb{D}$. Show that f is constant.

Exercise 5.2.2: Prove that if $f: \mathbb{D} \setminus \{0\} \to \mathbb{D} \setminus \{0\}$ is an automorphism, then $f(z) = e^{i\theta} z$ for some θ.

Exercise 5.2.3: Prove that if $f: \mathbb{D} \setminus \{0, 1/2\} \to \mathbb{D} \setminus \{0, 1/2\}$ is an automorphism, then $f(z) = z$ or $f(z) = \frac{1-2z}{2-z}$.

Exercise 5.2.4: Suppose $U \subset \mathbb{C}$ is open and $\{z_n\}$ is a sequence in U converging to $p \in U$. Let $S = \{z_n : n \in \mathbb{N}\} \cup \{p\}$ and let $f: U \setminus S \to \mathbb{C}$ be a bounded holomorphic function. Prove that f extends through S: There exists a holomorphic $F: U \to \mathbb{C}$ such that $F|_{U \setminus S} = f$.

Exercise 5.2.5: The Riemann extension theorem is (of course) not true for functions that are not holomorphic. Prove that $\frac{xy}{x^2+y^2}$ is a bounded infinitely (real) differentiable function on $\mathbb{R}^2 \setminus \{(0,0)\}$ with an isolated singularity, and this function does not extend through the singularity even continuously.

Exercise 5.2.6: Suppose that f is an entire holomorphic function such that $|f(z)| \leq e^{|\operatorname{Im} z|}$ for all z and such that $f'(0) = 1$ and $f(n\pi) = 0$ for all integers n. Prove that $f(z) = \sin z$.

Exercise 5.2.7: Suppose that f and g are entire functions and $|f| \leq |g|$ everywhere. Show that $f = cg$ for some $c \in \mathbb{C}$. Hint: Make sure to handle the zeros of f and g.

Exercise 5.2.8: Prove that there does not exist a holomorphic $f: \Delta_r(0) \to \mathbb{C}$ (for any $r > 0$) that is not identically zero and such that $f(z)e^{1/z}$ is bounded in $\Delta_r(0) \setminus \{0\}$.

Using the Riemann extension, we will show that at nonessential isolated singularities a holomorphic function blows up to a finite integral order. We obtain a criterion for poles, and we classify poles according to order just like zeros.

Corollary 5.2.3. *Suppose $U \subset \mathbb{C}$ is open, $p \in U$, and $f : U \setminus \{p\} \to \mathbb{C}$ holomorphic.*

(i) *If p is a pole, then there exists a $k \in \mathbb{N}$ such that*

$$g(z) = (z - p)^k f(z)$$

has a removable singularity at p.

(ii) *Conversely, if there exists a $k \in \mathbb{N}$ such that $g(z) = (z - p)^k f(z)$ is bounded near p (has a removable singularity), then f has a pole or a removable singularity at p.*

Proof. Suppose f has a pole at p. Then f is not zero in some punctured neighborhood $\Delta_r(p) \setminus \{p\}$ as it goes to infinity. As f goes to infinity, $1/f$ goes to zero, and so it is bounded in some $\Delta_r(p) \setminus \{p\}$. Thus, $1/f$ has a removable singularity at p by Riemann extension. Let h be holomorphic in $\Delta_r(p)$ such that $h(z) = 1/f(z)$ for $z \neq p$. By continuity, $h(p) = 0$. Hence $h(z) = (z - p)^k \psi(z)$ for some holomorphic ψ and $k \in \mathbb{N}$, where $\psi(p) \neq 0$. As h is not zero in $\Delta_r(p) \setminus \{p\}$, then ψ is not zero in $\Delta_r(p)$ and $1/\psi$ is holomorphic in $\Delta_r(p)$. First item follows as for $z \in \Delta_r(p) \setminus \{p\}$,

$$g(z) = (z - p)^k f(z) = (z - p)^k \frac{1}{(z - p)^k \psi(z)} = \frac{1}{\psi(z)}.$$

The converse statement follows by noting that if $g(z)$ has a removable singularity, then $g(z) = (z - p)^\ell \varphi(z)$ where $\varphi(p) \neq 0$ and $\ell \geq 0$ is an integer. Thus

$$f(z) = (z - p)^{\ell-k} \varphi(z),$$

and this expression either goes to ∞ if $k > \ell$ (f has a pole) or is bounded near p if $k \leq \ell$ (f has a removable singularity). $\qquad \square$

Definition 5.2.4. Given a holomorphic function f with a pole at p, the smallest $k \in \mathbb{N}$ such that $(z - p)^k f(z)$ is bounded near p is called the *order* of the pole. A pole of order 1 is called a *simple pole*.

What we have proved is that if f has a pole at p, f can be written as

$$f(z) = \frac{g(z)}{(z - p)^k},$$

where g is holomorphic, $g(p) \neq 0$, and k is the order of the pole. There is a symmetry between zeros and poles: If f has a zero of order k at p, then $1/f$ has a pole of order k at p. If f has a pole of order k at p, then $1/f$ has a removable singularity, and the extended function has a zero of order k at p. In other words, if f has a pole or a removable singularity, we can write $f(z) = (z - p)^\ell g(z)$ for some $\ell \in \mathbb{Z}$ and some holomorphic g such that $g(p) \neq 0$. The point p is a zero of order ℓ if $\ell > 0$, and it is a pole of order $-\ell$ if $\ell < 0$.

Exercise 5.2.9: *Suppose f has a pole of order $k \in \mathbb{N}$ at p. Show that there exists a holomorphic g defined near p such that $g(p) = 0$ and $g'(p) \neq 0$ and such that near p*

$$f(z) = \frac{1}{\left(g(z)\right)^k}.$$

Exercise 5.2.10: *Suppose $U \subset \mathbb{C}$ is open, $\overline{\mathbb{D}} \subset U$, and $f: U \setminus \{0\} \to \mathbb{C}$ is holomorphic. Suppose f has a simple pole at 0. Prove that for all $z \in \mathbb{D} \setminus \{0\}$,*

$$f(z) = \frac{1}{2\pi i} \int_{\partial \mathbb{D}} \frac{f(\zeta)}{\zeta - z} \, d\zeta + \frac{1}{z} \frac{1}{2\pi i} \int_{\partial \mathbb{D}} f(\zeta) \, d\zeta.$$

Exercise 5.2.11: *Suppose f has an isolated singularity at p. Suppose that $\{z_n\}$ and $\{\zeta_n\}$ are two sequences such that $\lim z_n = \lim \zeta_n = p$ and $\lim f(z_n) \neq \lim f(\zeta_n)$ (both limits exist). Show that f has an essential singularity at p.*

Exercise 5.2.12: *Suppose $f: \mathbb{D} \setminus \{0\} \to \mathbb{C}$ is holomorphic, not identically zero, and f has infinitely many zeros in $\Delta_{1/2}(0) \setminus \{0\}$. Prove that f has an essential singularity at 0.*

5.2.2i Singularities and the Laurent series

The terms of the Laurent series may be used to classify a singularity.

Proposition 5.2.5. *Suppose $f: \Delta_r(p) \setminus \{p\} \to \mathbb{C}$ is holomorphic, and*

$$f(z) = \sum_{n=-\infty}^{\infty} c_n (z - p)^n$$

is the corresponding Laurent series. The singularity at p is

(i) *removable if and only if $c_n = 0$ for all $n < 0$,*

(ii) *a pole of order $k \in \mathbb{N}$ if and only if $c_n = 0$ for all $n < -k$ and $c_{-k} \neq 0$,*

(iii) *essential if and only if $c_n \neq 0$ for infinitely many negative n.*

The fundamental point is that a Laurent series in a punctured disc $\Delta_r(p) \setminus \{p\}$ is unique, so if the singularity is removable, the power series for the extended function must equal the Laurent series.

Exercise 5.2.13: *Prove the proposition.*

Definition 5.2.6. At an isolated singularity, the negative part of the Laurent series

$$\sum_{n=-\infty}^{-1} c_n(z - p)^n$$

is called the *principal part*.

A singularity is removable if the principal part is zero, it is a pole if the principal part is finite, and it is essential if the principal part is infinite. Consider

$$e^{1/z} = \sum_{n=-\infty}^{0} \frac{1}{(-n)!} z^n.$$

The principal part is infinite and $e^{1/z}$ has an essential singularity at 0. This function is the first one anyone ever thinks of if asked for an example of an essential singularity.

Suppose $P(z)$ is the principal part of $f(z)$ at an isolated singularity. It is sometimes useful to consider $f(z) - P(z)$, which then has a removable singularity, as it is defined by a power series.

For entire functions, we can talk about the "singularity at infinity." If we think of $\mathbb{C} \subset \mathbb{C}_\infty$, then this phrase makes perfect sense. The mapping $z \mapsto 1/z$ is a self-mapping of the Riemann sphere $\mathbb{C}_\infty$ that takes infinity to zero. Given $f: \mathbb{C} \to \mathbb{C}$, the function $z \mapsto f(1/z)$ has an isolated singularity at the origin, and that is the singularity at infinity of f. That's exactly what happened with $e^{1/z}$ above. The function e^z has an essential singularity at infinity.

> *Exercise 5.2.14:* Prove that if f has a pole at the origin and g has an essential singularity at the origin, then $f + g$ has an essential singularity at the origin.
>
> *Exercise 5.2.15:* Find holomorphic functions f and g (different pairs for each part) with essential singularities at p, such that
>   a) $f + g$ has a removable singularity at p,
>   b) $f g$ has a removable singularity at p.
>
> *Exercise 5.2.16:* Suppose f is a nonconstant holomorphic function defined in an open neighborhood of the origin such that $f(0) = 0$ and g is holomorphic with an isolated singularity at the origin. Write $g \circ f$ for the composition where it is defined. Show that $g \circ f$ has an isolated singularity of the same type (removable, pole, essential) as g. Moreover, if $f'(0) \neq 0$ and g has a pole of order k, then $g \circ f$ has a pole of order k.
>
> *Exercise 5.2.17:* If f has a pole at p, then $e^{f(z)}$ has an essential singularity at p. Hint: First do it for a simple pole.
>
> *Exercise 5.2.18:* Show that an entire holomorphic $f: \mathbb{C} \to \mathbb{C}$ has a pole at infinity if and only if it is a nonconstant polynomial. The order of the pole is the degree of the polynomial.
>
> *Exercise 5.2.19:* Show that if $f: \mathbb{C} \to \mathbb{C}$ is an automorphism, then $f(z) = az + b$ for some constants $a \neq 0$ and b. Hint: Show that f has a simple pole at infinity.

5.2.3*i*   Wild world of essential singularities, Casorati–Weierstrass

Functions near an essential singularity achieve essentially every value arbitrarily close to the singularity. The function is very wild (and getting wilder and wilder) as it gets close to an essential singularity.

Theorem 5.2.7 (Casorati–Weierstrass*). *Suppose $U \subset \mathbb{C}$ is open and $f : U \setminus \{p\} \to \mathbb{C}$ is holomorphic with an essential singularity at $p \in U$. Then for every punctured disc $\Delta_r(p) \setminus \{p\} \subset U$, the image*

$$f\big(\Delta_r(p) \setminus \{p\}\big) = \big\{w \in \mathbb{C} : w = f(z), z \in \Delta_r(p) \setminus \{p\}\big\}$$

is dense in $\mathbb{C}$.

There is a stronger version of this theorem called the Picard theorem saying that in any punctured neighborhood, f achieves all values with at most one exception: $f\big(\Delta_r(p) \setminus \{p\}\big) = \mathbb{C}$ or $f\big(\Delta_r(p) \setminus \{p\}\big) = \mathbb{C} \setminus \{z_0\}$ for some z_0. But that is much harder to prove.

The intuitive idea of the proof of Casorati–Weierstrass is that if there is a whole disc $\Delta_s(q)$ missing from the image, then take q to ∞ by an LFT and $\Delta_s(q)$ will become the complement of a bounded closed disc, allowing one to use Riemann extension.

Proof. Suppose $f : U \setminus \{p\} \to \mathbb{C}$ is holomorphic, $\Delta_r(p) \subset U$, and that there is a $q \in \mathbb{C}$ and $s > 0$ such that $\Delta_s(q) \subset \mathbb{C} \setminus f\big(\Delta_r(p) \setminus \{p\}\big)$. Consider $g : \Delta_r(p) \setminus \{p\} \to \mathbb{C}$ given by

$$g(z) = \frac{1}{f(z) - q}.$$

By assumption, $|f(z) - q| \geq s$ for $z \in \Delta_r(p) \setminus \{p\}$. Hence $|g(z)| \leq 1/s$, and g has a removable singularity at p by Riemann extension. So assume that g is defined and holomorphic on all of $\Delta_r(p)$. For $z \in \Delta_r(p) \setminus \{p\}$,

$$f(z) = \frac{1}{g(z)} + q.$$

If g has a zero at p, then f has a pole at p. If g does not have a zero at p, then f has a removable singularity. In either case, f does not have an essential singularity at p.   □

> *Exercise 5.2.20: Prove the converse of Casorati–Weierstrass. Let $U \subset \mathbb{C}$ be open, $p \in U$, and $f : U \setminus \{p\} \to \mathbb{C}$ holomorphic. Prove that if $f\big(\Delta_r(p) \setminus \{p\}\big)$ is dense in $\mathbb{C}$ for all $r > 0$ such that $\Delta_r(p) \subset U$, then f has an essential singularity at p.*

---

*Some people say it should be called Casorati–Sochocki(–Weierstrass) theorem as Casorati and Sochocki both published it in 1868, (Casorati in Italian and Sochocki, who was Polish, in Russian) while Weierstrass published it in 1876 (in German). But really it first appeared in a book by Briot and Bouquet in 1859 (in French), so really it should be called the Briot–Bouquet theorem, no? If we all still published in Latin, we wouldn't be in this mess.

Exercise 5.2.21: *Suppose that* $g \colon \Delta_r(p) \setminus \{p\} \to \mathbb{C}$ *has an isolated singularity. Prove that* $f(z) = e^{g(z)}$ *has either a removable singularity, in which case* g *has a removable singularity, or* f *has an essential singularity. Remark: See also Exercise 5.2.17.*

Exercise 5.2.22: *Suppose* $f \colon \mathbb{C} \to \mathbb{C}$ *is holomorphic and nonconstant. Prove that* $f(\mathbb{C})$ *is dense in* $\mathbb{C}$*. Remark: The so-called "little Picard theorem" says that* $f(\mathbb{C})$ *is actually everything minus possibly one point, but that is much harder to prove.*

Exercise 5.2.23: *Suppose* $U \subset \mathbb{C}$ *is open,* $p \in U$*, and* $f \colon U \setminus \{p\} \to \mathbb{C}$ *holomorphic with an essential singularity at* p*.*
    a) Prove a "Picard for modulus" theorem: For every $r > 0$ *such that* $\Delta_r(p) \subset U$*, the set of all moduli of all the values of* f *on* $\Delta_r(p) \setminus \{p\}$*, that is,*

$$\big| f\big(\Delta_r(p) \setminus \{p\}\big) \big| = \big\{ |w| \in \mathbb{R} : w = f(z), z \in \Delta_r(p) \setminus \{p\} \big\},$$

    is $(0, \infty)$ *or* $[0, \infty)$*.*
    b) Show by example that both $(0, \infty)$ *and* $[0, \infty)$ *are possible.*

Exercise 5.2.24: *Suppose* $U \subset \mathbb{C}$ *is open,* $p \in U$*, and* $f \colon U \setminus \{p\} \to \mathbb{C}$ *is holomorphic with an essential singularity at* $p \in U$*. Then for every disc* $\Delta_r(p) \subset U$ *and every segment* $[a, b] \subset \mathbb{C}$*, we have* $f\big(\Delta_r(p) \setminus \{p\}\big) \cap [a, b] \neq \emptyset$*. Hint: See Exercise 2.2.17.*

5.2.4*i*   **Meromorphic functions**

Definition 5.2.8. A holomorphic function $f \colon U \setminus S \to \mathbb{C}$ with poles on a discrete set $S \subset U$ is said to be *meromorphic*.

A meromorphic function can be extended to a function

$$f \colon U \to \mathbb{C}_\infty$$

by simply setting $f(p) = \infty$ at all the poles. By the definition of a pole, the extended function is then continuous. In fact, a way to define a meromorphic function is as a "holomorphic function $f \colon U \to \mathbb{C}_\infty$." Holomorphicity at a pole can be rephrased as holomorphicity of the function $\frac{1}{f(z)}$ at p. There is a small technicality: Should one consider the function that is constantly ∞ as a meromorphic function or not? We will take the view that it is not a meromorphic function.

From now on, when we say that f is meromorphic, we mean that it is a holomorphic function with possible poles in U. We will assume that f is defined to be ∞ at the poles, and we will write $f \colon U \to \mathbb{C}_\infty$. That is, even though we could just say "$f \colon U \to \mathbb{C}_\infty$ is holomorphic," we will, for emphasis, usually say "$f \colon U \to \mathbb{C}_\infty$ is meromorphic."

Similarly, we can define a function on subsets of $\mathbb{C}_\infty$, just as we did with LFTs. We define holomorphicity at ∞ when $U \subset \mathbb{C}_\infty$ by saying that f is *holomorphic at* ∞ if $f(1/z)$ is holomorphic at 0. With this terminology, an LFT is a biholomorphic mapping

$f: \mathbb{C}_\infty \to \mathbb{C}_\infty$. It is left as an exercise that these are the only biholomorphisms of the Riemann sphere and so $\text{Aut}(\mathbb{C}_\infty)$ consists of all the LFTs.

> **Exercise 5.2.25:** *Show that a holomorphic $f: \mathbb{C}_\infty \to \mathbb{C}_\infty$ has at most finitely many poles and finitely many zeros.*
>
> **Exercise 5.2.26:** *Show that a holomorphic $f: \mathbb{C}_\infty \to \mathbb{C}_\infty$ is either constant or onto.*
>
> **Exercise 5.2.27:** *Show that a holomorphic $f: \mathbb{C}_\infty \to \mathbb{C}_\infty$ is a rational function (a polynomial divided by a polynomial).*
>
> **Exercise 5.2.28:** *Show that an injective holomorphic $f: \mathbb{C}_\infty \to \mathbb{C}_\infty$ is an LFT.*

5.3*i* Residue theorem

If f has an isolated singularity at p, expand f in the Laurent series on $\Delta_r(p) \setminus \{p\}$:

$$f(z) = \sum_{n=-\infty}^{\infty} c_n (z - p)^n. \tag{5.1}$$

The only power $(z - p)^n$ that does not have a primitive in $\Delta_r(p) \setminus \{p\}$ is $(z - p)^{-1}$. It is the only power a line integral of f around p "sees," that is, it is what's left, or the "residue"* of integrating $f(z)$ around a closed path.

Definition 5.3.1. Let f be a holomorphic function with an isolated singularity at p. Let the *residue* of f at p be

$$\text{Res}(f; p) \stackrel{\text{def}}{=} c_{-1},$$

where c_{-1} is the coefficient of $(z - p)^{-1}$ in the Laurent series expansion (5.1) in a punctured disc $\Delta_r(p) \setminus \{p\}$.

We know how to compute c_{-1}: For small enough $s > 0$,

$$\text{Res}(f; p) = \frac{1}{2\pi i} \int_{\partial \Delta_s(p)} f(z) \, dz.$$

Via Cauchy's theorem, we relate any integral around a cycle to the residues that lie inside the cycle. With that, we can state a theorem that is often used for computing integrals—even integrals that do not at all seem like line integrals or have any complex numbers in them.

---

*Q: Why did the mathematician name their dog Cauchy? A: Because it left a residue at every pole.

Theorem 5.3.2 (Residue theorem). *Suppose $U \subset \mathbb{C}$ is open, $S \subset U$ is a finite subset, and Γ is a cycle in $U \setminus S$ homologous to zero in U.** *Suppose $f : U \setminus S \to \mathbb{C}$ is holomorphic (isolated singularities on S). Then*

$$\frac{1}{2\pi i} \int_{\Gamma} f(z)\, dz = \sum_{p \in S} n(\Gamma; p) \operatorname{Res}(f; p).$$

Proof. Let $w_1, \ldots, w_\ell$ denote the elements of S. Let $r_1, \ldots, r_\ell$ be positive numbers such that the closed discs $\overline{\Delta_{r_1}(w_1)}, \ldots, \overline{\Delta_{r_\ell}(w_\ell)}$ are mutually disjoint (no pair of them intersects), and $\overline{\Delta_{r_j}(w_j)} \subset U$ for all j. See Figure 5.1.

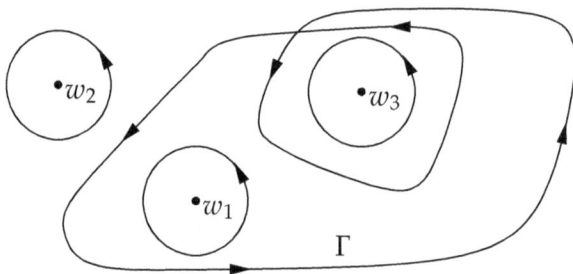

Figure 5.1: Proof of residue theorem by putting small discs around all singularities. Note that $n(\Gamma; w_1) = 1$, $n(\Gamma; w_2) = 0$, and $n(\Gamma; w_3) = 2$.

Define the cycle

$$\Lambda = \Gamma - n(\Gamma; w_1)\, \partial \Delta_{r_1}(w_1) - \cdots - n(\Gamma; w_\ell)\, \partial \Delta_{r_\ell}(w_\ell).$$

We claim that

$$n(\Lambda; p) = 0$$

for all $p \notin U \setminus S$. The winding number is defined by an integral and so

$$n(\Lambda; p) = n(\Gamma; p) - n(\Gamma; w_1)\, n(\partial \Delta_{r_1}(w_1); p) - \cdots - n(\Gamma; w_\ell)\, n(\partial \Delta_{r_\ell}(w_\ell); p).$$

If $p \notin U$, then $n(\Gamma; p) = 0$ as Γ is homologous to zero in U, and as $\overline{\Delta_{r_j}(w_j)} \subset U$ for all j, we get $n(\partial \Delta_{r_j}(w_j); p) = 0$, and the claim follows. If $p = w_k \in S$, then $n(\partial \Delta_{r_j}(w_j); p) = 0$ if $j \neq k$, and $n(\partial \Delta_{r_k}(w_k); p) = 1$. The claim again follows.

By the homology version of the Cauchy theorem, Theorem 4.2.3, we find

$$0 = \frac{1}{2\pi i} \int_{\Lambda} f(z)\, dz = \frac{1}{2\pi i} \int_{\Gamma} f(z)\, dz - \sum_{k=1}^{\ell} n(\Gamma; w_k) \frac{1}{2\pi i} \int_{\partial \Delta_{r_k}(w_k)} f(z)\, dz.$$

We recognize the formula for the c_{-1} term of the Laurent series at w_k, that is,

$$\frac{1}{2\pi i} \int_{\partial \Delta_{r_k}(w_k)} f(z)\, dz = \operatorname{Res}(f; w_k). \qquad \square$$

*As usual, this statement means that $n(\Gamma; z) = 0$ for all $z \in \mathbb{C} \setminus U$.

The residue theorem is supposed to be useful in computing line integrals. But at first glance it seems ridiculous. How does one compute c_{-1}? By an integral. Well how does that help then? It helps because there are easier ways to compute c_{-1} than by the line integral. The first one is almost criminally trivial, but it may be good to emphasize all of them by making them propositions.

Proposition 5.3.3. *Suppose f is holomorphic in an open neighborhood of p and g is holomorphic with an isolated singularity at p, then $\mathrm{Res}(f + g; p) = \mathrm{Res}(g; p)$.*

Proof. For a small enough $\epsilon > 0$,

$$\mathrm{Res}(f + g; p) = \frac{1}{2\pi i} \int_{\partial \Delta_\epsilon(p)} (f(z) + g(z))\, dz$$

$$= \frac{1}{2\pi i} \int_{\partial \Delta_\epsilon(p)} f(z)\, dz + \frac{1}{2\pi i} \int_{\partial \Delta_\epsilon(p)} g(z)\, dz = \mathrm{Res}(g; p). \qquad \square$$

Proposition 5.3.4. *Suppose a meromorphic f has a pole at p. If p is a simple pole of f, then*

$$\mathrm{Res}(f; p) = \lim_{z \to p} (z - p) f(z).$$

More generally, if p is a pole of f of order k, then

$$\mathrm{Res}(f; p) = \frac{1}{(k-1)!} \lim_{z \to p} \frac{d^{k-1}}{dz^{k-1}} \left[(z - p)^k f(z) \right].$$

Exercise **5.3.1:** *Prove the proposition.*

Proposition 5.3.5. *Suppose $f(z) = \frac{h(z)}{g(z)}$ where h and g are holomorphic at p and g has a simple zero at p (f has at most a simple pole at p). Then*

$$\mathrm{Res}(f; p) = \frac{h(p)}{g'(p)}.$$

Exercise **5.3.2:** *Prove the proposition.*

A common application of the residue theorem is to compute certain real integrals that are difficult by classical calculus. Let us work a couple of examples.

Example 5.3.6:

$$\int_{-\infty}^{\infty} \frac{1}{1 + x^2}\, dx.$$

OK, this one is easy to compute by classical calculus, but let us ignore that fact for the sake of the simplicity of the example.

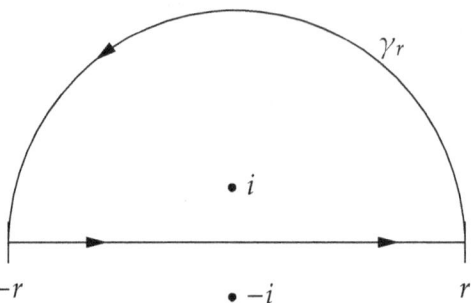

Figure 5.2: The cycle Γ_r.

Define the cycle $\Gamma_r = [-r, r] + \gamma_r$, where $\gamma_r(t) = re^{it}$ for $t \in [0, \pi]$, that is, γ_r is the upper semi-circle of the circle of radius r centered at the origin oriented counterclockwise. See Figure 5.2.

We "complexify" $\frac{1}{1+x^2}$ to make it $\frac{1}{1+z^2}$. That's just a fancy way of saying we are going to plug complex numbers into a real formula. By partial fractions,

$$\frac{1}{1+z^2} = \frac{1}{(z+i)(z-i)} = \frac{i}{2}\frac{1}{z+i} - \frac{i}{2}\frac{1}{z-i}.$$

There are isolated singularities at $\pm i$, both simple poles. The cycle Γ_r goes around i once, so $n(\Gamma_r; i) = 1$, but not around $-i$, that is, $n(\Gamma_r; -i) = 0$. So we only need to compute the residue around i. We can use any one of the three propositions:

$$\text{Res}\left(\frac{1}{1+z^2}; i\right) = \text{Res}\left(\frac{-i}{2}\frac{1}{z-i}; i\right) = \frac{-i}{2}, \qquad \text{Res}\left(\frac{1}{1+z^2}; i\right) = \lim_{z \to i}\frac{z-i}{1+z^2} = \frac{1}{2i} = \frac{-i}{2},$$

$$\text{Res}\left(\frac{1}{1+z^2}; i\right) = \frac{1}{2i} = \frac{-i}{2}.$$

The Residue theorem says

$$\pi = 2\pi i \text{ Res}\left(\frac{1}{1+z^2}; i\right) = \int_{\Gamma_r}\frac{1}{1+z^2}\,dz = \int_{-r}^{r}\frac{1}{1+x^2}\,dx + \int_{\gamma_r}\frac{1}{1+z^2}\,dz.$$

Let us find the limit as $r \to \infty$ of the second term. Assume $r > 1$. The length of γ_r is $r\pi$, and on γ_r, $|1+z^2| \geq r^2 - 1$. So

$$\left|\int_{\gamma_r}\frac{1}{1+z^2}\,dz\right| \leq r\pi\frac{1}{r^2-1} \underset{\text{as } r \to \infty}{\longrightarrow} 0.$$

Hence,

$$\int_{-\infty}^{\infty}\frac{1}{1+x^2}\,dx = \lim_{r \to \infty}\int_{-r}^{r}\frac{1}{1+x^2}\,dx = \pi.$$

Why taking the symmetric limit is sufficient to compute the double improper integral is left to the reader. After all, usually one has to take two independent limits.

Exercise 5.3.3: *Rigorously prove that in the example above, $n(\Gamma_r; i) = 1$ and $n(\Gamma_r; -i) = 0$.*

Another application to real integrals is to recognize a path integral. For example, integrals of trigonometric functions are often integrals over the unit circle. On the unit circle, $\bar{z} = 1/z$. So if $z = e^{i\theta}$, then $\cos\theta = \text{Re } z = \frac{z+1/z}{2}$ and $\sin\theta = \text{Im } z = \frac{z-1/z}{2i}$.

Example 5.3.7: If $c > 1$, then

$$\int_0^{2\pi} \frac{1}{c + \cos\theta}\, d\theta = \int_{\partial\mathbb{D}} \frac{1}{c + \frac{z+1/z}{2}}\, \frac{1}{iz}\, dz = -2i \int_{\partial\mathbb{D}} \frac{1}{z^2 + 2cz + 1}\, dz.$$

The function $\frac{1}{z^2+2cz+1}$ has two simple poles $-c \pm \sqrt{c^2-1}$, one inside and one outside the unit circle. Thus

$$\int_0^{2\pi} \frac{1}{c + \cos\theta}\, d\theta = (-2i)(2\pi i) \text{Res}\left(\frac{1}{z^2 + 2cz + 1}; -c + \sqrt{c^2-1} \right) = \frac{2\pi}{\sqrt{c^2-1}}.$$

Exercise 5.3.4: *For all integers $n \in \mathbb{Z}$, compute*

$$\int_{\partial\mathbb{D}} z^n e^{1/z}\, dz.$$

Exercise 5.3.5: *Compute using the residue theorem (hint: $\cos(3x) = \text{Re } e^{i3x}$):*

a) $\displaystyle\int_{-\infty}^{\infty} \frac{1}{(x^2 + 1)^2}\, dx,$

b) $\displaystyle\int_{-\infty}^{\infty} \frac{\cos(3x)}{x^4 + 1}\, dx.$

Exercise 5.3.6 (Inverse Laplace transform): *A common integral computed via the Residue theorem is the inverse Laplace transform via* Mellin's inversion formula. *Given $F(s)$,*

$$f(t) = \mathscr{L}^{-1}\big[F(s)\big] = \frac{1}{2\pi i} \lim_{r\to\infty} \int_{c-ir}^{c+ir} e^{st} F(s)\, ds, \qquad t \geq 0,$$

is the inverse as long as $c \in \mathbb{R}$ is bigger than the real part of all the singuarities of $F(s)$. Compute (using the residue theorem):

a) $\mathscr{L}^{-1}\left[\frac{1}{s(s+1)} \right],$

b) $\mathscr{L}^{-1}\left[\frac{s^2}{(s+2)^2(s^2+1)} \right].$

Hint: Pick a vertical line (pick a c) and an arc that goes around all the poles.

Exercise 5.3.7: *Compute using the residue theorem:*

a) $\displaystyle\int_0^{2\pi} \frac{\cos\theta}{2 + \cos\theta}\, d\theta,$

b) $\displaystyle\int_0^{\pi} \frac{\sin^2\theta}{2 + \cos\theta}\, d\theta.$

Exercise 5.3.8: *Suppose that $r > 1$, $f: \Delta_r(0) \setminus \{1\} \to \mathbb{C}$ is holomorphic, and suppose f has a simple pole with $\text{Res}(f; 1) = 1$. If the power series for f at 0 is $\sum_{n=0}^{\infty} c_n z^n$, show that $\lim_{n\to\infty} c_n$ exists and compute what it is. Hint: Try subtracting the pole away.*

Exercise 5.3.9: *Suppose f is holomorphic on $U = \{z \in \mathbb{C} : |z| > R\}$ for some $R > 0$. Define the residue of f at ∞, $\operatorname{Res}(f; \infty)$, to be the residue of $g(z) = -z^{-2} f(z^{-1})$ at 0.*
   a) Prove that for all $r > R$,

$$\operatorname{Res}(f; \infty) = \frac{-1}{2\pi i} \int_{\partial \Delta_r(0)} f(z)\, dz.$$

   That is, going around a circle in reverse is going around infinity rather than the center (if what we are "going around" is defined to be whatever is on our left).
   b) Suppose f is holomorphic on $\mathbb{C}$ except for finitely many isolated singularities. Prove that the sum of all residues of f including the residue at ∞ is zero.

Exercise 5.3.10: *Use the function $f(z) = \dfrac{e^{-z^2/2}}{1+e^{-\sqrt{\pi}(1+i)z}}$ and the rectangular path with vertices $-r, r, r + i\sqrt{\pi},$ and $-r + i\sqrt{\pi}$ to compute the integral $\int_{-\infty}^{\infty} e^{-x^2/2}\, dx.$*

5.4i Counting zeros and poles

5.4.1i The argument principle

Integration picks out singularities, and so it can count zeros and poles of a function.

Theorem 5.4.1 (Argument principle). *Suppose $U \subset \mathbb{C}$ is open and Γ is a cycle in U homologous to zero in U. Suppose $f : U \to \mathbb{C}_\infty$ is a meromorphic function with no zeros or poles on Γ. Let $z_1, \ldots, z_n$ denote the zeros of f counted with multiplicity, and let $p_1, \ldots, p_\ell$ denote the poles of f counted with multiplicity. Then*

$$\frac{1}{2\pi i} \int_\Gamma \frac{f'(z)}{f(z)}\, dz = \sum_{k=1}^{n} n(\Gamma; z_k) - \sum_{k=1}^{\ell} n(\Gamma; p_k).$$

Furthermore, if $h : U \to \mathbb{C}$ is holomorphic, then

$$\frac{1}{2\pi i} \int_\Gamma h(z)\frac{f'(z)}{f(z)}\, dz = \sum_{k=1}^{n} n(\Gamma; z_k)h(z_k) - \sum_{k=1}^{\ell} n(\Gamma; p_k)h(p_k).$$

By *zeros counted with multiplicity*, we mean that if a zero has multiplicity (order) m, we repeat it m times. For instance, $f(z) = z^2(z - 1)^3$ has the zeros $z_1, z_2, z_3, z_4, z_5 = 0, 0, 1, 1, 1$. Same with poles. The number of zeros or poles of a meromorphic function inside an open set is possibly countably infinite (unless f is identically zero). But there are only ever finitely many zeros or poles for which $n(\Gamma; z) \neq 0$ (see exercises

---

*This nifty solution is due to H. Kneser. The tricky bit with using the residue theorem is that $e^{-z^2/2}$ has no singularities itself, so one has to find a function that does.

below), as long as Γ is homologous to zero. One can even find a slightly smaller U that only includes finitely many zeros and poles and Γ is still homologous to zero in that smaller U, and so a theorem for finitely many zeros and poles is sufficient.

Why do we say that the theorem counts the number of zeros and poles? Suppose Γ only goes around every point in $z \in U$ at most once in the positive direction or not at all. That is, $n(\Gamma; z) = 1$ or 0 for all $z \in U$. We think of the "inside of Γ" as the points where $n(\Gamma; z) = 1$. If f has n zeros and ℓ poles (counting multiplicity) inside Γ, then

$$\frac{1}{2\pi i} \int_\Gamma \frac{f'(z)}{f(z)} \, dz = n - \ell.$$

The name "argument principle" comes from the fact that for a path γ, the integral $\int_\gamma \frac{f'(z)}{f(z)} \, dz$ computes i times the change in the argument of f as we traverse γ: The antiderivative of $\frac{f'(z)}{f(z)}$ is $\log f(z)$. We take some value of $\log f(z) = \log|f(z)| + i \arg f(z)$ at the beginning of γ, we follow it around γ, and subtract the value of $\log f(z)$ at the end. Another way to look at the integral is to write

$$\frac{1}{2\pi i} \int_\gamma \frac{f'(z)}{f(z)} \, dz = \frac{1}{2\pi i} \int_{f \circ \gamma} \frac{1}{\zeta} \, d\zeta = n(f \circ \gamma; 0).$$

The argument principle counts the number of times $f \circ \gamma$ winds around zero.

Proof of the argument principle. We prove the "Furthermore" as that proves the first part by considering $h \equiv 1$. The function $h(z)\frac{f'(z)}{f(z)}$ has isolated singularities at the zeros and poles of f. Let S be the set of zeros and poles of f, and apply the residue theorem:

$$\frac{1}{2\pi i} \int_\Gamma h(z)\frac{f'(z)}{f(z)} \, dz = \sum_{p \in S} n(\Gamma; p) \operatorname{Res}\left(h\frac{f'}{f}; p\right).$$

We simply compute the residues. Consider a zero of f of multiplicity m or a pole of order $-m$, and without loss of generality suppose it is the origin. Write $f(z) = z^m F(z)$ where F is holomorphic near 0, $F(0) \neq 0$, and $h(z) = h(0) + zH(z)$. Then $f'(z) = mz^{m-1}F(z) + z^m F'(z)$, and so

$$h(z)\frac{f'(z)}{f(z)} = \left(h(0) + zH(z)\right)\frac{mz^{m-1}F(z) + z^m F'(z)}{z^m F(z)}$$

$$= m\, h(0)\frac{1}{z} + h(0)\frac{F'(z)}{F(z)} + H(z)\frac{mF(z) + zF'(z)}{F(z)}.$$

Everything except $m\, h(0)\frac{1}{z}$ is holomorphic near 0. Hence, $\operatorname{Res}\left(h\frac{f'}{f}; 0\right) = m\, h(0)$. The theorem follows. $\qquad\square$

Besides the theoretical implications we will see, there are some immediate practical ones. The argument principle can be used to locate zeros of polynomials (or holomorphic functions more generally) by numerical computations. If we numerically estimate the integral to within a precision of at least 0.5 (no need to be extremely precise), then we know the number of zeros of the polynomial enclosed by the cycle. A related application is computing the power sums of the zeros. Given a cycle Γ going at most once around a certain region, such that $z_1, \ldots, z_n$ are the zeros of f inside Γ, then

$$\frac{1}{2\pi i} \int_\Gamma z^k \frac{f'(z)}{f(z)} \, dz = z_1^k + \cdots + z_n^k.$$

If there is one simple zero z_0 of f enclosed within Γ, then

$$\frac{1}{2\pi i} \int_\Gamma z \frac{f'(z)}{f(z)} \, dz = z_0.$$

One useful consequence is that zeros of a polynomial f vary continuously (interpreted in the right way) as the coefficients of f change.

Exercise 5.4.1: Suppose $U \subset \mathbb{C}$ is open, Γ is a cycle in U homologous to zero in U, and $f : U \to \mathbb{C}_\infty$ is meromorphic with no zeros or poles on Γ. Show that there are only finitely many zeros and poles z of f such that $n(\Gamma; z) \neq 0$.

Exercise 5.4.2: Suppose $U \subset \mathbb{C}$ is open, Γ is a cycle in U homologous to zero in U, and $f : U \to \mathbb{C}_\infty$ is meromorphic with no zeros or poles on Γ. Show that there exists an open $U' \subset U$ with Γ a cycle in U' homologous to zero in U', such that the only zeros or poles z of the restriction $f|_{U'}$ are such that $n(\Gamma; z) \neq 0$.

Exercise 5.4.3: Compute $\int_{\partial \mathbb{D}} \frac{z^3}{2z^2+1} \, dz$ with the argument principle. Hint: $\frac{z^3}{2z^2+1} = \frac{z^2}{4} \frac{4z}{2z^2+1}$.

Exercise 5.4.4: Let f be meromorphic on an open neighborhood of $\overline{\mathbb{D}}$ with no pole or zero on $\partial \mathbb{D}$. Suppose $\operatorname{Re} f(z) > 0$ for all $z \in \partial \mathbb{D}$, and $f(0) = 0$. Prove that f has a pole in $\mathbb{D}$.

Exercise 5.4.5: Suppose $f(z)$ is a degree 3 polynomial, $f(0) = 1$, and

$$\frac{1}{2\pi i} \int_{\partial \Delta_2(2i)} z \frac{f'(z)}{f(z)} \, dz = 2i + 1, \qquad \frac{1}{2\pi i} \int_{\partial \Delta_2(-2i)} z \frac{f'(z)}{f(z)} \, dz = -i,$$

$$\frac{1}{2\pi i} \int_{\partial \Delta_1(5)} z \frac{f'(z)}{f(z)} \, dz = 5.$$

Find f. You may assume f has no zeros on those 3 circles (the integrals exist after all).

Exercise 5.4.6: Let $P_t(z) = z^n + c_{n-1}(t)z^{n-1} + \cdots + c_1(t)z + c_0(t)$ be a polynomial where all the coefficients c_k are continuous functions of $[a, b]$.
    a) Prove that the power sums of the zeros of P_t (sums $z_1^k + \cdots + z_n^k$ where $z_1, \ldots, z_n$ are the zeros) are continuous functions of $t \in [a, b]$.
    b) Prove that if ξ_0 is a simple zero of P_{t_0} for some $t_0 \in (a, b)$, and is the unique zero in $\Delta_{2r}(\xi_0)$, then there exists an $\epsilon > 0$ and a continuous function $\xi : (t_0 - \epsilon, t_0 + \epsilon) \to \Delta_r(\xi_0)$ such that $\xi(t)$ is the unique (simple) zero of P_t in $\Delta_r(\xi_0)$.

> *Exercise 5.4.7:* Prove that if $U \subset \mathbb{C}$ is a domain, $p \notin U$, and there exists a holomorphic $g \colon U \to \mathbb{C}$ such that $g^2 = z - p$, then there does not exist a closed path γ such that $n(\gamma; p) = 1$. Hint: Use the motivation for the theorem, not the theorem itself. Note: Some more work (see Exercise 6.3.16) can show that this means that existence of square roots implies that U is simply connected.

5.4.2*i*   Rouché's theorem

The next theorem allows us to count zeros (or poles) of a function that is close to another function. In rough terms, *the number of zeros minus the number of poles (counting multiplicity) inside a curve does not change if the function does not change much on the curve.* For instance, the functions z^2 and $(z - \epsilon)(z + \epsilon)$ are close on $\partial \mathbb{D}$, and they have the same number of zeros in the disc. A nonzero point might "split" into a zero and a pole, that is, $\frac{z-\epsilon}{z+\epsilon}$ is very close to the function 1 on $\partial \mathbb{D}$. So poles are allowed to "cancel" zeros, but this balance is always maintained. If we do not allow any poles whatsoever, then the theorem says that the number of zeros does not change.

Theorem 5.4.2 (Rouché*). *Suppose $U \subset \mathbb{C}$ is open, Γ is a cycle in U homologous to zero in U, and $n(\Gamma; z)$ is either 0 or 1 for all $z \notin \Gamma$. Suppose $f \colon U \to \mathbb{C}_\infty$ and $g \colon U \to \mathbb{C}_\infty$ are meromorphic functions with no zeros or poles on Γ such that*

$$|f(z) - g(z)| < |f(z)| + |g(z)|$$

for all $z \in \Gamma$. Let $V = \{ z \in U \setminus \Gamma : n(\Gamma; z) = 1 \}$. Let N_f, N_g be the number of zeros in V and P_f, P_g the number of poles in V (both counting multiplicity) of f and g respectively. Then

$$N_f - P_f = N_g - P_g.$$

The condition on Γ means that the cycle is simple in the sense that it only goes around any particular point either once or not at all. We then count the number of zeros and poles inside Γ. The strictness of the inequality is the key point, the nonstrict inequality is always true for any f and g by the triangle inequality. Often the theorem is only applied for holomorphic functions, that is, functions without poles.

Corollary 5.4.3 (Rouché). *Let U, Γ and V be as in Theorem 5.4.2. Suppose $f \colon U \to \mathbb{C}$ and $g \colon U \to \mathbb{C}$ are holomorphic such that $|f(z) - g(z)| < |f(z)| + |g(z)|$ for all $z \in \Gamma$. Then f and g have the same number of zeros (counting multiplicity) in V.*

Observe that for holomorphic functions, the inequality precludes any zeros of f or g on Γ, so the statement of the hypotheses is simpler. This observation is actually quite convenient in the applications as it avoids having to show some technicalities.

The classical statement of the theorem uses the inequality

$$|f(z) - g(z)| < |f(z)|$$

---

*The stronger version we state was actually proved by Estermann in 1962.

as hypothesis. This weaker statement of the theorem is enough for vast majority of applications, and has a simpler geometric meaning. By the argument principle, the number of zeros (minus the number of poles) of f inside Γ corresponds to the number of times $f(z)$ winds around zero as z traverses Γ, that is, $n(f \circ \Gamma; 0)$. The classical visual "proof" using this weaker inequality is a dog on a leash with the master going around a tree. The master is at $f(z)$, the tree is at the origin, and the dog is at $g(z)$, so the length of the leash is $|f(z) - g(z)|$ and the distance of the master from the tree is $|f(z)|$. So the "proof" of the theorem is to observe that if the master walks around a tree k times, and the dog is never further from the master than the distance of the master to the tree, then the dog also walked around the tree k times. See Figure 5.3.

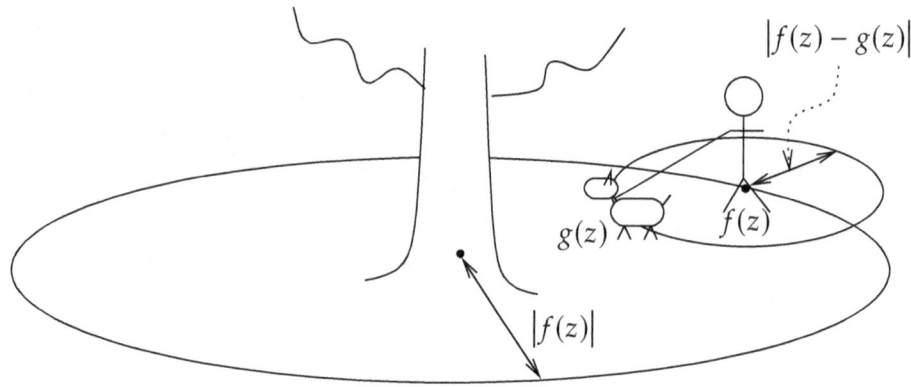

Figure 5.3: Dog and tree proof of Rouché's theorem.

Let us get to the rigorous proof of the symmetric version of the theorem.

Proof of Rouché. Write the inequality as

$$\left| \frac{f(z)}{g(z)} - 1 \right| < \left| \frac{f(z)}{g(z)} \right| + 1.$$

Let $\varphi(z) = \frac{f(z)}{g(z)}$. The inequality precludes $\varphi(z)$ ever being negative real on Γ (it is also never zero there), and hence it is so in a neighborhood of Γ. The function $\frac{\varphi'}{\varphi}$ has a well-defined antiderivative $\mathrm{Log} \circ \varphi$ on a neighborhood of Γ, where Log is the principal branch of the logarithm defined in $\mathbb{C} \setminus (-\infty, 0]$. Cauchy's theorem for derivatives (Corollary 3.2.6) together with the argument principle finishes the proof:

$$0 = \frac{1}{2\pi i} \int_\Gamma \frac{\varphi'(z)}{\varphi(z)} \, dz = \frac{1}{2\pi i} \int_\Gamma \left(\frac{f'(z)}{f(z)} - \frac{g'(z)}{g(z)} \right) dz$$

$$= \frac{1}{2\pi i} \int_\Gamma \frac{f'(z)}{f(z)} \, dz - \frac{1}{2\pi i} \int_\Gamma \frac{g'(z)}{g(z)} \, dz = (N_f - P_f) - (N_g - P_g). \quad \square$$

Example 5.4.4: A typical application of Rouché is to approximately locate zeros of polynomials. Consider $P(z) = z^n + 1$. We can explicitly find the roots, but let us forget we can do so, and show using Rouché that they are all on the unit circle.

First consider Γ to be the circle of radius $1 - \epsilon$ ($\epsilon > 0$) around the origin. On Γ,

$$|P(z) - 1| = |z|^n < 1 = |1|.$$

By Rouché ($P(z)$ is the dog and 1 is the master), $P(z)$ has the same number of zeros as the constant 1 (that is, no zeros) in $\Delta_{1-\epsilon}(0)$.

Second, take z^n instead of 1, and make Γ be the circle of radius $1 + \epsilon$ around the origin. On Γ,
$$|P(z) - z^n| = 1 < |z^n|.$$

By Rouché, $P(z)$ and z^n have the same number of zeros in $\Delta_{1+\epsilon}(0)$, that is, n zeros. As ϵ was arbitrary, all the zeros of $P(z)$ must be on the unit circle.

Example 5.4.5: Consider a more complicated polynomial $P(z) = z^4 + 12z^3 + 24z^2 + 4z + 6$. When $|z| = 1$,

$$\left|P(z) - (z^4 + 24z^2)\right| = \left|12z^3 + 4z + 6\right| \le \left|12z^3\right| + |4z| + |6|$$
$$= 22 < 23 = \left|\,|24z^2| - |z^4|\,\right| \le \left|z^4 + 24z^2\right|.$$

It is easy to see that $z^4 + 24z^2$ has zeros at $\pm\sqrt{24}i$ (outside the unit circle), and two zeros at the origin (inside the unit circle). Thus, $P(z)$ also has two zeros in $\mathbb{D}$.

On the other hand, if $|z| = 46 + \epsilon$ where $\epsilon > 0$, then

$$\left|P(z) - z^4\right| = \left|12z^3 + 24z^2 + 4z + 6\right| \le 46|z|^3 < |z|^4 = \left|z^4\right|.$$

So P has all four zeros in a disc $\Delta_{46+\epsilon}(0)$ for any $\epsilon > 0$, in other words, all zeros of P satisfy $|z| \le 46$. These are not the ideal estimates (largest zero of P has modulus less than 10), but they are explicit and they were easy to come by.

Exercise 5.4.8: Using Rouché's theorem, count the number of zeros of $z^7 - 4z^3 - 11$ in $\text{ann}(0; 1, 2)$.

Exercise 5.4.9: Suppose the monic polynomial $P(z) = z^n + a_{n-1}z^{n-1} + \cdots + a_0$ has no zeros on the unit circle and k zeros (counting multiplicity) in the unit disc ($k \le n$). Show that there exists an $\epsilon > 0$ such that if $|b_j - a_j| < \epsilon$ for $j = 0, \ldots, n - 1$, then $Q(z) = z^n + b_{n-1}z^{n-1} + \cdots + b_0$ has exactly k zeros (counting multiplicity) in $\mathbb{D}$.

Exercise 5.4.10: Suppose $P(z) = z^n + a_{n-1}z^{n-1} + \cdots + a_0$. If $M > \max\left\{1, \sum_{j=0}^{n-1}|a_j|\right\}$, prove that P has n zeros (counting multiplicity) in the disc $\Delta_M(0)$ and no zeros outside. Do this exercise without applying the fundamental theorem of algebra.

Exercise 5.4.11: Suppose $U \subset \mathbb{C}$ is open, $\overline{\mathbb{D}} \subset U$, and $f: U \to \mathbb{C}$ is a holomorphic function with no zeros on $\mathbb{D}$. Prove that there exists a $z \in \partial\mathbb{D}$ such that $|f(z) - z| \ge 1$.

Exercise 5.4.12: *Suppose $U \subset \mathbb{C}$ is open, $\overline{\mathbb{D}} \subset U$, and $f \colon U \to \mathbb{C}$ is a holomorphic function such that $|f(z)| \geq 1$ whenever $z \in \partial \mathbb{D}$, and such that for at least one $p \in \mathbb{D}$, we have $f(p) \in \mathbb{D}$. Prove that $\mathbb{D} \subset f(\mathbb{D})$.*

Exercise 5.4.13: *Suppose $U \subset \mathbb{C}$ is open, $\overline{\mathbb{D}} \subset U$, and $f \colon U \to \mathbb{C}$ is a holomorphic function such that $f\left(\overline{\mathbb{D}}\right) \subset \mathbb{D}$, then there exists exactly one $z_0 \in \mathbb{D}$ such that $f(z_0) = z_0$.*

5.4.3*i*   Hurwitz's theorem

Let us see what happens to zeros under limits of functions. That is, if we know the number of zeros of functions in a sequence, what can we tell about the number of zeros of the limit. Alternatively, if we have a limit function with k zeros, then what can we say about the number of zeros of the functions in the sequence. Recall that if a sequence of holomorphic functions converges uniformly on compact subsets, the limit is holomorphic (see Theorem 3.4.5).

Theorem 5.4.6 (Hurwitz). *Let $U \subset \mathbb{C}$ be open and $f_n \colon U \to \mathbb{C}$ a sequence of holomorphic functions converging uniformly on compact subsets to a holomorphic $f \colon U \to \mathbb{C}$. Suppose Γ is a cycle in U homologous to zero in U, such that $n(\Gamma; z)$ is 0 or 1 for all $z \notin \Gamma$. Suppose f has no zeros on Γ and k zeros (counting multiplicity) in $V = \{ z \in U \setminus \Gamma : n(\Gamma; z) = 1 \}$. Then there is an N such that for all $n \geq N$, f_n has k zeros (counting multiplicity) in V.*

Proof. As a set, Γ is compact, and so there is a $\delta > 0$ such that $\delta < |f(z)|$ for all $z \in \Gamma$. The functions f_n converge uniformly to f on Γ. So for all n large enough,

$$|f(z) - f_n(z)| < \delta < |f(z)|$$

for all $z \in \Gamma$. By Rouché's theorem, f and f_n have the same number of zeros in V. □

Note that it is necessary for f to not be zero on Γ. If $f(z) = z - 1$, then it is zero on the unit circle but not in the unit disc. The sequences of functions $z - (1 - 1/n)$ and $z - (1 + 1/n)$ both converge uniformly to $z - 1$, but $z - (1 - 1/n)$ has one zero in the unit disc and $z - (1 + 1/n)$ does not.

Example 5.4.7: For every integer $k > 0$, there is an N such that for all $d \geq N$, the polynomial

$$P_d(z) = \sum_{n=0}^{d} \frac{(-1)^n}{(2n)!} z^{2n}$$

has exactly $2k$ zeros in $\Delta_{\pi k}(0)$. This claim follows as $\cos(z)$ has exactly $2k$ zeros in that disc and the polynomials P_d are the partial sums of the power series of cosine, which converges uniformly on compact subsets.

The Γ in the theorem is there for defining the region V, a compact set with interior and nice boundary. Often the theorem is applied or stated with Γ being the boundary of a small disc: That is, suppose $\{f_n\}$ is a sequence of holomorphic functions converging uniformly on compact subsets to f on some open U. Suppose z_0 is a zero of f of order k. Then for a small enough disc $\Delta_r(z_0)$, making sure $\overline{\Delta_r(z_0)} \subset U$ and f is only zero at z_0 on $\overline{\Delta_r(z_0)}$, there exists an N such that for all $n \geq N$, f_n has k zeros counting multiplicity in $\Delta_r(z_0)$.

Example 5.4.8: Hurwitz's theorem does not work for real functions. Let $f(x) = x^2$ be a function on $\mathbb{R}$. Then f has a zero (actually a zero of order 2) at $x = 0$. The functions $f_n(x) = x^2 + 1/n$ converge uniformly to f, but f_n has no zero for any $x \in \mathbb{R}$.

On the other hand, consider $f(z) = z^2$ as a function of $\mathbb{C}$, and let $f_n(z) = z^2 + 1/n$. Again f_n goes to f uniformly. Now for any $\epsilon > 0$, $z^2 + 1/n$ has two zeros in $\Delta_\epsilon(0)$ for large enough n. In this case we can even compute them: $\pm i/\sqrt{n}$.

An interesting application of Hurwitz's theorem is that the limit of injective functions is either injective or constant. Injective holomorphic functions are sometimes called *univalent*.

Corollary 5.4.9. *Suppose $U \subset \mathbb{C}$ is a domain and $f_n \colon U \to \mathbb{C}$ are injective holomorphic functions that converge uniformly on compact subsets to $f \colon U \to \mathbb{C}$. Then f is either injective or constant.*

Proof. Assume f is nonconstant. Suppose there exist distinct z_1 and z_2 in U such that $f(z_1) = f(z_2) = w$. The function $f - w$ has isolated zeros at z_1 and z_2. Consider two disjoint small discs $\Delta_r(z_1)$ and $\Delta_r(z_2)$, whose closures are contained in U, and such that $f - w$ is not zero on $\overline{\Delta_r(z_1)} \setminus \{z_1\}$ or $\overline{\Delta_r(z_2)} \setminus \{z_2\}$. For a large enough n, Hurwitz says that $f_n - w$ has the same number of zeros in $\Delta_r(z_1)$ as $f - w$ and the same for $\Delta_r(z_2)$. So there are $z_1' \in \Delta_r(z_1)$ and $z_2' \in \Delta_r(z_2)$ such that $f_n(z_1') = f_n(z_2') = w$. In particular, f_n is not injective. $\qquad\square$

Exercise 5.4.14: *Suppose $U \subset \mathbb{C}$ is a domain, and suppose $f_n \colon U \to \mathbb{C}$ are holomorphic, nowhere zero, and converge uniformly on compact subsets to $f \colon U \to \mathbb{C}$. Show that either f is nowhere zero, or f is identically zero. Give examples of both possible conclusions.*

Exercise 5.4.15:
  a) *Suppose $f_n \colon \mathbb{D} \to \mathbb{C}$ is a sequence of holomorphic functions converging to $f \colon \mathbb{D} \to \mathbb{C}$ uniformly on compact subsets such that for each $0 < r < 1$ the number of zeros (counting multiplicity) of f_n in $\Delta_r(0)$ goes to infinity as $n \to \infty$. Prove that $f \equiv 0$.*
  b) *Find an example sequence of such maps such that $f_n(\mathbb{D}) = \mathbb{D}$ for all n.*

Exercise 5.4.16: *Suppose $f_n \colon \mathbb{C} \to \mathbb{C}$ is a sequence of holomorphic functions converging uniformly on compact subsets to $f \colon \mathbb{C} \to \mathbb{C}$, which is not identically zero. Suppose that all the zeros of f_n are real for all n. Prove that all the zeros of f are real.*

Exercise 5.4.17: *Suppose $U \subset \mathbb{C}$ is a domain, $f_n : U \to \mathbb{C}$ are holomorphic converging uniformly on compact subsets to a nonconstant $f : U \to \mathbb{C}$, and $f(z_0) = w_0$ for some z_0, w_0. Prove that there exists a sequence $\{z_n\}$ in U such that $\lim z_n = z_0$ and $f_n(z_n) = w_0$ for all large enough n.*

Exercise 5.4.18: *Suppose $P_t(z) = z^n + \sum_{k=0}^{n-1} a_k(t) z^k$ is a polynomial with continuous coefficients $a_k : [0,1] \to \mathbb{C}$. Suppose P_t has no zeros on $\partial \mathbb{D}$ for all $t \in [0,1]$. Then the number of zeros of P_t (counting multiplicity) in $\mathbb{D}$ is constant as a function of t.*

Exercise 5.4.19:
　　a) *Find an example sequence of automorphisms of $\mathbb{D}$ converging uniformly on compact subsets of $\mathbb{D}$ to a constant.*
　　b) *Automorphisms of $\mathbb{D}$ extend to be continuous on $\overline{\mathbb{D}}$. Prove that if a sequence of automorphisms converges uniformly on $\overline{\mathbb{D}}$, then the limit is an automorphism. Hint: Prove it is injective and its derivative is never zero.*

Exercise 5.4.20: *Let $f_n(x) = \frac{x(x-1/n)(x+1/n)}{(1/n)^2 + x^2}$ and $f(x) = x$.*
　　a) *Show that $\{f_n\}$ converges uniformly on compact subsets of the real line to f.*
　　b) *Show that on every interval $(-\epsilon, \epsilon)$, f_n has three distinct zeros for large enough n, while $f(x)$ has a simple zero.*
　　c) *Plug in complex values, $f_n(z)$, and show that $\{f_n\}$ does not converge uniformly to anything on any disc $\Delta_\epsilon(0)$.*

5.5i　　The open mapping theorem

A continuous function from a domain $U \subset \mathbb{R}^2$ to $\mathbb{R}^2$ can do all sorts of things to the topology. The surprisingly famous* map $(x, y) \mapsto (x, xy)$ takes all of $\mathbb{R}^2$, which is both an open and a closed set, to the set $\{(x, y) : x \neq 0 \text{ or } y = 0\}$, which is neither open nor closed. Holomorphic functions are always nice to your topology. Recall that for a continuous map, $f^{-1}(V)$ is open whenever V is open. Nonconstant holomorphic functions have this property also in reverse. So while not every holomorphic function is invertible, at least it behaves as if it were as far as the topology is concerned.

Theorem 5.5.1 (Open mapping). *Let $U \subset \mathbb{C}$ be a domain and $f : U \to \mathbb{C}$ be holomorphic and nonconstant. Then $f(V)$ is an open set for every open set $V \subset U$.*

Proof. Suppose f is nonconstant. As U is connected, f is nonconstant near every point. Let $V \subset U$ be open and $p \in V$. As f is holomorphic and nonconstant near p, there is a closed disc $\overline{\Delta_r(p)} \subset V$ small enough such that $f(z) \neq f(p)$ for $z \in \partial \Delta_r(p)$. There is a

---

*It is called a "blow-up" and it is used in obtaining a Fields Medal. Alas, the medal has already been obtained by Hironaka, and you have to find your own map if you want your own medal.

$\delta > 0$ such that $|f(z) - f(p)| > \delta$ for all $z \in \partial\Delta_r(p)$. The function $z \mapsto f(z) - f(p)$ has at least one zero in $\Delta_r(p)$ (at p). Take any $w \in \Delta_\delta(f(p))$. Then for all $z \in \partial\Delta_r(p)$,

$$\left|(f(z) - w) - (f(z) - f(p))\right| = |f(p) - w| < \delta < |f(z) - f(p)|.$$

By Rouché, $z \mapsto f(z) - w$ has at least one zero in $\Delta_r(p)$. In other words,

$$\Delta_\delta(f(p)) \subset f(\Delta_r(p)) \subset f(V). \qquad\qquad \square$$

The open mapping theorem is really a stronger version of the maximum modulus principle: If for any $p \in U$, $f(p)$ is always in the interior of $f(V)$ for any neighborhood V of p, then $|f(z)|$ cannot achieve a maximum at p. Notice also that the proof says something stronger than the theorem statement. It gives an explicit bound. It says that if $|f(z) - f(p)| > \delta$ for $z \in \partial\Delta_r(p)$, then $\Delta_\delta(f(p)) \subset f(\Delta_r(p))$.

Exercise 5.5.1 (Easy)*: Suppose $U, V \subset \mathbb{C}$ are open and $f: U \to \mathbb{C}$ is holomorphic and bijective. Prove that $f^{-1}: V \to U$ is continuous.*

Exercise 5.5.2 (Easy)*: Let $U \subset \mathbb{C}$ be a domain and let $f: U \to \mathbb{C}$ be holomorphic and let $f(U) \subset V$ for some closed set V. Suppose $f(z) \in \partial V$ for some $z \in U$. Prove f is constant.*

Exercise 5.5.3: Let $U \subset \mathbb{C}$ be a domain and let $f: U \to \mathbb{C}$ be holomorphic. Prove that if $|\operatorname{Im} f(z)| = |\operatorname{Re} f(z)|$ for all $z \in U$, then f is constant.

Exercise 5.5.4: Let $U \subset \mathbb{C}$ be a domain and let $f: U \to \mathbb{C}$ be holomorphic. Suppose $|f''(z)| \leq M$ for all $z \in U$ for some $M > 0$. Suppose $p \in U$ where $f'(p) \neq 0$ and $r > 0$ is such that $\frac{M}{2}r < |f'(p)|$ and $\overline{\Delta_r(p)} \subset U$. Let $\delta = |f'(p)|r - \frac{M}{2}r^2 = r\left(|f'(p)| - \frac{M}{2}r\right) > 0$. Prove that $\Delta_\delta(f(p)) \subset f(\Delta_r(p))$.

Exercise 5.5.5: Let $U \subset \mathbb{C}$ be a nonempty bounded domain and let $f: U \to \mathbb{C}$ be holomorphic. Suppose that f is nonconstant and for every $p \in \partial U$,

$$\lim_{z \to p} |f(z)| = 1.$$

Prove that $f(U) = \mathbb{D}$.

Exercise 5.5.6: Let $f: \mathbb{C} \to \mathbb{C}$ be an entire holomorphic function that is real-valued on the unit circle $\partial\mathbb{D}$. Show that f is constant. What if $\partial\mathbb{D}$ is replaced by the real line?

5.6i Inverses of holomorphic functions

The standard local relationship between derivative and injectivity of a function is the inverse function theorem (Theorem 2.2.8). It says that if $f'(z)$ is nonzero somewhere, then in some neighborhood of that point, f is injective and has an inverse g such that

$$g'(w) = \frac{1}{f'(g(w))}.$$

You now have enough machinery to give a simple proof of the inverse function theorem for holomorphic functions without needing the real inverse function.

> *Exercise 5.6.1:* Prove the inverse function theorem (Theorem 2.2.8) using the following outline, without appealing to the real inverse function theorem (Theorem B.3.16).
> a) Show that for some neighborhood V of p, $f|_V$ is injective. Hint: $f(z) - f(p)$ has a simple zero at p.
> b) Show that $W = f(V)$ is open and the inverse $g \colon W \to V$ is continuous.
> c) By looking directly at the difference quotient $\frac{g(w)-g(w_0)}{w-w_0}$, show that g is complex differentiable at all $w_0 \in W$.

So far, nothing really new with holomorphic functions. It is the next result that is surprising: Being injective implies that the derivative is nonzero. A priori, this statement should make no sense at all and it is not true for real functions. A function $f \colon \mathbb{R} \to \mathbb{R}$ such as $f(x) = x^3$ is injective, but its derivative is zero at $x = 0$, and $f^{-1}(x) = \sqrt[3]{x}$ is not differentiable at $x = 0$. On the other hand, intuitively, a holomorphic function is locally like z^k for some k, and the only way that z^k is injective for complex z is if $k = 1$, in which case the derivative is nonzero. You can make that argument into a proof, but it is a bit more complicated than what we will do.

Lemma 5.6.1. *If $U \subset \mathbb{C}$ is open and $f \colon U \to \mathbb{C}$ is holomorphic and injective, then f' is never zero.*

Proof. We prove the contrapositive. Suppose $f'(p) = 0$ for some p and suppose f is nonconstant (near p). So f' has an isolated zero at p. Let $\overline{\Delta_r(p)} \subset U$ be small enough so that $f'(z) \neq 0$ for all $z \in \Delta_r(p) \setminus \{p\}$, and such that $|f(z) - f(p)| > \delta > 0$ for all $z \in \partial\Delta_r(p)$. The function $z \mapsto f(z) - f(p)$ has a zero at p of multiplicity at least two. For $w \in \Delta_\delta(f(p)) \setminus \{f(p)\}$, we have that $z \mapsto f(z) - w$ has at least two zeros in $\Delta_r(p)$ counting multiplicity via Rouché as before. As the derivative of $f(z)$, hence also of $f(z) - w$, is nonzero in the punctured disc, all these zeros are of multiplicity one. Ergo, $f(z) - w$ has more than one distinct zero, and consequently, f is not injective. $\square$

Lemma 5.6.2. *Suppose $U \subset \mathbb{C}$ is open, $f \colon U \to \mathbb{C}$ is holomorphic and injective, and $\overline{\Delta_r(p)} \subset U$. Then for all $w \in f(\Delta_r(p))$,*

$$f^{-1}(w) = \frac{1}{2\pi i} \int_{\partial\Delta_r(p)} \frac{f'(z)z}{f(z) - w} \, dz.$$

Note that $f(\Delta_r(p))$ is a neighborhood of $f(p)$ by the open mapping theorem.

Proof. Fix $w \in f(\Delta_r(p))$ and suppose $\zeta \in \Delta_r(p)$ is such that $f(\zeta) = w$. The derivative of f is never zero by Lemma 5.6.1, and so $z \mapsto f(z) - w$ has a simple zero at $z = \zeta$. By the residue theorem and Proposition 5.3.5,

$$\frac{1}{2\pi i} \int_{\partial \Delta_r(p)} \frac{f'(z)z}{f(z) - w} \, dz = \operatorname{Res}\left(\frac{f'(z)z}{f(z) - w}; \zeta \right) = \frac{f'(\zeta)\zeta}{f'(\zeta)} = \zeta = f^{-1}(w). \qquad \square$$

Alternatively, you could note that the formula is just applying the argument principle with $h(z) = z$. In fact, that is a way that you can guess the formula.

Putting the two lemmas together obtains the main result of this section.

Theorem 5.6.3. *If $U \subset \mathbb{C}$ is open and $f : U \to \mathbb{C}$ is holomorphic and injective, then $f(U)$ is open, f' is never zero on U, and $f^{-1} : f(U) \to U$ is holomorphic.*

Proof. If f is injective, then it is not constant anywhere, and so $f(U)$ is open by the open mapping theorem. By Lemma 5.6.1, f' is never zero on U, by Lemma 5.6.2 the inverse is locally defined by an integral, and by Lemma 3.4.1, f^{-1} is holomorphic. $\square$

The inverse function theorem could be used to prove that f^{-1} is holomorphic instead of Lemma 3.4.1, although the fact that there is an integral formula for f^{-1} is quite interesting on its own.

Exercise 5.6.2: Suppose f is holomorphic in an open neighborhood of the closed unit disc $\overline{\mathbb{D}}$ such that for every $z_0 \in \mathbb{D}$,

$$\int_{\partial \mathbb{D}} \frac{f'(z)}{f(z) - f(z_0)} \, dz = 2\pi i.$$

Prove that $f|_{\mathbb{D}}$ is a biholomorphism of $\mathbb{D}$ and $f(\mathbb{D})$.

Exercise 5.6.3: Suppose $U, V \subset \mathbb{C}$ are domains. Suppose $f_n : U \to V$ are bijective holomorphic mappings that converge uniformly on compact subsets to a nonconstant holomorphic f (which is injective by Corollary 5.4.9). Show that $f(U) = V$ and that $\{f_n^{-1}\}$ converges uniformly on compact subsets to f^{-1}.

Exercise 5.6.4: Suppose $U, V \subset \mathbb{C}$ are open, $k \in \mathbb{N}$, and $f : U \to V$ is an onto k-to-1 holomorphic map, that is, for each $w \in V$, $f^{-1}(w)$ is k distinct points. Prove that f' is never zero on U.

Exercise 5.6.5: Suppose $U \subset \mathbb{C}$ is open, $f : U \to \mathbb{C}$ is holomorphic, $V = f(U)$, such that $z \mapsto f(z) - w$ has two zeros counting multiplicity for every $w \in V$. Call the zeros $z_1(w)$ and $z_2(w)$ (given in some unspecified order).
   a) Prove that $w \mapsto z_1(w) + z_2(w)$ and $w \mapsto z_1(w)z_2(w)$ are holomorphic functions on V. Hint: Argument principle.
   b) Show that $z_1(w)$ and $z_2(w)$ are solutions of $z^2 + b(w)z + c(w) = 0$ for some functions b and c holomorphic on V.
   c) For $f(z) = z^2$, $U = V = \mathbb{C}$, show that neither z_1 nor z_2 is a continuous function of w, no matter how one orders the zeros.

6i Montel and Riemann

A round man cannot be expected to fit in a square hole right away. He must have time to modify his shape.

—*Mark Twain*

6.1*i* Equicontinuity and the Arzelà–Ascoli theorem

6.1.1*i* Convergence of subsequences

The point of Montel's theorem is to find a simple criterion for relatively compact subsets of the set of holomorphic functions. That is, we will try to figure out when does a sequence of holomorphic functions contain a convergent subsequence. We would really like something like the Bolzano–Weierstrass for sequences of numbers: *If $\{z_n\}$ is a bounded sequence in $\mathbb{C}$, then it has a convergent subsequence.* Interestingly, Montel provides just that kind of theorem for holomorphic functions. However, before we get to Montel, we must seek a weaker result of this kind for continuous functions, the Arzelà–Ascoli theorem. For continuous functions, we need something more than just boundedness to get the convergent subsequence, we need some sort of uniformity in the continuity. But let us start with boundedness.

Definition 6.1.1. Let X be any set. Consider a sequence of functions $f_n \colon X \to \mathbb{C}$. We say that $\{f_n\}$ is *pointwise bounded* if for every $x \in X$, there is an $M_x \in \mathbb{R}$ such that

$$|f_n(x)| \leq M_x \qquad \text{for all } n \in \mathbb{N}.$$

We say that $\{f_n\}$ is *uniformly bounded* if there is an $M \in \mathbb{R}$ such that

$$|f_n(x)| \leq M \qquad \text{for all } n \in \mathbb{N} \text{ and all } x \in X.$$

A sequence of functions that converges pointwise is pointwise bounded. The sequence of bounded functions $\left\{\frac{n^2 x}{1+n^2 x^2}\right\}$ for $x \in \mathbb{R}$ is not uniformly bounded, but it is pointwise bounded as it converges pointwise (exercise). On the other hand, a uniformly bounded sequence of functions may not contain any subsequence that converges even pointwise. For instance, $\sin(nx)$ on the real line is one such example.* Below we show that for such a sequence there must always exist a subsequence converging at countably many points, but $\mathbb{C}$ (or any open subset) is uncountable. Moreover, the functions x^n are uniformly bounded and converge pointwise to a function on the unit interval $[0, 1]$, but the limit is discontinuous. We desire continuous functions, ergo we must require better convergence than pointwise. For now, we ignore continuity and show that we get a pointwise converging subsequence on a countable set if we start with pointwise bounded functions. The proof is a nice example of a diagonalization argument.

Proposition 6.1.2. *Let X be a countable set and $\{f_n\}$ a pointwise bounded sequence of functions $f_n \colon X \to \mathbb{C}$. Then $\{f_n\}$ has a subsequence that converges pointwise.*

Proof. Let $x_1, x_2, x_3, \ldots$ be an enumeration of the elements of X. The sequence $\{f_n(x_1)\}_{n=1}^{\infty}$ is bounded and hence there exists a subsequence of $\{f_n\}_{n=1}^{\infty}$, which we denote by $\{f_{1,k}\}_{k=1}^{\infty}$, such that $\{f_{1,k}(x_1)\}_{k=1}^{\infty}$ converges. Suppose we already defined $\{f_{m,k}\}_{k=1}^{\infty}$, a subsequence of $\{f_{m-1,k}\}_{k=1}^{\infty}$, such that $\{f_{m,k}(x_j)\}_{k=1}^{\infty}$ converges for $j = 1, 2, \ldots, m$. Let $\{f_{m+1,k}\}_{k=1}^{\infty}$ be a subsequence of $\{f_{m,k}\}_{k=1}^{\infty}$ such that $\{f_{m+1,k}(x_{m+1})\}_{k=1}^{\infty}$ converges (and hence it converges for all x_j for $j = 1, 2, \ldots, m + 1$). Rinse and repeat.

If X is finite, we are done as the process stops at some point. If X is countably infinite, we pick the sequence $\{f_{k,k}\}_{k=1}^{\infty}$, which is a subsequence of the original sequence $\{f_n\}_{n=1}^{\infty}$. For any m, the tail $\{f_{k,k}\}_{k=m}^{\infty}$ is a subsequence of $\{f_{m,k}\}_{k=1}^{\infty}$ and hence for any m the sequence $\{f_{k,k}(x_m)\}_{k=1}^{\infty}$ converges. $\qquad\square$

Exercise 6.1.1: *Show that the sequence of functions $\left\{\frac{n^2 x}{1+n^2 x^2}\right\}$ for $x \in \mathbb{R}$ is not uniformly bounded, but it is pointwise bounded (in fact it converges pointwise).*

Exercise 6.1.2: *Prove that a uniformly convergent sequence of bounded functions converging to a bounded function is uniformly bounded.*

Exercise 6.1.3: *Define a sequence of continuous functions $f_n \colon \mathbb{R} \to [0, 1]$ such that $\{f_n(x)\}$ converges to 1 on a dense set of x and it converges to 0 on another dense set. Hint: Do it piecewise.*

Exercise 6.1.4 (Requires measure theory): *Prove that on no interval $[a, b] \subset \mathbb{R}$ does $\sin(nx)$ have a pointwise convergent subsequence. First, if f is the pointwise limit of a subsequence on $[a, b]$, use Riemann–Lebesgue lemma to show that $f = 0$ almost everywhere. Second, consider $\int_{[a,b]} f^2 \, dx$ and the dominated convergence theorem to find a contradiction.*

---

*A proof of this fact requires some measure theory, see the exercises.

6.1.2*i*   Equicontinuity

For larger than countable sets, in order to find convergent subsequences of continuous functions we need some uniformity of continuity across the sequence.

Definition 6.1.3. Let (X, d) be a metric space. A set S of functions $f \colon X \to \mathbb{C}$ is *equicontinuous* at $x \in X$ when for every $\epsilon > 0$, there is a $\delta > 0$ such that if $y \in X$ with $d(x, y) < \delta$, then

$$|f(x) - f(y)| < \epsilon \qquad \text{for all } f \in S.$$

We say S is *equicontinuous* if it is equicontinuous at every $x \in X$.

The set S is *uniformly equicontinuous* when for every $\epsilon > 0$, there is a $\delta > 0$ such that if $x, y \in X$ with $d(x, y) < \delta$, then

$$|f(x) - f(y)| < \epsilon \qquad \text{for all } f \in S.$$

For finite sets S, equicontinuity and uniform equicontinuity is the same as continuity and uniform continuity. The notion is interesting for infinite sets.

Exercise 6.1.5: Prove that a finite set of functions continuous at x is equicontinuous at x, and a finite set of uniformly continuous functions is uniformly equicontinuous.

Proposition 6.1.4. Let (X, d) be a compact metric space. Consider an equicontinuous sequence of functions $f_n \colon X \to \mathbb{C}$. Then the sequence $\{f_n\}$ is uniformly equicontinuous.

Proof. Argue by contrapositive. Suppose that $\{f_n\}$ is not uniformly equicontinuous. Then there exists an $\epsilon > 0$ such that for every $k \in \mathbb{N}$, there are $x_k, y_k \in X$ with $d(x_k, y_k) < 1/k$ such that $|f_{n_k}(x_k) - f_{n_k}(y_k)| \geq \epsilon$ for some n_k. By compactness, $\{x_k\}$ and $\{y_k\}$ have convergent subsequences, so without loss of generality, suppose that they converge, in which case they converge to the same $x \in X$. For any $\delta > 0$, take k such that $d(x, x_k) < \delta$ and $d(x, y_k) < \delta$. Then

$$\epsilon \leq |f_{n_k}(x_k) - f_{n_k}(y_k)| \leq |f_{n_k}(x_k) - f_{n_k}(x)| + |f_{n_k}(x) - f_{n_k}(y_k)|.$$

So either $|f_{n_k}(x_k) - f_{n_k}(x)|$ or $|f_{n_k}(x) - f_{n_k}(y_k)|$ is bigger than or equal to a fixed $\epsilon/2$. As we can do that for any δ, $\{f_n\}$ is not equicontinuous at x. $\square$

Exercise 6.1.6: Suppose (X, d) is a compact metric space, and a sequence of continuous functions $f_n \colon X \to \mathbb{C}$ converges uniformly. Prove that $\{f_n\}$ is uniformly equicontinuous.

Exercise 6.1.7: Suppose S is a set of differentiable functions $f \colon [0, 1] \to \mathbb{R}$ such that $|f'(x)| \leq 1$ for all $x \in [0, 1]$. Prove that S is uniformly equicontinuous.

6.1.3*i*   Arzelà–Ascoli

For continuous functions, our analogue of Bolzano–Weierstrass is the Arzelà–Ascoli theorem. Unlike Bolzano–Weierstrass, Arzelà–Ascoli requires equicontinuity in addition to boundedness. We will start with the theorem on compact metric spaces, and then move to open sets. We start with a lemma showing that a countable dense set exists in any compact metric space. We will then be able to apply our result about countable sets.

Proposition 6.1.5. *A compact metric space (X, d) contains a countable dense subset, that is, there exists a countable $D \subset X$ such that $\overline{D} = X$.*

Denote by $B(x, \delta) = \{y \in X : d(x, y) < \delta\}$ the ball of radius δ.

Proof. As X is compact, for every $n \in \mathbb{N}$ there exist $x_{n,1}, x_{n,2}, \dots, x_{n,k_n} \in X$ such that

$$X = B(x_{n,1}, 1/n) \cup \cdots \cup B(x_{n,k_n}, 1/n).$$

Let $D = \bigcup_{n=1}^{\infty} \{x_{n,1}, x_{n,2}, \dots, x_{n,k_n}\}$. The set D is countable as it is a countable union of finite sets. For every $x \in X$ and every $\epsilon > 0$, there exists an n such that $1/n < \epsilon$ and an $x_{n,\ell} \in D$ such that

$$x \in B(x_{n,\ell}, 1/n) \subset B(x_{n,\ell}, \epsilon).$$

Hence $x \in \overline{D}$, so $\overline{D} = X$, and D is dense. $\qquad\square$

Theorem 6.1.6 (Arzelà–Ascoli). *Let (X, d) be a compact metric space, and let $\{f_n\}$ be a pointwise bounded and equicontinuous sequence of functions $f_n \colon X \to \mathbb{C}$. Then $\{f_n\}$ is uniformly bounded and has a uniformly convergent subsequence.*

Proof. First, we show that the sequence is uniformly bounded. As X is compact, the sequence $\{f_n\}$ is uniformly equicontinuous. Hence, there is a $\delta > 0$ such that for all $x \in X$ and all $n \in \mathbb{N}$,

$$B(x, \delta) \subset f_n^{-1}\big(B(f_n(x), 1)\big).$$

By compactness, there exist $x_1, \dots, x_k$ such that

$$X = B(x_1, \delta) \cup \cdots \cup B(x_k, \delta).$$

As $\{f_n\}$ is pointwise bounded, there exists an M such that for $\ell = 1, \dots, k$,

$$|f_n(x_\ell)| \leq M \qquad \text{for all } n.$$

Given any $x \in X$, $x \in B(x_\ell, \delta)$ for some ℓ, and hence $x \in f_n^{-1}\big(B(f_n(x_\ell), 1)\big)$ for all n. In other words, $|f_n(x) - f_n(x_\ell)| < 1$. So $\{f_n\}$ is uniformly bounded as for all n,

$$|f_n(x)| < 1 + |f_n(x_\ell)| \leq 1 + M.$$

Next, pick a countable dense subset $D \subset X$. By Proposition 6.1.2, find a subsequence $\{f_{n_j}\}$ that converges pointwise on D. Write $g_j = f_{n_j}$ for simplicity. The sequence $\{g_n\}$ is uniformly equicontinuous. Let $\epsilon > 0$ be given. There exists a $\delta > 0$ such that for all $x \in X$ and all $n \in \mathbb{N}$,

$$B(x, \delta) \subset g_n^{-1}\big(B(g_n(x), \epsilon/3)\big).$$

By density of D, every $x \in X$ is in $B(y, \delta)$ for some $y \in D$. By compactness of X, there is a finite subset $\{x_1, \ldots, x_k\} \subset D$ such that

$$X = B(x_1, \delta) \cup \cdots \cup B(x_k, \delta).$$

As there are finitely many points and $\{g_n\}$ converges pointwise on D, there exists a single N such that for all $n, m \geq N$,

$$|g_n(x_\ell) - g_m(x_\ell)| < \epsilon/3 \qquad \text{for } \ell = 1, \ldots, k.$$

Let $x \in X$ be arbitrary. There is some ℓ such that $x \in B(x_\ell, \delta)$ and so for all $j \in \mathbb{N}$,

$$|g_j(x) - g_j(x_\ell)| < \epsilon/3.$$

So for $n, m \geq N$,

$$|g_n(x) - g_m(x)| \leq |g_n(x) - g_n(x_\ell)| + |g_n(x_\ell) - g_m(x_\ell)| + |g_m(x_\ell) - g_m(x)| < \epsilon.$$

Hence, the sequence is uniformly Cauchy. By completeness of $\mathbb{C}$, it is uniformly convergent. $\qquad \square$

Before we prove Arzelà–Ascoli for open sets in $\mathbb{C}$, we need a useful lemma, which deserves to be stated separately. It is sometimes called an *exhaustion by compact sets*.

Lemma 6.1.7. *Let $U \subset \mathbb{C}$ be open. Then there exists a sequence K_n of compact subsets of U such that $K_n \subset K_{n+1}^\circ$ (each set is contained in the interior of the next), $\bigcup_{n=1}^\infty K_n = U$, and for every compact $K \subset U$, there is an n such that $K \subset K_n$.*

Proof. Let $d(z, \partial U)$ denote the distance to the boundary of U. Define

$$K_n = \big\{z \in U : d(z, \partial U) \geq 1/n \text{ and } |z| \leq n\big\}.$$

The set K_n is compact: It is closed (in $\mathbb{C}$) by Proposition A.5.5 and obviously bounded. It is also easy to see that $U = \bigcup K_n$. The interior of K_n is given (exercise) by

$$K_n^\circ = \big\{z \in U : d(z, \partial U) > 1/n \text{ and } |z| < n\big\}.$$

It is then clear that $K_n \subset K_{n+1}^\circ$, and $U = \bigcup K_n^\circ$ (it is an open cover). Therefore, any compact $K \subset U$ is contained in some K_n°, and hence in K_n. $\qquad \square$

Exercise **6.1.8:** *Prove the formula of K_n°. That is, prove that the interior of K_n is $K_n^\circ = \big\{z \in U : d(z, \partial U) > 1/n \text{ and } |z| < n\big\}$. Then prove that $K_n \subset K_{n+1}^\circ$ for all n.*

We next prove a version of Arzelà–Ascoli for open subsets of $\mathbb{C}$, that is, a version that gives us uniform convergence on every compact subset of an open set.

Corollary 6.1.8 (Arzelà–Ascoli). *Let $U \subset \mathbb{C}$ be open and let $\{f_n\}$ be a pointwise bounded and equicontinuous sequence of functions $f_n \colon U \to \mathbb{C}$. Then $\{f_n\}$ contains a subsequence that converges uniformly on compact subsets.*

Proof. Find the exhaustion by compact sets $\{K_\ell\}$ from the lemma. Using the Arzelà–Ascoli theorem on compact sets, find a subsequence $\{f_{1,n}\}$ of $\{f_n\}$ that converges uniformly on K_1. Then find a subsequence $\{f_{2,n}\}$ of $\{f_{1,n}\}$ that converges uniformly on K_2, and so on. Finally take the diagonal sequence $\{f_{n,n}\}$. Any compact $K \subset U$ is contained in some K_ℓ. The ℓ-tail of the sequence $\{f_{n,n}\}$ is a subsequence of $\{f_{\ell,n}\}$ and hence uniformly convergent on K_ℓ and thus on K. □

Exercise 6.1.9: *Suppose that $f_n \colon [0,1] \to \mathbb{C}$ are functions that are pointwise bounded, (real) differentiable, and for some $M > 0$, we have $|f_n'(t)| \leq M$ for all $t \in [0,1]$ and all n. Prove that there exists a subsequence that converges uniformly on $[0,1]$.*

Exercise 6.1.10: *Let $f_n \colon [-1,1] \to \mathbb{R}$ be given by $f_n(x) = \frac{nx}{1+(nx)^2}$. Prove that the sequence is uniformly bounded, converges pointwise to 0, but does not converge uniformly to 0. Which hypothesis of Arzelà–Ascoli is not satisfied? Prove your assertion.*

Exercise 6.1.11: *Suppose $f_n \colon \partial \mathbb{D} \to \mathbb{C}$ are uniformly bounded continuous functions. Let $g(z,w)$ be a continuous function on $\overline{\mathbb{D}} \times \partial \mathbb{D}$. Define $F_n \colon \overline{\mathbb{D}} \to \mathbb{C}$ by*

$$F_n(z) = \int_{\partial \mathbb{D}} f_n(w)\, g(z,w)\, dw.$$

Show that $\{F_n\}$ has a uniformly convergent subsequence.

Exercise 6.1.12: *Suppose (X,d) is a compact metric space and $\{f_n\}$ an equicontinuous sequence of functions on X. If $\{f_n\}$ converges pointwise, show that it converges uniformly.*

Exercise 6.1.13: *Define $f_n \colon [0,1] \to \mathbb{C}$ by $f_n(t) = e^{i(2\pi t + n)}$.*
  a) *Prove that $\{f_n\}$ is a uniformly equicontinuous uniformly bounded sequence.*
  b) *Let $\delta \in \mathbb{R}$ be given, and define $g(t) = e^{i(2\pi t + \delta)}$. Prove that there exists a subsequence of $\{f_n\}$ converging uniformly to g.*
Feel free to use the Kronecker density theorem: $\{e^{in}\}_{n=1}^\infty$ is dense in the unit circle.

6.2*i*   Montel's theorem

For holomorphic functions, a bound on the function means a bound on the derivative using the Cauchy integral formula. A uniform bound on the derivative gives equicontinuity, and so it comes as no surprise that a uniformly bounded set of holomorphic functions is equicontinuous. Let us use some of the traditional language.*

---

*We wouldn't want the reader to miss out on all the jokes about how Montel had a "normal family."

Definition 6.2.1. Let $U \subset \mathbb{C}$ be open. A set $\mathcal{F}$ of holomorphic functions $f : U \to \mathbb{C}$ is a *normal family* if every sequence in $\mathcal{F}$ has a subsequence that converges uniformly on compact subsets (the limit need not be in $\mathcal{F}$).

A set $\mathcal{F}$ of functions on U is *locally bounded* if for every $p \in U$, there is a disc $\Delta_r(p) \subset U$ and $M > 0$, such that $\|f\|_{\Delta_r(p)} \leq M$ for all $f \in \mathcal{F}$.

In more modern language, a set $\mathcal{F}$ is a normal family if it is precompact, or relatively compact, in the space of holomorphic functions on U. But we didn't define an actual topology or metric on this space in this book, so we will just use the traditional verbiage.

Theorem 6.2.2 (Montel). *Let $U \subset \mathbb{C}$ be open and let $\mathcal{F}$ be a locally bounded set of holomorphic functions on U. Then $\mathcal{F}$ is a normal family (every sequence has a subsequence that converges uniformly on compact subsets).*

The theorem allows quite incredible applications. The thing is, using just the definitions, it is often easy to show that a sequence is bounded, but it is hard to show that it converges (or has a subsequence that does). One can generate "approximate" solutions to a problem without worrying about them being close to something. The reader has seen applications of this idea (using Bolzano–Weierstrass) when working with sequences in $\mathbb{C}$ or $\mathbb{R}$ before (several times in this book already in fact).

Proof. The point of the proof is to apply Arzelà–Ascoli, so let us go through the hypotheses. Clearly $\mathcal{F}$ is pointwise bounded as it is bounded on discs around every point. We need to show that it is equicontinuous at every point.

Consider $p \in U$ and suppose $\overline{\Delta_r(p)} \subset U$. The family $\mathcal{F}$ is bounded on this disc, say $\|f\|_{\overline{\Delta_r(p)}} \leq M$ for all $f \in \mathcal{F}$. For $z \in \overline{\Delta_{r/2}(p)}$,

$$|f'(z)| = \left| \frac{1}{2\pi i} \int_{\partial \Delta_r(p)} \frac{f(\zeta)}{(\zeta - z)^2} \, d\zeta \right| \leq \frac{1}{2\pi} \int_{\partial \Delta_r(p)} \frac{|f(\zeta)|}{|\zeta - z|^2} |d\zeta|$$

$$\leq \frac{1}{2\pi} \int_{\partial \Delta_r(p)} \frac{M}{(r/2)^2} |d\zeta| = \frac{4M}{r}.$$

And so

$$|f(z) - f(p)| = \left| \int_{[p,z]} f'(\zeta) \, d\zeta \right| \leq \int_{[p,z]} |f'(\zeta)| \, |d\zeta| \leq \frac{4M}{r} |z - p|.$$

As M does not depend on the particular f, we get that $\mathcal{F}$ is equicontinuous at p (it is in fact Lipschitz at p with the same Lipschitz constant for every $f \in \mathcal{F}$). Therefore, we may apply Arzelà–Ascoli, Corollary 6.1.8, to any sequence in $\mathcal{F}$ to find a convergent subsequence and hence $\mathcal{F}$ is a normal family.   □

Montel's theorem is very useful for solving extremal problems: finding a holomorphic function that satisfies a certain extremal condition such as maximizing the derivative. There are several examples of this usage in the exercises below, and in fact our main application will be to prove the Riemann mapping theorem, which is proved by solving an extremal problem using Montel's theorem.

Another common application of Montel is to use it via Vitali's theorem, which is an exercise below. Given a locally bounded sequence of holomorphic functions, one only needs to prove pointwise convergence at "enough" points to get that the sequence actually converges uniformly on compact subsets.

Exercise 6.2.1: Prove the converse to Montel: If $\mathcal{F}$ is a normal family of holomorphic functions on an open set $U \subset \mathbb{C}$, then $\mathcal{F}$ is locally bounded.

Exercise 6.2.2: For open $U \subset \mathbb{C}$, prove that "locally bounded" means "bounded on compact subsets," that is, $\mathcal{F}$ is locally bounded if and only if for every compact $K \subset U$ there is an $M > 0$ such that $\|f\|_K \leq M$ for all $f \in \mathcal{F}$.

Exercise 6.2.3 (Vitali's theorem)*: Suppose $U \subset \mathbb{C}$ is a domain, $\{f_n\}$ is a locally bounded sequence of holomorphic functions $f_n : U \to \mathbb{C}$ converging pointwise on a set $E \subset U$, and E has a limit point in U. Prove that $\{f_n\}$ converges uniformly on compact subsets in U.*

Exercise 6.2.4: Let $U \subset \mathbb{C}$ be open and $\mathcal{F}$ a normal family of holomorphic functions on U. Show that $\{f' : f \in \mathcal{F}\}$ is a normal family.

Exercise 6.2.5: Let $U \subset \mathbb{C}$ be open and $\mathcal{F}$ a set of holomorphic functions such that $\{f' : f \in \mathcal{F}\}$ is a normal family.
  a) Show that $\mathcal{F}$ need not be normal family.
  b) Add a simple hypothesis (one that is weaker than "$\mathcal{F}$ is locally bounded") that would make it a normal family.

Exercise 6.2.6: Given $c \in [0,1)$ let $\mathcal{F}_c$ be the set of holomorphic $f : \mathbb{D} \to \mathbb{D}$ such that $f(0) = 0$ and $f(1/2) = c$.
  a) Prove that $\mathcal{F}_c = \emptyset$ if $c \in (1/2, 1)$ and $\mathcal{F}_c \neq \emptyset$ if $c \in [0, 1/2]$.
  b) Prove that for each $c \in [0, 1/2]$, there exists an $f \in \mathcal{F}_c$ such that $|f'(0)|$ is minimal (among the functions in $\mathcal{F}_c$), and let $m_c = \inf\{|f'(0)| : f \in \mathcal{F}_c\}$.
  c) Prove that $m_c > 0$ if $c > 1/4$ and $m_c = 0$ if $c \leq 1/4$.

Exercise 6.2.7: Let $U \subset \mathbb{C}$ be a domain, $p \in U$, and suppose there exists a nonconstant bounded holomorphic function on U.
  a) Prove that there exists a holomorphic $F : U \to \mathbb{D}$ such that $F'(p) \neq 0$, and if $f : U \to \mathbb{D}$ is holomorphic, then $|f'(p)| \leq |F'(p)|$.
  b) Prove that necessarily $F(p) = 0$.

Exercise 6.2.8: Let $U \subset \mathbb{C}$ be open. Show that the set of holomorphic $f : U \to \mathbb{C}$ such that

$$\int_U |f(x + iy)|^2 \, dx \, dy < 1$$

is a normal family. Hint: Prove $|f(p)|^2 \leq \frac{1}{2\pi} \int_0^{2\pi} |f(p + re^{it})|^2 dt$ for $p \in U$ and small r.

Exercise 6.2.9: *Find the largest domain $U \subset \mathbb{C}$ such that the family of functions defined by $z \mapsto e^{cz}$, $c \geq 0$, is a normal family and, of course, prove your assertion.*

Exercise 6.2.10: *Show that if the partial sums of a power series centered at p are uniformly bounded on $\Delta_r(p)$ for some $r > 0$, then the power series converges in $\Delta_r(p)$.*

6.3*i*   Riemann mapping theorem

6.3.1*i*   The theorem

The Riemann mapping theorem says that every simply connected domain in $\mathbb{C}$ (except $\mathbb{C}$ itself) is really equivalent (biholomorphically, conformally) to $\mathbb{D}$. It is a theorem that gets cited a lot in all sorts of branches of mathematics.*

Theorem 6.3.1 (Riemann mapping). *Let $U \subset \mathbb{C}$ be a simply connected domain such that $U \neq \mathbb{C}$. Let $p \in U$ be given. Then there exists a unique biholomorphic (conformal) map $f : U \to \mathbb{D}$ such that $f(p) = 0$ and $f'(p) > 0$ (real and positive).*

See Figure 6.1 for the mapping that takes the upper half-disc to the unit disc. You will explicitly construct this map in Exercise 6.3.2.

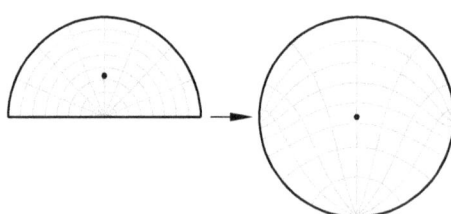

Figure 6.1: The Riemann map for the upper half-disc with $p = (\sqrt{2} - 1)i$.

The proof of the theorem is a wonderful example of solving a problem by formulating a related extremal problem. The f in the theorem maximizes $|f'(p)|$ among injective holomorphic maps taking p to 0. We will prove that maximizing $|f'(p)|$ is equivalent to f being onto: For any map that is not onto, we find one with a bigger $|f'(p)|$ that still goes into the disc. Such maps are bounded, so Montel gives us a convergent sequence with $|f'(p)|$ going to the supremum, and the limit then has to be onto. Why would one think of this extremal problem? Well, maximizing the derivative seems like a good way to spread out the values. We wish to make the image as large as possible, and we get furthest if the velocity is largest, no?

---

*For example, a beginning course on topology might cite the theorem to say that any simply connected domain in $\mathbb{R}^2$ is homeomorphic to the disc, and that's despite the topologists only needing the mapping to be continuous.

Proof. Consider $\mathcal{F}$ to be the family of injective (univalent) holomorphic $f: U \to \mathbb{D}$ such that $f(p) = 0$. We're trying to find an $f \in \mathcal{F}$ that is onto. Before we rush off to Montel, however, we must show that some injective $f: U \to \mathbb{D}$ actually exists, that is, that $\mathcal{F}$ is not empty. Here is where we use that U is not equal to the complex plane (by Liouville, $\mathcal{F}$ would be empty if $U = \mathbb{C}$). If the complement of U contained an open set, things would be easy, but we can only use that the complement of U contains at least one point, and that U is simply connected. We will use the simply-connectedness to construct a square root, as a square root squishes things together.

Suppose $q \in \mathbb{C} \setminus U$. The function $z \mapsto z - q$ is never zero and U is simply connected, so $z - q$ has a holomorphic square root—there exists a $g: U \to \mathbb{C}$ such that $(g(z))^2 = z - q$. The set $g(U)$ is open by the open mapping theorem. If $g(z) = g(\zeta)$, then $(g(z))^2 = (g(\zeta))^2$ and so $z = \zeta$. Thus g is injective. What's more, $g(z) = -g(\zeta)$ also implies $z = \zeta$. Thus, $g(z) = -g(\zeta)$ never happens since g is never zero. In other words, $g(U) \cap (-g(U)) = \emptyset$, where $-g(U) = \{w : -w \in g(U)\}$. The set $-g(U)$ is also open, and so the complement of $g(U)$ contains an open disc $\Delta_r(\xi)$. Hence,

$$z \mapsto \frac{r}{g(z) - \xi}$$

takes U to $\mathbb{D}$ and is injective. By composing with the correct automorphism of the disc, we find a map that takes p to 0. Hence, $\mathcal{F}$ is nonempty.

OK. Now suppose that $f: U \to \mathbb{D}$ is an injective holomorphic map such that $f(p) = 0$, but that f is not onto: There is a $q \in \mathbb{D} \setminus f(U)$. Recall that $\varphi_q(z) = \frac{z-q}{1-\bar{q}z}$ is an automorphism of $\mathbb{D}$ that takes q to 0, and consider $\varphi_q \circ f$. This function is not zero on U and thus there exists a holomorphic square root g on U, that is, $(g(z))^2 = \varphi_q(f(z))$. The square root of a number in $\mathbb{D}$ is still in $\mathbb{D}$, so g takes U to $\mathbb{D}$. If $g(z) = g(\zeta)$, then $\varphi_q(f(z)) = \varphi_q(f(\zeta))$ and as $\varphi_q \circ f$ is injective, $z = \zeta$. So g is injective. The function g takes p to one of the roots of $-q$. Define

$$h = \varphi_{g(p)} \circ g.$$

In particular, $h(p) = 0$, and $h \in \mathcal{F}$. The inverse of φ_a is φ_{-a}, so $g = \varphi_{-g(p)} \circ h$. Next, differentiate $\varphi_q \circ f = g^2$ at p, noting that $\varphi_a'(0) = 1 - |a|^2$:

$$(1 - |q|^2)f'(p) = \varphi_q'(f(p))f'(p) = 2g(p)g'(p)$$
$$= 2g(p)\varphi_{-g(p)}'(h(p))h'(p) = 2g(p)(1 - |g(p)|^2)h'(p).$$

As $(g(p))^2 = -q$, then

$$|f'(p)| = \frac{2|g(p)|(1 - |g(p)|^2)}{1 - |q|^2}|h'(p)| = \frac{2\sqrt{|q|}}{1 + |q|}|h'(p)|.$$

If $|q| < 1$, then $\frac{2\sqrt{|q|}}{1+|q|} < 1$ (calculus exercise). In other words,

$$|f'(p)| < |h'(p)|.$$

Take a sequence $\{f_n\}$ in $\mathscr{F}$ such that

$$\lim_{n \to \infty} |f_n'(p)| = \sup_{f \in \mathscr{F}} |f'(p)|.$$

As all functions in $\mathscr{F}$ are bounded by 1, Montel says that there exists a subsequence that converges uniformly on compact subsets to some holomorphic f. Assume $\{f_n\}$ is that subsequence. By the corollary to Hurwitz, Corollary 5.4.9, the function f is injective or constant. It cannot be constant, as $f_n'(p)$ converges to $f'(p)$, which cannot be zero as $0 < |f_n'(p)| \leq |f'(p)|$ for any n. We must have $f(p) = 0$ by taking the limit. Similarly $|f(z)| \leq 1$ for all $z \in U$ by taking limits, so $f(U) \subset \overline{\mathbb{D}}$. By the open mapping theorem, $f(U) \subset \mathbb{D}$, so $f \in \mathscr{F}$. If f was not onto, then it would not be the one that achieves the supremum of $|f'(p)|$, see the construction of h above. Thus $f(U) = \mathbb{D}$, and f is the desired map. Multiplying by $e^{i\theta}$ we can make $f'(p) > 0$.

The uniqueness is left as an exercise.   □

Exercise **6.3.1:** *Finish the proof of the theorem: Given an open $U \subset \mathbb{C}$ and $p \in U$, prove that a biholomorphic $f : U \to \mathbb{D}$ such that $f(p) = 0$ and $f'(p) > 0$ is unique if it exists.*

Remark 6.3.2. The map from the theorem can be useful, but the theorem itself doesn't tell you how to construct it. There are entire books written on the subject, collecting the techniques for constructing these maps for various types of domains. For instance, there is an explicit formula for the map given any polygon called the Schwarz–Christoffel mapping. Let us not worry about these constructions here.

Remark 6.3.3. Another interesting question we will not address is the boundary regularity of the map. That is, does the map extend to the closure $\overline{U}$, and how "nice" it is. If U is bounded by a Jordan curve, it is known that f extends to be continuous on $\overline{U}$. If U has smooth boundary (locally a graph of a smooth function), then f extends smoothly to $\overline{U}$. If U has real-analytic boundary (locally a graph of a real-analytic function), then f extends holomorphically a bit past the boundary. Again, we refer the reader to more advanced literature.

Remark 6.3.4. In the proof we only used that U was simply connected to obtain square roots of nowhere zero functions. So the result we actually proved is that existence of square roots on $U \neq \mathbb{C}$ implies that U is biholomorphic to $\mathbb{D}$. In other words, U being simply connected is equivalent to the existence of square roots.

Exercise **6.3.2:** *Find the unique biholomorphic map (the Riemann map) $f : U \to \mathbb{D}$ explicitly for the following U and p, that is, such that $f(p) = 0$ and $f'(p) > 0$:*
   a) The strip $U = \{z : -1 < \operatorname{Im} z < 1\}$, $p = 0$.
   b) The quadrant $U = \{z : \operatorname{Re} z > 0 \text{ and } \operatorname{Im} z > 0\}$, $p = \frac{1+i}{\sqrt{2}}$.
   c) The upper half-disc $U = \{z : |z| < 1 \text{ and } \operatorname{Im} z > 0\}$, $p = (\sqrt{2} - 1)i$. Hint: Start with the inverse of the Cayley map, and don't worry about p at first. See Figure 6.1.

Exercise 6.3.3: *Suppose $V \subset \mathbb{C}$ is a simply connected domain, and $V \neq \mathbb{C}$. Show that every holomorphic $f: \mathbb{C} \to V$ is constant.*

Exercise 6.3.4: *Suppose $U \subset \mathbb{C}$ is a simply connected domain. Show that for every two points $z, w \in U$, there exists an automorphism $\psi \in \mathrm{Aut}(U)$ such that $\psi(z) = w$.*

Exercise 6.3.5:
  a) *Suppose $U \subset \mathbb{C}$ is a simply connected domain, $U \neq \mathbb{C}$, $p, q \in U$ are distinct points, and $f: U \to U$ is holomorphic such that $f(p) = p$ and $f(q) = q$. Prove that f is the identity, that is, $f(z) = z$ for all $z \in U$.*
  b) *Find a counterexample if $U = \mathbb{C}$.*

Exercise 6.3.6: *A Riemann-mapping-like theorem for multiply connected domains (domains with holes) is not true (at least not in the most obvious way): Show that the punctured disc $\mathrm{ann}(0; 0, 1) = \mathbb{D} \setminus \{0\}$ and the annulus $\mathrm{ann}(0; 1, 2)$ are not biholomorphic.*

Exercise 6.3.7: *Suppose $U \subset \mathbb{C}$ is a domain. Suppose one connected component of $\mathbb{C}_\infty \setminus U$ is more than one point.*
  a) *Prove that U is biholomorphic to a subset of $\mathbb{D}$.*
  b) *If $\mathbb{C}_\infty \setminus U$ has finitely many connected components, then U is biholomorphic to $\mathbb{D} \setminus K$ for some (possibly empty) compact set $K \subset \mathbb{D}$, where K has finitely many components.*
Hint: What if the component contained ∞?

Exercise 6.3.8: *Let $S \subset \mathbb{C}$ be a countable closed subset. Prove that $U = \mathbb{C} \setminus S$ is not biholomorphic to any subset of $\mathbb{D}$. Hint: A countable closed contains isolated points since every nonempty perfect set is uncountable (feel free to assume this fact).*

Exercise 6.3.9: *Prove that if $f: \mathbb{C} \to \mathbb{C}$ is entire holomorphic and injective, then f is onto.*

6.3.2*i* Simply connected is simply connected ★

Let us finally prove that simply connected (in the sense of homology, Definition 4.3.1 that we have been using all this time), is equivalent to simply connected in the sense of homotopy. The proof of this corollary is a wonderful example of something that would be quite difficult without the Riemann mapping theorem, and it is almost trivial with the mapping theorem. The key idea is that any path is trivially homotopic to the zero path in the disc just by scaling. See also Example 4.5.2.

Corollary 6.3.5. *A domain $U \subset \mathbb{C}$ is simply connected in the sense of homotopy if and only if it is simply connected in the sense of Definition 4.3.1.*

Proof. Proposition 4.5.10 says that if U is simply connected in the sense of homotopy, then U is simply connected in the sense of homology. So let us prove the converse.

Suppose U is simply connected in the sense of homology. Let $\gamma: [a, b] \to U$ be a continuous function with $\gamma(a) = \gamma(b)$. We wish to show that γ is homotopic to a constant in U. The Riemann mapping theorem says that either $U = \mathbb{C}$ or there is a biholomorphism $f \colon U \to \mathbb{D}$.

If $U = \mathbb{C}$, then define the homotopy $H \colon [a,b] \times [0,1] \to \mathbb{C}$ as

$$H(t,s) = (1-s)\gamma(t).$$

If $U \neq \mathbb{C}$, define $H \colon [a,b] \times [0,1] \to U$ as

$$H(t,s) = f^{-1}\Big((1-s)f\big(\gamma(t)\big)\Big).$$

In other words, we map U to $\mathbb{D}$ and consider $f \circ \gamma$. Then we define H in the same way as we did for $\mathbb{C}$, and then we take the whole thing back to U. $\qquad\square$

Remark 6.3.6. Let us again emphasize that the definition "simply connected in terms of homology" is not standard. It is just a shortcut we took in case one wants to skip homotopy on first reading. In the wild (outside of this book), "simply connected" is always in terms of homotopy. Furthermore, while for domains in $\mathbb{C}$ the two concepts happen to be the same, they are not the same for more general topological spaces.

Exercise 6.3.10: *Without using the Riemann mapping theorem, Corollary 6.3.5, or mapping to the disc in any way, prove by constructing an explicit homotopy that the slit plane $\mathbb{C} \setminus (-\infty, 0]$ is simply connected in the sense of homotopy, see the proof of the corollary.*

6.3.3i   Cycles around compacts and simply-connectedness

As a second and perhaps much less obvious application of the Riemann mapping theorem, we will prove a lemma that around any compact set in some domain U we can find a cycle that goes around this compact set and is homologous to zero in U. What's interesting is that there is no simply connected domain in sight in this problem, but we can still use the Riemann mapping theorem to greatly simplify the topology of the situation. There are more direct and constructive (but no less technical) ways of proving the lemma below,[*] but I have an irrational affinity to using the mapping theorem, and this is my book after all.

This lemma will allow us to finish the proof that simply connected domains in $\mathbb{C}$ are precisely those where the complement in $\mathbb{C}_\infty$ is connected. That is, we will prove the converse of Proposition 4.3.7. First the lemma.

Lemma 6.3.7. *Let $U \subset \mathbb{C}$ be open and suppose that $K \subset U$ is compact and nonempty. Then there exists a cycle Γ in $U \setminus K$ such that $n(\Gamma; z) = 1$ for all $z \in K$ and $n(\Gamma; z) = 0$ for all $z \in \mathbb{C} \setminus U$ (Γ is homologous to zero in U) and such that $n(\Gamma; z)$ is 0 or 1 for all $z \notin \Gamma$.*

---

[*]The typical proof will put a fine enough square grid on $\mathbb{C}$ and then show that if we add up all these small cycles whose squares happen to intersect the compact set and remove the doubled sides, we get the cycle we want.

The intuitive idea of the proof is rather simple. Suppose K is connected. Take one point of it to infinity by an LFT (an inversion). Then use the Riemann mapping theorem to make the complement of K go to the disc (or several discs). Then the path is a backwards (clockwise) circle very close to the boundary of the disc to "go around" (after reversing the inversion) what's "outside the disc," that is, it will go around K. The complement of U corresponds to a compact set inside the disc so this way we will not "go around" the complement of U. Then we repeat the procedure for all components of K. See Figure 6.2. As expected, we hit a bunch of technicalities such as K possibly having infinitely many connected components, that the complement of a connected K can have multiple components, and of course that vague intuitive ideas are nice, but we need to actually do some grubby computation to prove anything.

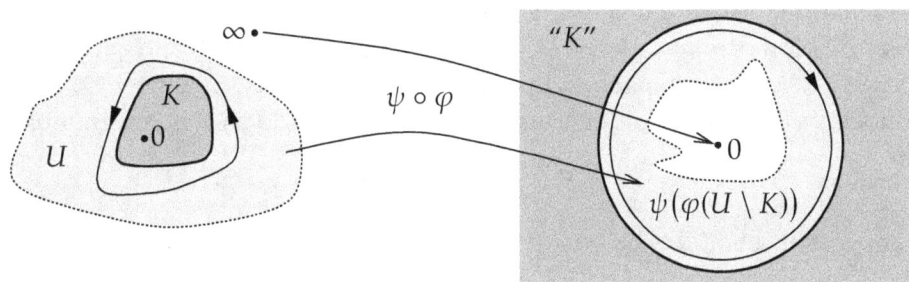

Figure 6.2: The idea of the proof: The complement of K with infinity goes to the inside of the disc by Riemann mapping theorem. The inversion is the φ and ψ is the mapping from Riemann's theorem. The outside of the disc on the right is not really the image of K, but morally one could think of it that way (hence the quotes).

Proof. We first enlarge K so that it has only finitely many components. For a small enough $r > 0$, we cover K by finitely many discs $\Delta_r(z)$ such that $\overline{\Delta_r(z)} \subset U$. In particular, for some $z_1, \ldots, z_m$,

$$K' = \overline{\Delta_r(z_1)} \cup \cdots \cup \overline{\Delta_r(z_m)}$$

is a compact subset of U and $K \subset K'$. As K' is a union of finitely many closed discs, it has finitely many topological components. If we find a Γ in U around K', then we are done as $K \subset K'$.

Let $K_1, \ldots, K_n$ be the components of K'. The components are closed, and as there are finitely many, $K_2 \cup \cdots \cup K_n$ is also closed. If we prove the lemma for the connected compact set K_1 and the open set $U \setminus (K_2 \cup \cdots \cup K_n)$ to find a cycle Γ_1, then we claim we are done: We could repeat the procedure for each K_j to find Γ_j and let $\Gamma = \Gamma_1 + \cdots + \Gamma_n$. As Γ_j winds exactly once around every point of K_j and does not wind around any point of K_ℓ for $\ell \neq j$, then Γ will wind exactly once around any point of K', and it is still homologous to zero in U.

So, without loss of generality, assume that K is connected. Also assume that $0 \in K$. If $K = \{0\}$, we are done trivially, so assume K is more than one point. Consider $\varphi : \mathbb{C}_\infty \to \mathbb{C}_\infty$ be the inversion LFT: $\varphi(z) = 1/z$ for $z \in \mathbb{C} \setminus \{0\}$, $\varphi(0) = \infty$, and $\varphi(\infty) = 0$. Being an LFT, φ is a biholomorphic mapping of $\mathbb{C}_\infty$ to itself, where $\varphi^{-1} = \varphi$. Let

$$V = \varphi(\mathbb{C}_\infty \setminus K).$$

Note that $\infty \notin V$, $0 \in V$, $V \neq \mathbb{C}$ (K is more than one point), and $\mathbb{C}_\infty \setminus V = \varphi(K)$ is connected. So each connected component of V is a simply connected domain by Proposition 4.3.7 (exercise). As we can assume that K is a finite union of closed discs, $\mathbb{C}_\infty \setminus K$ and therefore V have finitely many connected components. Let $V_1, \ldots, V_m$ be the connected components of V. By the Riemann mapping theorem, there exists a biholomorphic map from V_j to $\Delta_1(q_j)$, where $q_1, \ldots, q_m$ are some points far enough apart so that the discs are disjoint. Write $D = \Delta_1(q_1) \cup \cdots \cup \Delta_1(q_m)$. In other words, there is a biholomorphic $\psi : V \to D$. We can arrange that $q_1 = 0$ and $\psi(0) = 0$.

The set $\mathbb{C}_\infty \setminus U$ is compact: It is a closed subset of a compact set $\mathbb{C}_\infty$. As φ is continuous, $\varphi(\mathbb{C}_\infty \setminus U)$ is compact, and a subset of V. And as ψ is continuous,

$$S = \psi\big(\varphi(\mathbb{C}_\infty \setminus U)\big)$$

is a compact subset of D. There is an $r < 1$ such that

$$S \subset \Delta_r(q_1) \cup \cdots \cup \Delta_r(q_m).$$

Consider the paths $\gamma_j(t) = q_j + re^{-it}$ for $t \in [0, 2\pi]$, that is, $\gamma_j = -\partial\Delta_r(q_j)$. Let $\Gamma_j = \varphi^{-1} \circ \psi^{-1} \circ \gamma_j$, and $\Gamma = \Gamma_1 + \cdots + \Gamma_m$.

Let us compute the winding numbers. For any p not on Γ, compute the winding number (see Proposition 3.1.7):

$$n(\Gamma; p) = \sum_{j=1}^{m} \frac{1}{2\pi i} \int_{\varphi^{-1} \circ \psi^{-1} \circ \gamma_j} \frac{1}{z - p}\, dz = \sum_{j=1}^{m} \frac{1}{2\pi i} \int_{\psi^{-1} \circ \gamma_j} \frac{-1}{(1 - \zeta p)\zeta}\, d\zeta$$

$$= \sum_{j=1}^{m} \frac{1}{2\pi i} \int_{\gamma_j} \frac{-1}{\big(1 - \psi^{-1}(\xi)p\big)\, \psi^{-1}(\xi)\, \psi'\big(\psi^{-1}(\xi)\big)}\, d\xi.$$

First suppose that $p \in \mathbb{C} \setminus U$. The function

$$h(\xi) = \frac{-1}{\big(1 - \psi^{-1}(\xi)p\big)\, \psi^{-1}(\xi)\, \psi'\big(\psi^{-1}(\xi)\big)}$$

defined on D has two poles: one at $\psi(1/p)$ and one at $q_1 = 0$ (the third factor in the denominator is never zero). They are both simple poles as is easy to check and the residues are (using Proposition 5.3.5)

$$\mathrm{Res}(h; 0) = \frac{-1}{\big(1 - \psi^{-1}(0)p\big)\, \psi'\big(\psi^{-1}(0)\big)} \frac{1}{\frac{1}{\psi'(\psi^{-1}(0))}} = -1$$

and

$$\operatorname{Res}\big(h;\psi(1/p)\big) = \frac{-1}{\psi^{-1}\big(\psi(1/p)\big)\,\psi'\big(\psi^{-1}(\psi(1/p))\big)}\,\frac{1}{\frac{-1}{\psi'(\psi^{-1}(\psi(1/p)))}\,p} = 1.$$

The path γ_1 goes around $q_1 = 0$. Some γ_j goes around $\psi(1/p)$, as r was picked sufficiently large precisely so that the circles γ_j go around S, that is, points that are the image of $\mathbb{C} \setminus U$, in particular, $\psi(1/p) = \psi\big(\varphi(p)\big)$. The sum of the residues is zero, so by the residue theorem,

$$n(\Gamma; p) = 0.$$

Next suppose $p \in K$. As before $n(\Gamma; p) = \sum_j \frac{1}{2\pi i}\int_{\gamma_j} h(\xi)\,d\xi$. As $p \in K$, $\psi^{-1}(\xi) \neq 1/p$ for all $\xi \in D$, and so h has just the pole at 0. Since γ_1 traverses the circle backwards,

$$n(\Gamma; p) = \sum_{j=1}^m \frac{1}{2\pi i}\int_{\gamma_j} h(\xi)\,d\xi = \frac{1}{2\pi i}\int_{\gamma_1} h(\xi)\,d\xi = -\operatorname{Res}(h;0) = 1.$$

If p is any other point not on Γ, then $n(\Gamma; p)$ is either 0 or 1, depending on if there is a pole at $\psi(1/p)$ and if Γ goes around it or not. $\qquad\square$

Exercise 6.3.11: Prove that if $K \subset \mathbb{C}_\infty$ is compact and connected, then every component of $\mathbb{C}_\infty \setminus K$ is a simply connected domain. Hint: Prove that the complement of each one of these components is connected.

We can now prove the simplest topological characterization of simply connected domains in $\mathbb{C}$.

Theorem 6.3.8. *Let $U \subset \mathbb{C}$ be a domain. Then $\mathbb{C}_\infty \setminus U$ is connected if and only if U is simply connected.*

Proof. The forward direction is Proposition 4.3.7. Let's do the backwards direction by contrapositive. Suppose $\mathbb{C}_\infty \setminus U$ is disconnected. Then there are two nonempty disjoint closed sets S and K such that $S \cup K = \mathbb{C}_\infty \setminus U$. Assume $\infty \in S$. The set $U' = U \cup K$ is open as S is closed, $U' \subset \mathbb{C}$, and $K \subset U'$ is compact. Apply the lemma to find a cycle Γ in $U = U' \setminus K$ such that $n(\Gamma; z) = 1$ for all $z \in K$. In other words, Γ is not homologous to zero in U. $\qquad\square$

Exercise 6.3.12: Suppose $K \subset \mathbb{C}$ is compact and connected, $\mathbb{C} \setminus K$ is connected, and K is more than one point. Prove that there exists a biholomorphic map $\psi \colon \mathbb{C} \setminus K \to \mathbb{C} \setminus \overline{\mathbb{D}}$.

Exercise 6.3.13: Construct an example compact set $K \subset \mathbb{C}$ with a connected component K_1 with the following property. For every cycle Γ in $\mathbb{C} \setminus K$ such that $n(\Gamma; z) = 1$ for all $z \in K_1$, there exists a $\zeta \in K \setminus K_1$ where $n(\Gamma; \zeta) = 1$. Why does this example not contradict the construction in the proof of the lemma?

***Exercise* 6.3.14:** *Suppose* $\{f_n\}$ *is a sequence of holomorphic functions on an open set* $U \subset \mathbb{C}$ *that converges uniformly on compact subsets to a nonconstant* $f: U \to \mathbb{C}$. *Let* $K \subset U$ *be a compact set. Prove that for every open neighborhood* V *of* K *in* U *(so* $K \subset V \subset U$*) there exists a smaller open neighborhood* W *(so* $K \subset W \subset V$*) and an* $N \in \mathbb{N}$ *such that* f *and* f_n *have the same number of zeros in* W *for all* $n \geq N$.

***Exercise* 6.3.15:** *Given an open* $U \subset \mathbb{C}$, *a compact nonempty* $K \subset U$, *and a* $\delta > 0$, *prove there exists a cycle* Γ *in* $U \setminus K$ *homologous to zero in* U, *such that* $n(\Gamma; z)$ *is either 0 or 1 for all* $z \notin \Gamma$, *such that* $n(\Gamma; z) = 1$ *for all* $z \in K$, *and such that for every* $p \in \Gamma$ *there is a* $q \in K$ *such that* $|p - q| < \delta$ *(*Γ *is within* δ *of* K*).*

***Exercise* 6.3.16:** *Let* $U \subset \mathbb{C}$ *be a domain.*
   a) *Prove that if* U *is not simply connected, then there exists a* $p \in \mathbb{C} \setminus U$ *and a cycle* Γ *in* U *such that* $n(\Gamma; p) = 1$.
   b) *Fix an integer* $k \geq 2$. *Suppose that for every nowhere zero holomorphic* $f: U \to \mathbb{C}$, *there exists a holomorphic* $g: U \to \mathbb{C}$ *such that* $g^k = f$. *Prove that* U *is simply connected. Hint: See Exercise 5.4.7 for a hint.*

7*i* \ Harmonic Functions

If you cannot get rid of the family skeleton, you may as well make it dance.

—George Bernard Shaw

7.1*i* \ Harmonic functions

Hitherto, we examined holomorphic functions as complex-valued functions. However, to do analysis one needs inequalities, and complex numbers are not ordered. Let us consider the real and imaginary parts of holomorphic functions, the so-called harmonic functions. Interestingly, harmonic functions come up often in applied mathematics, for example as steady state heat or the distribution of electrostatic potential in a region without charge. Harmonic functions are used to study holomorphic functions and vice versa, holomorphic functions are used to study harmonic functions.

Most of the results we will prove for harmonic functions are analogues of the results for holomorphic functions. The reader is encouraged to look for these connections. Just as it is best to study animals in their natural habitat, many results we proved for holomorphic functions are better understood as results for harmonic functions. Nevertheless, the results for harmonic functions are not always a simple application of what has already been proved for holomorphic functions, and even the statements or proofs of the analogous results may be quite different. Harmonic functions are somewhat more general than real and imaginary parts of holomorphic functions. They are real and imaginary parts of holomorphic functions only locally, but perhaps not globally.

7.1.1*i* \ Real and imaginary parts of holomorphic functions

Definition 7.1.1. Let $U \subset \mathbb{C}$ be open. A twice continuously (real) differentiable (C^2 for short) function $f : U \to \mathbb{R}$ is *harmonic* if

$$\nabla^2 f = \frac{\partial^2 f}{\partial x^2} + \frac{\partial^2 f}{\partial y^2} = 0 \quad \text{on } U.$$

The operator ∇^2, sometimes written Δ, is the *Laplacian*.* It is the trace of the Hessian matrix. It is convenient to note that

$$4\frac{\partial^2}{\partial\bar{z}\partial z}f = 4\left[\frac{1}{2}\left(\frac{\partial}{\partial x} + i\frac{\partial}{\partial y}\right)\right]\left[\frac{1}{2}\left(\frac{\partial}{\partial x} - i\frac{\partial}{\partial y}\right)\right]f = \left[\frac{\partial^2}{\partial x^2} + \frac{\partial^2}{\partial y^2}\right]f = \nabla^2 f.$$

Namely, f is harmonic if and only if $\frac{\partial f}{\partial \bar{z}}$ is holomorphic. So suppose f is harmonic. Locally (in some neighborhood), we find a primitive g of $\frac{\partial f}{\partial \bar{z}}$, and we write

$$f(z) = g(z) + c(z),$$

where g is holomorphic, c is at least C^2, and $\frac{\partial c}{\partial z} \equiv 0$. Let $h = \bar{c}$ be the complex conjugate of c. Then

$$\frac{\partial h}{\partial \bar{z}} = \frac{\partial \bar{c}}{\partial \bar{z}} = \overline{\frac{\partial c}{\partial z}} = 0,$$

so h is holomorphic. Thus, locally, we found holomorphic g and h so that

$$f(z) = g(z) + \overline{h(z)}.$$

Consider the holomorphic $\varphi(z) = g(z) + h(z)$. As f is real-valued,

$$f(z) = \text{Re}\, f(z) = \frac{g(z) + \overline{h(z)} + \overline{g(z)} + h(z)}{2} = \frac{g(z) + h(z) + \overline{g(z) + h(z)}}{2} = \text{Re}\, \varphi(z).$$

So any harmonic function f is locally the real part of a holomorphic function. Similarly f is locally the imaginary part of a holomorphic function. This all works only in some neighborhood. We cannot necessarily find a single φ (the g and h above) in the entire domain U, unless U is simply connected. If it is not, we can always pick a simply connected neighborhood, such as a disc. If U is simply connected, then $\frac{\partial f}{\partial \bar{z}}$ has a primitive in U, and the computation above leads to the following proposition.

Proposition 7.1.2. *Let $U \subset \mathbb{C}$ be a simply connected domain and $f: U \to \mathbb{R}$ a harmonic function. Then there exists a holomorphic $\varphi: U \to \mathbb{C}$ such that $f = \text{Re}\, \varphi$.*

Conversely, suppose that f is the real-part of a holomorphic function φ:

$$f(z) = \text{Re}\, \varphi(z) = \frac{1}{2}\left(\varphi(z) + \overline{\varphi(z)}\right).$$

Notice that

$$\nabla^2 = 4\frac{\partial^2}{\partial\bar{z}\partial z} = 4\frac{\partial^2}{\partial z\partial\bar{z}}.$$

Then

$$\nabla^2 f = 4\frac{\partial^2}{\partial\bar{z}\partial z}\left(\frac{1}{2}\left(\varphi(z) + \overline{\varphi(z)}\right)\right) = 2\left(\frac{\partial}{\partial z}\left(\frac{\partial}{\partial\bar{z}}\varphi(z)\right) + \frac{\partial}{\partial\bar{z}}\left(\frac{\partial}{\partial z}\overline{\varphi(z)}\right)\right) = 0.$$

We have thus proved the following characterization of harmonic functions.

---

*The Laplacian is defined in $\mathbb{R}^n$ for any n by $\nabla^2 f = \frac{\partial^2 f}{\partial x_1^2} + \cdots + \frac{\partial^2 f}{\partial x_n^2}$, and so there are harmonic functions in any dimension. We are interested in the complex plane, $n = 2$, which is surprisingly different from the $n \geq 3$ case. When $n \geq 3$, the theory has far less to do with complex analysis.

Proposition 7.1.3. *Let $U \subset \mathbb{C}$ be open and $f : U \to \mathbb{R}$ a function.*

(i) *The function f is harmonic if and only if for every point $p \in U$ there exists an open neighborhood V of p and a holomorphic $\varphi : V \to \mathbb{C}$ such that $f = \operatorname{Re} \varphi$ on V.*

(ii) *The function f is harmonic if and only if for every $p \in U$, there exists a power series expansion*

$$f(z) = c_0 + \sum_{n=1}^{\infty} c_n(z - p)^n + \bar{c}_n(\overline{z - p})^n$$

converging uniformly absolutely on every closed disc $\overline{\Delta_r(p)} \subset U$.

As holomorphic functions are infinitely differentiable, harmonic functions are as well. Actually, we see above that harmonic functions have a real power series (a power series in x and y, or equivalently in z and $\bar{z}$) and so they are what is called real-analytic.

Proposition 7.1.4. *If $U \subset \mathbb{C}$ is open and $f : U \to \mathbb{R}$ is harmonic, then f is infinitely (real) differentiable.*

Starting with a harmonic function f, finding the holomorphic function whose real part is f means finding another harmonic function g such that $f + ig$ is holomorphic.

Definition 7.1.5. *Let $U \subset \mathbb{C}$ be open and $f : U \to \mathbb{R}$ harmonic. If $g : U \to \mathbb{R}$ is harmonic and $f + ig$ is holomorphic, then g is called the *harmonic conjugate* of f.*

Proposition 7.1.2 says that every harmonic function on a simply connected domain has a harmonic conjugate. On the other hand, on the punctured plane $\mathbb{C} \setminus \{0\}$, the harmonic function $\log|z|$ fails to have a harmonic conjugate. If it did have a harmonic conjugate, then $\log$ would have a branch in $\mathbb{C} \setminus \{0\}$, which it does not. See Figure 4.1, the graph of the real part on the left is continuous, but the corresponding imaginary part is not a function. That we cannot find a different conjugate for $\log|z|$ follows from the following proposition.

Proposition 7.1.6. *Suppose $U \subset \mathbb{C}$ is a domain and $f : U \to \mathbb{R}$ is harmonic. If g_1 and g_2 are two harmonic conjugates of f, then $g_1 = g_2 + C$ for some $C \in \mathbb{R}$.*

The proof is trivial: The hypothesis implies that

$$\frac{(f + ig_1) - (f + ig_2)}{i} = g_1 - g_2$$

is holomorphic and it is real-valued on U, and thus constant.

The real and imaginary parts of a holomorphic function are harmonic; however, the modulus $|f(z)|$ is not. Not to fear, $\log|f(z)|$ is harmonic, at least where f is nonzero. The fact that $\log|f(z)|$ is harmonic is just as useful as the fact that the real and imaginary parts of f are harmonic. The proof is left as an exercise.

Proposition 7.1.7. *Suppose $U \subset \mathbb{C}$ is open, $f \colon U \to \mathbb{C}$ is holomorphic and never zero. Then*

$$z \mapsto \log|f(z)|$$

is harmonic.

Exercise 7.1.1: Prove Proposition 7.1.7.

Exercise 7.1.2: Show that the following functions of x and y (where $z = x + iy$) are harmonic (either on $\mathbb{C}$ or on the set given) and find their harmonic conjugate.

    a) y b) xy c) $\arctan(y/x)$ on $x \neq 0$ d) $\frac{x}{x^2+y^2}$ on $z \neq 0$

Exercise 7.1.3: Suppose that $U \subset \mathbb{C}$ is a simply connected domain and $f \colon U \to \mathbb{R}$ a harmonic function. Prove that there exists a holomorphic function $\varphi \colon U \to \mathbb{C}$ such that $f(z) = \log|\varphi(z)|$.

Exercise 7.1.4: Let $U, V \subset \mathbb{C}$ be open sets and $f \colon U \to V$ be holomorphic. Prove:
    a) If $g \colon V \to \mathbb{R}$ is harmonic, then $g \circ f$ is harmonic.
    b) Let f be a biholomorphism of U and V. Then $g \colon V \to \mathbb{R}$ is harmonic if and only if $g \circ f$ is harmonic.

Exercise 7.1.5: Prove that if $f \colon \mathbb{D} \to \mathbb{R}$ is harmonic, then $f(z/|z|^2)$ is harmonic in $\mathbb{C} \setminus \overline{\mathbb{D}}$.

Exercise 7.1.6: Suppose $U \subset \mathbb{C}$ is a domain and $f \colon U \to \mathbb{C}$ is holomorphic. Prove that if $z \mapsto |f(z)|^2$ is harmonic, then f is constant.

Exercise 7.1.7: Suppose $f \colon \mathbb{C} \to \mathbb{R}$ is harmonic.
    a) Show that there exists a holomorphic $F \colon \mathbb{C} \setminus \{0\} \to \mathbb{C}$ such that $F = f$ on $\partial \mathbb{D}$.
    b) Show that if F from part a) has a pole at the origin, then f is the real part of a holomorphic polynomial.

Exercise 7.1.8: Suppose $U \subset \mathbb{C}$ is open, $\overline{\mathbb{D}} \subset U$, and $f \colon U \to \mathbb{R}$ is harmonic. Expand f as a real power series at the origin as in Proposition 7.1.3, and find a formula for the c_n in terms of an integral around $\partial \mathbb{D}$.

Exercise 7.1.9: Prove the Liouville* theorem for harmonic functions: If $f \colon \mathbb{C} \to \mathbb{R}$ is harmonic and nonnegative, then f is constant.

Remark 7.1.8. As in Exercise 7.1.9, the analogue of "bounded" for holomorphic functions is "nonnegative" for harmonic functions. After all, if f is a bounded holomorphic function, then $\log|f(z) + M|$ or $\operatorname{Re} f(z) + M$ is nonnegative for large

---

*See Nelson, Edward *A proof of Liouville's theorem.* Proc. Amer. Math. Soc. **12** (1961), 995 (one of the shortest published papers) for an elegant proof for bounded functions. But you can't use it, you don't have the tools for it yet.

enough M. Conversely, if $\log|f(z)| \geq 0$, then $\frac{1}{f(z)}$ is bounded, and if $\operatorname{Re} f(z) \geq 0$, then $\frac{f(z)-1}{f(z)+1}$ is bounded (composing f with an LFT taking the right half-plane to the disc).

Remark 7.1.9. The procedure above, writing

$$\frac{\partial^2}{\partial x^2} + \frac{\partial^2}{\partial y^2} = 4\frac{\partial^2}{\partial \bar{z}\partial z},$$

that is, a sum of derivatives as a composition of different derivatives, so that we could integrate in these two new variables, may sound familiar. In this case, we found that a harmonic f is a sum of a function of z (a holomorphic function) and a function of $\bar{z}$ (an antiholomorphic function).

The procedure is analogous to the D'Alembert solution of the one-dimensional wave equation, which you may have seen in undergraduate differential equations. The wave operator is $\frac{\partial^2}{\partial t^2} - \frac{\partial^2}{\partial x^2}$ with a minus sign, where we use x and t as variables to be traditional. The wave operator decomposes as

$$\frac{\partial^2}{\partial t^2} - \frac{\partial^2}{\partial x^2} = \left[\frac{\partial}{\partial t} - \frac{\partial}{\partial x}\right]\left[\frac{\partial}{\partial t} + \frac{\partial}{\partial x}\right].$$

If we write $\mu = x + t$ and $\eta = x - t$ (the so-called characteristic coordinates), then

$$\frac{\partial^2}{\partial t^2} - \frac{\partial^2}{\partial x^2} = -4\frac{\partial^2}{\partial \eta \partial \mu}.$$

As before, a solution f to the wave equation is a function of μ plus a function of η. That is, $f(x,t) = A(\mu) + B(\eta) = A(x + t) + B(x - t)$, two waves travelling in opposite directions. The functions A and B need not be nice at all, any twice real differentiable functions. It is interesting that one puny minus sign makes such a huge difference.

7.1.2i Identity and the maximum principle

A consequence of the propositions above is the identity theorem for harmonic functions. The zero set of a harmonic function is allowed to have limit points. For instance, $\operatorname{Re} z$ is zero on the entire imaginary axis. However, we are still not allowed open sets for nonconstant harmonic functions. It is really a property of real-analytic functions, that is, functions that have a power series representation in terms of x and y or z and $\bar{z}$, but we do not wish to get far into power series in two variables.

Theorem 7.1.10 (Identity). *Let $U \subset \mathbb{C}$ be a domain and $f : U \to \mathbb{R}$ a harmonic function. Suppose $V \subset U$ is a nonempty open subset and $f = 0$ on V. Then $f \equiv 0$.*

Proof. Let Z_f be the zero set of f and let Z be the closure of the interior of Z_f in the subspace topology of U. The set Z is nonempty by hypothesis, so consider some $p \in Z$. Consider $\Delta_r(p) \subset U$. There exists a holomorphic $h : \Delta_r(p) \to \mathbb{C}$ such that $f = \operatorname{Re} h$ on $\Delta_r(p)$. On some open subset of the disc, f is zero. So the holomorphic function h is purely imaginary on an open subset of $\Delta_r(p)$, and hence it is constant on that open subset. By the identity theorem for holomorphic functions, h, and hence f, is constant on $\Delta_r(p)$. Since f is zero somewhere on the disc and constant, it is zero on the entire disc. Thus Z is open. As Z is also closed and U is connected, $Z = U$.   □

The maximum principle is really a theorem about harmonic functions rather than holomorphic functions. We will prove it using holomorphic functions and the open mapping theorem (Theorem 5.5.1)* although there is a more natural proof using the mean value property, which we will see later.

Theorem 7.1.11 (Maximum principle). *Suppose $U \subset \mathbb{C}$ is a domain and $f : U \to \mathbb{R}$ is harmonic. If f attains a local maximum (or a local minimum) in U, then f is constant.*

Proof. Suppose that f attains a local maximum at $p \in U$. The statement for a minimum follows by considering $-f$. Let $\Delta_r(p) \subset U$ be a disc such that p is the maximum of f on $\Delta_r(p)$. There exists a holomorphic $h : \Delta_r(p) \to \mathbb{C}$ such that $f = \operatorname{Re} h$. Then h takes $\Delta_r(p)$ to a subset of $X = \{ w \in \mathbb{C} : \operatorname{Re} w \leq f(p) \}$. The point $h(p)$ is on the boundary of X as $\operatorname{Re} h(p) = f(p)$. Hence, $h\big(\Delta_r(p)\big)$ is not open, which can only happen if h is constant by the open mapping theorem. As f is constant on $\Delta_r(p)$, it is constant on U by the identity theorem.   □

Exercise 7.1.10: *Prove that the maximum principle for harmonic functions implies the maximum modulus principle for holomorphic functions. Hint: Consider $\log|f(z)|$.*

Exercise 7.1.11: *Prove the second version of the maximum principle: If $U \subset \mathbb{C}$ is a bounded domain and $f : \overline{U} \to \mathbb{R}$ is continuous and harmonic on U, then f achieves both its maximum and its minimum on the boundary ∂U.*

Exercise 7.1.12: *Suppose $U \subset \mathbb{C}$ is open such that $\mathbb{R} \cap U \neq \emptyset$ and $\mathbb{R} \cap U$ is connected. Suppose $f : U \to \mathbb{R}$ is harmonic and the zero set of the restriction $f|_{\mathbb{R} \cap U}$ has a limit point in $\mathbb{R} \cap U$. Prove that $f|_{\mathbb{R} \cap U} \equiv 0$.*

Exercise 7.1.13: *Suppose $U \subset \mathbb{C}$ is a domain and $f : U \to \mathbb{R}$ is harmonic. Prove that $f(U)$ is an open interval or a single point.*

7.2i \ The Dirichlet problem in a disc and applications

7.2.1i \ The Dirichlet problem in a disc and the Poisson kernel

It is useful to find a harmonic function given boundary values: Given an open $U \subset \mathbb{C}$ and a continuous $f : \partial U \to \mathbb{R}$, find a continuous $g : \overline{U} \to \mathbb{R}$, harmonic on U, such that $g|_{\partial U} = f$. This problem is called the *Dirichlet problem*, and it is solvable for many

---

*The open mapping theorem is, after all, a stronger version of the maximum modulus principle.

(though not all) open sets. If the solution exists on a bounded domain, then it is unique. On unbounded domains, the solution need not be unique, see Exercise 7.2.5.

Proposition 7.2.1. *Suppose $U \subset \mathbb{C}$ is a bounded domain, $f, g \colon \overline{U} \to \mathbb{R}$ are continuous functions, harmonic on U, such that $f = g$ on ∂U. Then $f = g$ on U.*

Proof. Apply the maximum principle (second version, Exercise 7.1.11) to $f - g$. □

The solution of the problem in a disc is rather useful and rather explicit. It is achieved by integration against the so-called *Poisson kernel*. The Poisson kernel for the unit disc $\mathbb{D} \subset \mathbb{C}$, is

$$P_r(\theta) = \frac{1}{2\pi} \frac{1 - r^2}{1 + r^2 - 2r \cos \theta} = \frac{1}{2\pi} \operatorname{Re} \left(\frac{1 + re^{i\theta}}{1 - re^{i\theta}} \right), \qquad \text{for } 0 \le r < 1.$$

As a function of $z = re^{i\theta} \in \mathbb{D}$, the Poisson kernel is (see Figure 7.1)

$$z \mapsto \frac{1}{2\pi} \operatorname{Re} \left(\frac{1 + z}{1 - z} \right).$$

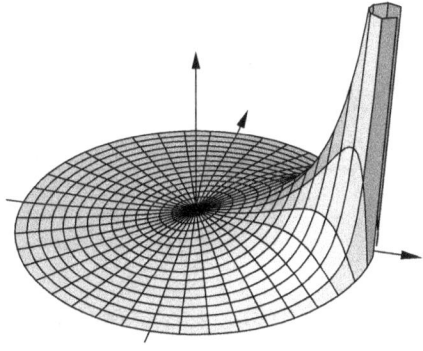

Figure 7.1: Graph of the Poisson kernel on $\mathbb{D}$, the pole at $z = 1$ is cut off at 2.

Proposition 7.2.2.

(i) $P_r(\theta) > 0$ for all $0 \le r < 1$ and all θ.

(ii) $\int_{-\pi}^{\pi} P_r(\theta) \, d\theta = 1$ for all $0 \le r < 1$.

(iii) For any given $\delta > 0$, $\sup\{P_r(\theta) : \delta \le |\theta| \le \pi\} \to 0$ as $r \uparrow 1$.

In the proof, it is useful to visualize the graph of P_r as a function of θ for a fixed r. See Figure 7.2.

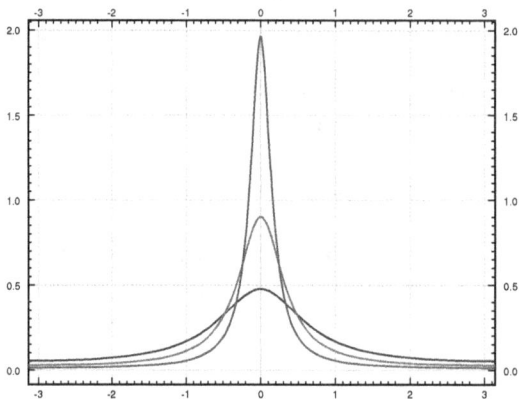

Figure 7.2: The graph of P_r as a function of θ on $[-\pi, \pi]$ for $r = 0.5$, $r = 0.7$, and $r = 0.85$.

Proof. The first item follows because for $0 \le r < 1$, we find $1 - r^2 > 0$ and

$$1 + r^2 - 2r \cos \theta \ge 1 + r^2 - 2r = (1 - r)^2 > 0.$$

For the second item,

$$\int_{-\pi}^{\pi} P_r(\theta) \, d\theta = \frac{1}{2\pi} \int_{-\pi}^{\pi} \mathrm{Re} \left(\frac{1 + re^{i\theta}}{1 - re^{i\theta}} \right) d\theta$$

$$= \mathrm{Re} \, \frac{1}{2\pi i} \int_{-\pi}^{\pi} \frac{1 + re^{i\theta}}{1 - re^{i\theta}} \frac{1}{re^{i\theta}} \, ire^{i\theta} \, d\theta$$

$$= \mathrm{Re} \, \frac{1}{2\pi i} \int_{\partial \Delta_r(0)} \frac{(1 + z)/(1 - z)}{z} \, dz = \mathrm{Re} \, \frac{1 + 0}{1 - 0} = 1.$$

The equality on the third line follows by the Cauchy integral formula using the function $\frac{1+z}{1-z}$ evaluated at 0.

For the third item, we only need to prove the result for $\delta \le \theta \le \pi$ by symmetry (P_r is even). On $(0, \pi)$, P_r is strictly decreasing as $\cos \theta$ is strictly decreasing. So we only need to show that $P_r(\delta)$ goes to 0 as $r \to 1$ if $\delta > 0$. This fact follows as $r \mapsto \frac{1+re^{i\delta}}{1-re^{i\delta}}$ is continuous at $r = 1$ and

$$\frac{1 + e^{i\delta}}{1 - e^{i\delta}} = \frac{(1 + e^{i\delta})(1 - e^{-i\delta})}{(1 - e^{i\delta})(1 - e^{-i\delta})} = \frac{e^{i\delta} - e^{-i\delta}}{|1 - e^{i\delta}|^2} = i \, \frac{2 \, \mathrm{Im} \, e^{i\delta}}{|1 - e^{i\delta}|^2}$$

is purely imaginary.   □

Theorem 7.2.3. *Let $f : \partial \mathbb{D} \to \mathbb{R}$ be continuous. Then $Pf : \overline{\mathbb{D}} \to \mathbb{R}$, defined by*

$$Pf(re^{i\theta}) = \begin{cases} \int_{-\pi}^{\pi} f(e^{it}) P_r(\theta - t) \, dt & \text{if } r < 1, \\ f(e^{i\theta}) & \text{if } r = 1, \end{cases}$$

is harmonic in $\mathbb{D}$ and continuous on $\overline{\mathbb{D}}$.

Proof. First, we prove Pf is harmonic in $\mathbb{D}$. Let $z = re^{i\theta}$. Then for any fixed t,

$$P_r(\theta - t) = \frac{1}{2\pi} \operatorname{Re}\left(\frac{1 + re^{i(\theta-t)}}{1 - re^{i(\theta-t)}}\right) = \frac{1}{2\pi} \operatorname{Re}\left(\frac{1 + ze^{-it}}{1 - ze^{-it}}\right)$$

is harmonic as a function of $z = re^{i\theta}$. By differentiating under the integral,

$$Pf(z) = Pf(re^{i\theta}) = \int_{-\pi}^{\pi} f(e^{it})P_r(\theta - t)\,dt = \frac{1}{2\pi}\int_{-\pi}^{\pi} f(e^{it})\operatorname{Re}\left(\frac{1 + ze^{-it}}{1 - ze^{-it}}\right) dt$$

is harmonic at $z = re^{i\theta} \in \mathbb{D}$.

Next we prove continuity at points in $\partial\mathbb{D}$. As both P_r and $f(e^{it})$ are 2π-periodic, we change variables:

$$Pf(re^{i\theta}) = \int_{-\pi}^{\pi} f(e^{it})P_r(\theta - t)\,dt = \int_{-\pi}^{\pi} f(e^{i(\theta-t)})P_r(t)\,dt.$$

Let M be the supremum of $|f|$ on $\partial\mathbb{D}$. Suppose $\epsilon > 0$ is given. As f is uniformly continuous on $\partial\mathbb{D}$, consider $\delta > 0$ small enough so that $\left|f(e^{i(\theta-t)}) - f(e^{i\theta})\right| < \epsilon/2$ whenever $|t| < \delta$. Proposition 7.2.2 says there exists a $\delta' > 0$ such that if $1 - \delta' < r < 1$, then $0 < P_r(t) < \frac{\epsilon}{8M\pi}$ whenever $\delta \le |t| \le \pi$.

Since $\int_{-\pi}^{\pi} P_r(t)\,dt = 1$,*

$$f(e^{i\theta}) = \int_{-\pi}^{\pi} f(e^{i\theta})P_r(t)\,dt.$$

So

$$|Pf(re^{i\theta}) - f(e^{i\theta})| = \left|\int_{-\pi}^{\pi} \left(f(e^{i(\theta-t)}) - f(e^{i\theta})\right)P_r(t)\,dt\right|$$

$$\le \left|\int_{-\pi}^{-\delta} \cdots dt\right| + \left|\int_{-\delta}^{\delta} \cdots dt\right| + \left|\int_{\delta}^{\pi} \cdots dt\right|.$$

Let us estimate the three integrals. First,

$$\left|\int_{-\pi}^{-\delta} \left(f(e^{i(\theta-t)}) - f(e^{i\theta})\right)P_r(t)\,dt\right| \le \int_{-\pi}^{-\delta} \left|f(e^{i(\theta-t)}) - f(e^{i\theta})\right|P_r(t)\,dt$$

$$\le (\pi - \delta)2M\frac{\epsilon}{8M\pi} < \frac{\epsilon}{4}.$$

The integral from δ to π is exactly the same. Next the middle integral,

$$\left|\int_{-\delta}^{\delta} \left(f(e^{i(\theta-t)}) - f(e^{i\theta})\right)P_r(t)\,dt\right| \le \int_{-\delta}^{\delta} \left|f(e^{i(\theta-t)}) - f(e^{i\theta})\right|P_r(t)\,dt$$

$$\le \int_{-\delta}^{\delta} \frac{\epsilon}{2}P_r(t)\,dt \le \int_{-\pi}^{\pi} \frac{\epsilon}{2}P_r(t)\,dt = \frac{\epsilon}{2}.$$

---

*This is a trick you see all the time in analysis, it is good to remember it.

Putting it all together, as long as $1 - \delta' < r < 1$,

$$|Pf(re^{i\theta}) - f(e^{i\theta})| < \frac{\epsilon}{4} + \frac{\epsilon}{2} + \frac{\epsilon}{4} = \epsilon.$$

So $Pf(re^{i\theta}) \to f(e^{i\theta})$ uniformly in θ as $r \uparrow 1$. The uniformity is important below.

Finally, for any $z_0 = e^{i\theta_0} \in \partial\mathbb{D}$, we must show that $Pf(z)$ tends to $Pf(z_0) = f(z_0)$ as $z \in \overline{\mathbb{D}}$ tends to z_0. Let $\epsilon > 0$ be given. As $f = Pf|_{\partial\mathbb{D}}$ is continuous, pick a $\delta > 0$ such that $|Pf(e^{i\theta}) - Pf(e^{i\theta_0})| < \epsilon/2$ whenever $|\theta - \theta_0| < \delta$. Also make δ small enough so that $|Pf(re^{i\theta}) - Pf(e^{i\theta})| < \epsilon/2$ when $1 - \delta < r \leq 1$ for all θ. Putting the two estimates together, we get

$$|Pf(re^{i\theta}) - Pf(e^{i\theta_0})| \leq |Pf(re^{i\theta}) - Pf(e^{i\theta})| + |Pf(e^{i\theta}) - Pf(e^{i\theta_0})| < \epsilon$$

whenever $z = re^{i\theta}$ satisfies $1 - \delta < r \leq 1$ and $|\theta - \theta_0| < \delta$. See Figure 7.3. Therefore, Pf is continuous at z_0. $\qquad\square$

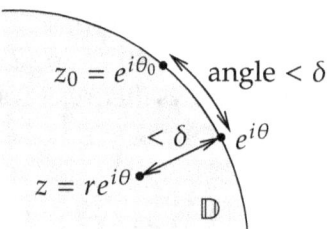

Figure 7.3: Continuity of Pf at z_0.

We remark that in the proof we used the topology on $\overline{\mathbb{D}}$ given by the polar coordinates, and we estimated the coordinates separately. Polar coordinates give a nice local homeomorphism (a continuous bijective map with a continuous inverse) outside of the origin, which is sufficient for us as we only worried about points on or near the boundary of $\mathbb{D}$. The reader who is still unconvinced should write out the details as an exercise.

Exercise 7.2.1: *We proved that given $\epsilon > 0$, there exists a $\delta > 0$ such that $|Pf(re^{i\theta}) - Pf(e^{i\theta_0})| < \epsilon$ when $|\theta - \theta_0| < \delta$ and $1 - \delta < r \leq 1$. Prove that this statement really does mean that $\lim_{z \to z_0} Pf(z) = f(z_0) = Pf(z_0)$.*

Translation and scaling give the more general version for any disc.

Corollary 7.2.4. *Let $f \colon \partial\Delta_R(p) \to \mathbb{R}$ be continuous. Then $Pf \colon \overline{\Delta_R(p)} \to \mathbb{R}$, defined by*

$$Pf(p + re^{i\theta}) = \begin{cases} \int_{-\pi}^{\pi} f(p + Re^{it}) P_{r/R}(\theta - t)\,dt & \text{if } r < R, \\ f(p + Re^{i\theta}) & \text{if } r = R, \end{cases}$$

is harmonic in $\Delta_R(p)$ and continuous on $\overline{\Delta_R(p)}$.

Exercise 7.2.2: *Prove the corollary.*

As the Poisson integral gives a solution of the Dirichlet problem on a disc, and as the solution to the Dirichlet problem is unique, the Poisson integral gives a representation of harmonic functions in terms of boundary values, just like the Cauchy integral formula does for holomorphic functions. That is, if $\overline{\Delta_R(p)} \subset U$ and f is harmonic in U, then $P[f|_{\partial\Delta_R(p)}] = f|_{\overline{\Delta_R(p)}}$. In particular, for $z = p + re^{i\theta} \in \Delta_R(p)$,

$$f(z) = \int_{-\pi}^{\pi} f(p + Re^{it})P_{r/R}(\theta - t)\,dt.$$

A difference with the Cauchy integral formula is that the Poisson kernel changes based on the domain. The Poisson kernel exists for other domains than the disc (as long as the boundary is nice enough), although in general we do not have an explicit formula. In the Cauchy integral formula the kernel is $\frac{1}{\zeta-z}$ no matter the path that we were integrating around, that is, no matter what domain we were solving in.

The Poisson kernel is also a reproducing kernel for holomorphic functions, as holomorphic functions are harmonic (their real and imaginary parts are). If f gives the boundary values for a holomorphic function, then Pf is holomorphic and it equals the Cauchy transform Cf inside the disc. Unlike the Cauchy transform, however, Pf is always continuous up to the boundary given continuous data on the disc. Thus, if f is not the boundary value of a holomorphic function, Cf and Pf are different in the disc. For example, if $f = z + \bar{z}$ on the circle, then $Pf = z + \bar{z}$ in $\mathbb{D}$ but $Cf = z$ in $\mathbb{D}$.

It is particularly useful to notice that in the corollary if we plug in $r = 0$, we get

$$Pf(p) = \frac{1}{2\pi}\int_{-\pi}^{\pi} f(p + Re^{it})\,dt. \tag{7.1}$$

The value at the center of the disc, $Pf(p)$, is the average value of f on $\partial\Delta_R(p)$. In the next section, we will see that this property actually characterizes harmonic functions.

Exercise 7.2.3 (Easy): *Prove that given any continuous $f: \partial\mathbb{D} \to \mathbb{C}$, there exists a holomorphic $F: \mathbb{D} \to \mathbb{C}$ such that $\operatorname{Re} F$ extends continuously to $\overline{\mathbb{D}}$ (agrees with a continuous function on $\overline{\mathbb{D}}$) and such that $\operatorname{Re} F = \operatorname{Re} f$ on $\partial\mathbb{D}$. That is, given arbitrary boundary data, we cannot in general find a holomorphic function with those boundary values, but we can do it at least for the real part.*

Exercise 7.2.4: *State and prove a version of Theorem 7.2.3 for a function that is bounded on $\partial\mathbb{D}$ and continuous at all but finitely many points on $\partial\mathbb{D}$. The conclusion should of course be then that $Pf(z)$ ($z \in \mathbb{D}$) tends to $f(z_0)$ ($z_0 \in \partial\mathbb{D}$) only if f is continuous at z_0. Note: More advanced students should note that one does not need boundedness, just $f \in L^1(\partial\mathbb{D})$.*

Exercise 7.2.5: *The Dirichlet problem on the upper half-plane $\mathbb{H} = \{z \in \mathbb{C} : \operatorname{Im} z > 0\}$ does not have a unique solution. Hint: Find two distinct harmonic functions that are zero on $\mathbb{R}$.*

Exercise 7.2.6: *Given a bounded continuous* $f : \mathbb{R} \to \mathbb{R}$, *prove that* $Pf : \overline{\mathbb{H}} \to \mathbb{R}$,

$$Pf(x + iy) = \begin{cases} \frac{1}{\pi} \int_{-\infty}^{\infty} f(t) \frac{y}{(x-t)^2 + y^2} \, dt & \text{if } y > 0, \\ f(x) & \text{if } y = 0, \end{cases}$$

is harmonic in $\mathbb{H}$ *and continuous on* $\overline{\mathbb{H}}$.

Exercise 7.2.7: *Given an open* $U \subset \mathbb{C}$, *a continuous function* $f : \overline{U} \to \mathbb{R}$, *positive and harmonic on* U, *and zero on* ∂U *is called a* **Martin function**.
   a) *Find a Martin function on the upper half-plane* $\mathbb{H}$.
   b) *Find a Martin function on* $\{z \in \mathbb{C} : \mathrm{Re}\, z > 0, 0 < \mathrm{Im}\, z < 1\}$. *Hint: Hyperbolic sine.*
   c) *Prove that if* U *is bounded, then there are no Martin functions on* U.

Exercise 7.2.8: *Explicitly solve the following Dirichlet problem: Let* $0 < r < R$ *and* $a, b \in \mathbb{R}$ *be given. Find a continuous* $f : \overline{\mathrm{ann}(0; r, R)} \to \mathbb{R}$, *harmonic on* $\mathrm{ann}(0; r, R)$, *such that* $f = a$ *on* $|z| = r$ *and* $f = b$ *on* $|z| = R$.

Exercise 7.2.9: *Derive the* **Schwarz integral formula**, *which recovers a holomorphic function out of the real parts of the boundary values and the value of the imaginary part at one point. If* $f : \overline{\mathbb{D}} \to \mathbb{C}$ *is continuous and holomorphic on* $\mathbb{D}$, *then for all* $z \in \mathbb{D}$,

$$f(z) = \frac{1}{2\pi i} \int_{\partial \mathbb{D}} \frac{\zeta + z}{\zeta - z} \frac{\mathrm{Re}\, f(\zeta)}{\zeta} \, d\zeta + i \, \mathrm{Im}\, f(0).$$

Exercise 7.2.10: *Let* $q_1, q_2, \ldots$ *be an enumeration of rational numbers in* $[0, 1]$.
   a) *Define* $\varphi : [0, 1] \to \mathbb{R}$ *by* $\varphi(t) = \sum_{j=1}^{\infty} 2^{-j} \chi_{[q_j, 1]}(t)$, *where* $\chi_{[q_j, 1]}(t) = 1$ *if* $t \in [q_j, 1]$ *and zero otherwise (the indicator function). Show that* φ *is discontinuous at every rational number in* $(0, 1]$, *nondecreasing, and bounded (hence Riemann integrable).*
   b) *Define* $\Phi(t) = \int_0^t \varphi(s) \, ds$, *show that* Φ *is increasing, continuous, but not differentiable on a dense set in* $[0, 1]$. *Use it to construct a* $\psi(t)$ *that is* 2π*-periodic, continuous, and not differentiable on a dense subset of* $\mathbb{R}$.
   c) *Find a continuous* $u : \overline{\mathbb{D}} \to \mathbb{R}$ *such that* $u|_{\mathbb{D}}$ *is harmonic and* $u(e^{it}) = \psi(t)$, *then find a holomorphic* $h : \mathbb{D} \to \mathbb{C}$ *such that* $\mathrm{Re}\, h = u$.
   d) *Show that* h *does not extend through any point of the boundary, that is, for every* $z_0 \in \partial \mathbb{D}$ *and every open neighborhood* U *of* z_0, *there exists no holomorphic* $f : U \to \mathbb{C}$ *such that* $f = h$ *on* $\mathbb{D} \cap U$.

7.2.2*i*   Mean-value property

We can define harmonic functions in one real variable by saying f is harmonic if $\nabla^2 f = \frac{\partial^2}{\partial x^2} f = f'' = 0$, that is, $f(x) = Ax + B$, an affine linear function. It is quite useful to think of harmonic functions on $\mathbb{C}$ as one particular analogue of affine linear functions to $\mathbb{C}$, although we ought not to take this analogy too far, of course. One property of affine linear functions is a mean-value property, that is, given $a < b$ then $f\left(\frac{a+b}{2}\right) = \frac{f(a)+f(b)}{2}$. The value at the center of an interval is equal to the average of values at the ends. In fact, if a continuous function satisfies this equality for all intervals $[a, b]$, then it is affine linear (exercise). This mean-value property completely characterizes affine linear (that is, harmonic) functions in $\mathbb{R}$. It is rather interesting that the same kind of property characterizes harmonic functions in $\mathbb{C}$ as well, although we have to replace an interval with a disc.

Theorem 7.2.5 (Mean-value property). *Suppose $U \subset \mathbb{C}$ is open. A continuous $f : U \to \mathbb{R}$ is harmonic if and only if for every $p \in U$, there exists an $R_p > 0$ such that $\Delta_{R_p}(p) \subset U$ and*

$$f(p) = \frac{1}{2\pi} \int_{-\pi}^{\pi} f(p + re^{i\theta}) \, d\theta \qquad \text{for all } r < R_p.$$

Moreover, if f is harmonic, then we may choose any $R_p > 0$ such that $\Delta_{R_p}(p) \subset U$.

Proof. One direction (and the "Moreover") follows quickly. Suppose f is harmonic. Take $p \in U$ and any $R_p > 0$ such that $\Delta_{R_p}(p) \subset U$. For any $r < R_p$, solve the Dirichlet problem in $\Delta_r(p)$ using the Poisson kernel given the boundary values $f|_{\partial \Delta_r(p)}$. Using (7.1) and the uniqueness of the solution of the Dirichlet problem, we find

$$f(p) = P\big[f|_{\partial\Delta_r(p)}\big](p) = \frac{1}{2\pi} \int_{-\pi}^{\pi} f(p + re^{it}) \, dt.$$

Conversely, suppose f is continuous and satisfies the mean-value property for all p and all $r < R_p$. Let $\overline{\Delta_s(q)} \subset U$ be an arbitrary closed disc. Let $h = P\big[f|_{\partial\Delta_s(q)}\big]$ be the solution of the Dirichlet problem in $\Delta_s(q)$ with boundary values given by f. Consider $\varphi = f - h$, which is continuous, identically zero on $\partial\Delta_s(q)$ and satisfies the mean-value property on the same circles as f (as long as they lie in $\Delta_s(q)$). Suppose for contradiction that φ is positive somewhere on $\Delta_s(q)$. Let φ achieve a maximum at $p \in \Delta_s(q)$, so $\varphi(p) > 0$. The set $X \subset \Delta_s(q)$ where $\varphi(z) = \varphi(p)$ is compact. Assume p is the point on X closest to $\partial\Delta_s(q)$. For some small $r < R_p$, the circle $\partial\Delta_r(p) \subset \Delta_s(q)$ and there is a constant C so that $\varphi(z) \leq C < \varphi(p)$ for z on a nonempty open subset of $\partial\Delta_r(p)$. See Figure 7.4. In particular, as φ is supposed to satisfy the mean-value property on $\Delta_r(p)$, we get a contradiction

$$\frac{1}{2\pi} \int_{-\pi}^{\pi} \varphi(p + re^{i\theta}) \, d\theta < \varphi(p).$$

We have proved that $\varphi \leq 0$ on $\Delta_s(q)$. Applying the same logic to $-\varphi$, we find that $\varphi = 0$ on $\Delta_s(q)$. Namely, $f = h$ and h is harmonic, so f is harmonic on $\Delta_s(q)$ (and thus on U). $\qquad\square$

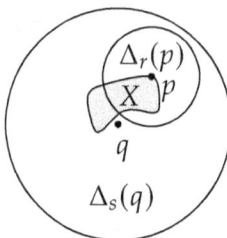

Figure 7.4: The discs $\Delta_s(q)$ and $\Delta_r(p)$ and the set X.

One immediate consequence of the mean value property is that a uniform limit on compact subsets of harmonic functions is harmonic. Just like for holomorphic functions, this result would be hard to prove using the definition of harmonic functions. Given a sequence $\{f_n\}$ of any old C^2 functions with uniform limit f, the limit of $\nabla^2 f_n$ is not necessarily $\nabla^2 f$. But uniform limits do go under the integral. The following result is one part of what is called *Harnack's first theorem* (several results about harmonic functions are named for Harnack).

Theorem 7.2.6 (Harnack's first). *Let $U \subset \mathbb{C}$ be open, and let $f_n \colon U \to \mathbb{R}$ be a sequence of harmonic functions converging uniformly on compact subsets to $f \colon U \to \mathbb{R}$. Then f is harmonic.*

Proof. First, f is continuous. Given any disc $\overline{\Delta_r(p)} \subset U$, the sequence $\{f_n\}$ converges uniformly on the boundary of the disc and at p and hence

$$f(p) = \lim_{n\to\infty} f_n(p) = \lim_{n\to\infty} \frac{1}{2\pi} \int_{-\pi}^{\pi} f_n(p + re^{i\theta})\, d\theta = \frac{1}{2\pi} \int_{-\pi}^{\pi} f(p + re^{i\theta})\, d\theta.$$

The result follows by the mean-value property. $\square$

Exercise 7.2.11: Suppose $f \colon \mathbb{R} \to \mathbb{R}$ is a continuous function such that $f\left(\frac{a+b}{2}\right) = \frac{f(a)+f(b)}{2}$ whenever $a < b$. Prove that $f(x) = Ax + B$ for some constants A, B.

Exercise 7.2.12: Prove the maximum principle for harmonic functions directly from the mean-value property.

Exercise 7.2.13: Suppose $U \subset \mathbb{C}$ is open, $f \colon U \to \mathbb{R}$ is continuous, $p \in U$ and f is harmonic on $U \setminus \{p\}$. Prove that f is, in fact, harmonic on all of U.

Exercise 7.2.14: Suppose f is harmonic in a neighborhood of $\overline{\Delta_r(0)}$ and $f(0) = 0$. Prove that

$$\frac{1}{2} \int_{-\pi}^{\pi} \left| f(re^{it}) \right| dt = \int_{-\pi}^{\pi} \max\{f(re^{it}), 0\}\, dt.$$

Exercise 7.2.15: *Suppose f is harmonic in a neighborhood of $\overline{\mathbb{D}}$, $f(0) = 0$, and $\int_{-\pi}^{\pi} |f(e^{it})| \, dt = 4\pi$. Prove that there exists a t such that $f(e^{it}) = 1$ and an s such that $f(e^{is}) = -1$.*

Exercise 7.2.16: *Suppose $U \subset \mathbb{C}$ is open and $f : U \times [0,1] \to \mathbb{R}$ is continuous such that for every fixed $t \in [0,1]$, $z \mapsto f(z,t)$ is harmonic. Prove that $g : U \to \mathbb{R}$ defined by*

$$g(z) = \int_0^1 f(z,t) \, dt$$

is harmonic. Hint: Fubini not Leibniz.

Exercise 7.2.17: *Let $U \subset \mathbb{C}$ be open. Prove that a continuous $f : U \to \mathbb{R}$ is harmonic if and only if it satisfies the* disc mean-value property *for every $\overline{\Delta_r(p)} \subset U$:*

$$f(p) = \frac{1}{\pi r^2} \int_{\Delta_r(p)} f(z) \, dA.$$

Exercise 7.2.18: *With a little care, it is not necessary to assume the mean-value property for all small enough discs. Suppose $f : \overline{\mathbb{D}} \to \mathbb{R}$ is continuous and such that for every $p \in \mathbb{D}$, there exists an r such that $\overline{\Delta_r(p)} \subset \mathbb{D}$ and*

$$f(p) = \frac{1}{2\pi} \int_{-\pi}^{\pi} f(p + re^{i\theta}) \, d\theta.$$

Prove that f is harmonic in $\mathbb{D}$.

7.2.3*i*   Harnack's inequality

Like holomorphic functions, harmonic functions defined in a disc cannot just do whatever they want inside the disc. Their behavior is somewhat controlled by the size of the disc: The further "inside" their domain of definition the disc is, the more control we have. The basic statement of this control is the *Harnack's inequality* in the disc. For holomorphic functions, an analogous result is Schwarz's lemma, where we require that the functions are bounded (they are valued in a disc). For harmonic functions, the analogue of boundedness is nonnegativity.

Theorem 7.2.7 (Harnack's inequality). *Suppose $f : \Delta_R(p) \to \mathbb{R}$ is harmonic and nonnegative, and suppose $0 < r < R$. Then for all $z \in \overline{\Delta_r(p)}$,*

$$\frac{R-r}{R+r} f(p) \leq f(z) \leq \frac{R+r}{R-r} f(p).$$

Proof. If we prove the inequality for z on $\partial \Delta_r(p)$, i.e., $|z - p| = r$, we are done as $\frac{R+r}{R-r}$ is increasing in r and $\frac{R-r}{R+r}$ is decreasing in r. So assume $z = p + re^{i\theta}$.

Let S be such that $0 < r < S < R$. Using Corollary 7.2.4 and the uniqueness of the solution of the Dirichlet problem,

$$f(z) = f(p + re^{i\theta}) = \int_{-\pi}^{\pi} f(p + Se^{it}) P_{r/S}(\theta - t)\, dt$$

$$= \frac{1}{2\pi} \int_{-\pi}^{\pi} f(p + Se^{it}) \frac{S^2 - r^2}{S^2 + r^2 - 2Sr \cos(\theta - t)}\, dt.$$

We estimate

$$\frac{S - r}{S + r} = \frac{S^2 - r^2}{S^2 + r^2 + 2Sr} \leq \frac{S^2 - r^2}{S^2 + r^2 - 2Sr \cos(\theta - t)} \leq \frac{S^2 - r^2}{S^2 + r^2 - 2Sr} = \frac{S + r}{S - r}.$$

For $z = p + re^{i\theta}$, using that f is nonnegative,

$$f(z) = \int_{-\pi}^{\pi} f(p + Se^{it}) P_{r/S}(\theta - t)\, dt \leq \frac{S + r}{S - r} \left(\frac{1}{2\pi} \int_{-\pi}^{\pi} f(p + Se^{it})\, dt \right) = \frac{S + r}{S - r} f(p).$$

The lower inequality follows in the same way. As $S < R$ was arbitrary, the inequality in the theorem follows by taking a limit. $\qquad\square$

The inequalities are optimal. In the unit disc $\mathbb{D}$, the theorem says

$$\frac{1 - r}{1 + r} f(0) \leq f(z) \leq \frac{1 + r}{1 - r} f(0).$$

Consider the function $f(z) = \mathrm{Re}\, \frac{1+z}{1-z}$. Except for the $\frac{1}{2\pi}$ it is the Poisson kernel and so $f(z) > 0$ on $\mathbb{D}$. Note that $f(0) = 1$. Plugging in $z = r$, we get equality in the right-hand inequality above, and plugging in $z = -r$ we get equality in the left-hand inequality above. In other words, the two constants $\frac{R-r}{R+r}$ and $\frac{R+r}{R-r}$ are optimal.

There is also a general version of Harnack's inequality on any domain.

Corollary 7.2.8 (Harnack's inequality). *Suppose $U \subset \mathbb{C}$ is a domain and $K \subset U$ is compact. Then there exists a $C > 0$ such that*

$$\sup_{z \in K} f(z) \leq C \inf_{z \in K} f(z),$$

for every harmonic and nonnegative function f defined on U.

Proof. If we prove the theorem for a larger K we are done, so we replace K by a larger connected compact subset of U. First, make K have only finitely many components by replacing it by finitely many closed discs (see the proof of Lemma 6.3.7). Then, U is path connected and so adding finitely many paths to K, we can connect the discs.

Suppose $r > 0$ is less than half the distance from K to ∂U. There exist N discs $\Delta_r(z_1), \ldots, \Delta_r(z_N)$ that cover K, where $z_j \in K$, and so $\Delta_{2r}(z_j) \subset U$ for every j. Fix $\zeta, \xi \in K$. After relabeling the discs, $\zeta \in \Delta_r(z_1)$ and $\xi \in \Delta_r(z_n)$ for some $n \leq N$. As K

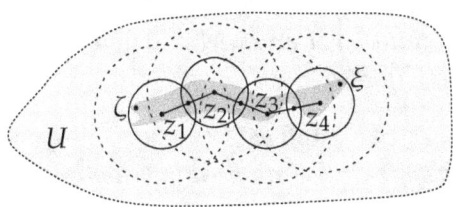

Figure 7.5: A chain of discs with centers z_1, z_2, z_3, z_4 connecting ζ to ξ in K (marked in darker shade). Midpoints of the segments between z_j and z_{j+1} are marked as well. The solid circles are discs of radius r and dashed circles are discs of radius $2r$.

is connected, we also arrange that $\Delta_r(z_j) \cap \Delta_r(z_{j+1})$ is nonempty for all $j = 1, \ldots, n-1$. Not all the discs need to be used. The proof is illustrated in Figure 7.5.

Let f be an arbitrary nonnegative harmonic function on U. For any j, as $\Delta_{2r}(z_j) \subset U$, if we take any $w \in \Delta_r(z_j)$, we find that

$$\frac{1}{3}f(z_j) = \frac{2r-r}{2r+r}f(z_j) \le f(w) \le \frac{2r+r}{2r-r}f(z_j) = 3f(z_j).$$

In other words, $f(w) \le 3f(z_j)$ and $f(z_j) \le 3f(w)$.

We now follow the chain of discs. First, as $\zeta \in \Delta_r(z_1)$, we get

$$f(\zeta) \le 3f(z_1).$$

Second, let q be the midpoint between z_j and z_{j+1}. Simple geometry dictates that $q \in \Delta_r(z_j) \cap \Delta_r(z_{j+1})$. Thus,

$$f(z_j) \le 3f(q) \le 3\big(3f(z_{j+1})\big) = 3^2 f(z_{j+1}).$$

Third, as $\xi \in \Delta_r(z_n)$, we get

$$f(z_n) \le 3f(\xi).$$

All in all,

$$f(\zeta) \le 3^{2n} f(\xi) \le 3^{2N} f(\xi).$$

The number N only depends on K, not on ζ, ξ, or f. As ζ and ξ were arbitrary, the theorem follows. □

The constant we get in the proof is not optimal, but it is explicit. Assuming K is connected, we can compute a specific C by knowing the distance of K to the boundary and the number of the r-discs needed to cover K.

Exercise 7.2.19 (Easy): *Show by example that Harnack's general inequality need not hold if U is not assumed to be connected.*

Exercise 7.2.20 (Easy): *Find the following counterexample of Harnack's inequality if f is not assumed to be nonnegative. For every $M > 0$ find a harmonic function $f : \mathbb{D} \to \mathbb{R}$ such that $f(0) = 1$ and $f(1/2) \geq M$.*

Exercise 7.2.21: Fix some $s > t > 0$. Let $U = \{z \in \mathbb{C} : -s < \operatorname{Re} z < s, -1 < \operatorname{Im} z < 1\}$. Compute an explicit constant C (doesn't need to be optimal) for the following K for the general Harnack's inequality:

    a) $K = [-t, t]$.   *b) $K = \{-t, t\}$.*

Exercise 7.2.22: Affine linear functions $Ax + B$ are the one-real-variable versions of harmonic functions. State and prove Harnack's inequality (analogue of Theorem 7.2.7) in the affine linear setting for an interval $[a, b]$ instead of a disc. Find the optimal constants in the two inequalities just like we got for a disc, and prove that the constants are optimal.

Exercise 7.2.23: Let $U = \mathbb{D}$ and $K = \overline{\Delta_r(0)}$, $r < 1$, in the general Harnack's inequality. Prove that the C from the theorem must necessarily go to infinity as $r \uparrow 1$.

Exercise 7.2.24: Use Harnack's inequality to prove Liouville's theorem for harmonic functions: If $f : \mathbb{C} \to \mathbb{R}$ is harmonic and nonnegative, then f is constant.

7.2.4*i*   Harnack's principle

Harnack's inequality yields that increasing sequences of harmonic functions converge to harmonic functions. This theorem is variously called Harnack's principle or Harnack's second theorem.

Theorem 7.2.9 (Harnack's principle). *Let $U \subset \mathbb{C}$ be a domain and $\{f_n\}$ a sequence of harmonic functions on U such that $f_1 \leq f_2 \leq f_3 \leq \cdots$. Then either $f_n \to +\infty$ uniformly on compact subsets, or $f_n \to f$ for a harmonic $f : U \to \mathbb{R}$ uniformly on compact subsets.*

Proof. Without loss of generality assume that $f_n \geq 0$ for all n. If not, apply the theorem to the functions $f_n - f_1$, which are all nonnegative.

By monotonicity, $\{f_n\}$ converges pointwise (possibly to $+\infty$). If $\lim f_n(p) = +\infty$ for some $p \in U$, let $K \subset U$ be compact, and let $K' = K \cup \{p\}$. Harnack's inequality (for K') says there is a C such that

$$f_n(p) \leq \sup_{z \in K'} f_n(z) \leq C \inf_{z \in K'} f_n(z) \leq C \inf_{z \in K} f_n(z).$$

Thus, $f_n(z) \to +\infty$ uniformly on K.

Therefore, suppose the limit of $\{f_n(z)\}$ is finite for every $z \in U$. Let $f : U \to \mathbb{R}$ be the limit. Let $K \subset U$ be any compact subset, C the constant from Harnack's inequality, and $p \in K$ any point. Given $\epsilon > 0$, there is an N such that whenever $m > n \geq N$, we get $f_m(p) - f_n(p) < \epsilon/C$. Then

$$\sup_{z \in K} \bigl(f_m(z) - f_n(z)\bigr) \leq C \inf_{z \in K} \bigl(f_m(z) - f_n(z)\bigr) \leq C \bigl(f_m(p) - f_n(p)\bigr) < \epsilon.$$

In other words, $\{f_n\}$ is uniformly Cauchy on K and hence converges uniformly on K to f. The function f is harmonic by Harnack's first theorem, Theorem 7.2.6. $\qquad\square$

Another way of stating Harnack's principle is to make it about a sequence of nonnegative functions less than some fixed harmonic function f. If the sequence converges to f at one point, it converges uniformly on compact subsets. Moreover, as nonnegative harmonic functions are the harmonic analogues of bounded holomorphic functions, we expect a version of Montel's theorem for nonnegative harmonic functions, and Harnack delivers that as well. We leave both proofs as exercises.

Exercise 7.2.25: *Prove yet another version of Harnack's principle. Suppose $U \subset \mathbb{C}$ is a domain, $\{f_n\}$ is a sequence of nonnegative harmonic functions on U, and $p \in U$ is fixed.*
   a) If $f_n(p) \to +\infty$, then $\{f_n\}$ converges to $+\infty$ uniformly on compact subsets.
   b) If $f : U \to \mathbb{R}$ is harmonic, $f_n(z) \leq f(z)$ for all $z \in U$, and $f_n(p) \to f(p)$, then $\{f_n\}$ converges to f uniformly on compact subsets.

Exercise 7.2.26: *Prove a Montel-like theorem for harmonic functions. Suppose $U \subset \mathbb{C}$ is a domain and $\{f_n\}$ is a sequence of nonnegative harmonic functions. Show that at least one (or both) of the following are true:*

   (i) There exists a subsequence converging to $+\infty$ uniformly on compact subsets.

   (ii) There exists a subsequence converging to a harmonic function uniformly on compact subsets.

7.3i   Extending harmonic functions

7.3.1i   Isolated singularities

For harmonic functions, we get the following classification of removable singularities, which is sharp, that is, best possible. The harmonic function $\log|z|$ has a nonremovable singularity at the origin. Any function that blows up any slower than that, doesn't actually blow up and, in fact, extends to be harmonic at the origin.

Theorem 7.3.1. *Suppose $U \subset \mathbb{C}$ is open, $p \in U$, and $f : U \setminus \{p\} \to \mathbb{R}$ is harmonic such that*

$$\lim_{z \to p} \frac{f(z)}{\log|z - p|} = 0.$$

Then there exists a harmonic $F : U \to \mathbb{R}$ such that $f = F|_{U \setminus \{p\}}$.

Proof. By considering $f(az + b)$, we may assume, without loss of generality, that $p = 0$ and $\overline{\mathbb{D}} \subset U$. Solve the Dirichlet problem to find a continuous $u : \overline{\mathbb{D}} \to \mathbb{R}$, harmonic in

$\mathbb{D}$, such that $u|_{\partial\mathbb{D}} = f|_{\partial\mathbb{D}}$. We wish to show that u equals f in $\mathbb{D} \setminus \{0\}$. The function $g = f - u$ is harmonic in $\mathbb{D} \setminus \{0\}$ and it is zero on $\partial\mathbb{D}$. Furthermore,

$$\lim_{z \to 0} \frac{g(z)}{-\log|z|} = 0.$$

Equivalently, given any $\epsilon > 0$, there is a $\delta > 0$ (we can assume $\delta < 1$) such that for all $z \in \overline{\Delta_\delta(0)} \setminus \{0\}$,

$$-\epsilon(-\log|z|) \leq g(z) \leq \epsilon(-\log|z|). \tag{7.2}$$

The estimate (7.2) holds also when $|z| = 1$ as $g = 0$ there. The functions $-\log|z|$ and g are harmonic outside of the origin, so the maximum principle (the version in Exercise 7.1.11) implies that (7.2) holds also for $\delta < |z| < 1$, and thus for all $z \in \mathbb{D} \setminus \{0\}$. As the estimate holds for all $\epsilon > 0$, we have $g(z) = 0$ for all $z \in \mathbb{D} \setminus \{0\}$. So u is the extension near 0 that we are looking for. □

An isolated singularity of a harmonic function g could be very wild, for example $\operatorname{Re} e^{1/z}$ or similar. But if g is $\log|f(z)|$ for a holomorphic f that has either a pole or a zero at the origin, then near the origin f behaves like z^n for $n \in \mathbb{Z}$ and

$$\log|z^n| = n \log|z|.$$

In other words, the function g behaves like $\log|z|$. The expression is positive near the origin if $n < 0$ and negative if $n > 0$. *Bôcher's theorem* is the converse of this reasoning: A nonnegative harmonic function at an isolated singularity at the origin can be written as $g(z) - C \log|z|$, where g is harmonic at the origin.

Theorem 7.3.2 (Bôcher). *Suppose $U \subset \mathbb{C}$ is open, $p \in U$, and $f: U \setminus \{p\} \to \mathbb{R}$ is harmonic and nonnegative. Then there exists a harmonic function $g: U \to \mathbb{R}$ and a $C \geq 0$ such that for all $z \in U \setminus \{p\}$,*

$$f(z) = g(z) - C \log|z - p|.$$

Proof. Without loss of generality suppose $p = 0$ and $U = \mathbb{D}$. We would like to use the theory of holomorphic functions, but $\mathbb{D} \setminus \{0\}$ is not simply connected. We cannot simply find a harmonic conjugate in the entire punctured disc. However, we can find a harmonic conjugate locally, and any two harmonic conjugates differ by a constant (Proposition 7.1.6). Thus if Φ is locally a holomorphic function such that $\operatorname{Re} \Phi = f$, then Φ' is well-defined in the entire punctured disc $\mathbb{D} \setminus \{0\}$. Fix some $q \in \mathbb{D} \setminus \{0\}$ and some Φ defined near q such that $\operatorname{Re} \Phi = f$. Then there exists a holomorphic $\varphi: \mathbb{D} \setminus \{0\} \to \mathbb{C}$ such that $\Phi' = \varphi$ near q.

Expand φ in $\mathbb{D} \setminus \{0\}$ using Laurent series,

$$\varphi(z) = \sum_{n=-\infty}^{\infty} c_n z^n.$$

Using Proposition 4.4.3, antidifferentiating $1/z$ separately, we find that locally near q,

$$\Phi(z) = A + c_{-1} \log z + \sum_{n=-\infty, n \neq -1}^{\infty} \frac{c_n}{n+1} z^{n+1} = c_{-1} \log z + \psi(z),$$

for some branch of the log, where ψ is a holomorphic function on $\mathbb{D} \setminus \{0\}$. Taking the real part we get f, a well-defined function on $\mathbb{D} \setminus \{0\}$. Therefore, $c_{-1} \log z$ has real part that is well-defined in $\mathbb{D} \setminus \{0\}$, meaning $c_{-1} \in \mathbb{R}$. So

$$f(z) = \operatorname{Re} \Phi(z) = c_{-1} \log|z| + \operatorname{Re} \psi(z).$$

Nonnegativity of $f(z)$ says that for some integer $k \leq c_{-1}$,

$$-k \log|z| \leq -c_{-1} \log|z| \leq \operatorname{Re} \psi(z).$$

Then

$$|z^{-k}| \leq e^{\operatorname{Re} \psi(z)} = \left| e^{\psi(z)} \right| \quad \text{or} \quad |z^{-k} e^{-\psi(z)}| \leq 1.$$

That is, $z^{-k} e^{-\psi(z)}$ has a removable singularity at the origin. By Exercise 5.2.21, $e^{-\psi(z)}$ cannot have a pole. So $e^{-\psi(z)}$ and thus $\psi(z)$ has a removable singularity. As $-c_{-1} \log|z| \leq \operatorname{Re} \psi(z)$, we find $-c_{-1} \log|z|$ is bounded from above near the origin and so $-c_{-1} \geq 0$. We are done, ψ extends through the origin and

$$f(z) = \operatorname{Re} \psi(z) - (-c_{-1}) \log|z|. \qquad \square$$

Exercise 7.3.1: Prove that a holomorphic $f : \Delta_r(p) \setminus \{p\} \to \mathbb{C}$ such that the real part of f is bounded has a removable singularity at p. Prove it using harmonic functions.

Exercise 7.3.2: Prove that the Dirichlet problem is not necessarily solvable in the punctured disc $\mathbb{D} \setminus \{0\}$.

Exercise 7.3.3: Prove that given f, the function g and the constant C in Bôcher's theorem are unique.

Exercise 7.3.4: Suppose $f : \mathbb{D} \setminus \{0\} \to \mathbb{R}$ is nonnegative and harmonic. Let $z = x + iy$. Prove that the C from Bôcher's theorem can be computed by

$$C = \frac{-r}{2\pi} \int_0^{2\pi} \left((\cos t) f_x(r \cos t, r \sin t) + (\sin t) f_y(r \cos t, r \sin t) \right) dt.$$

7.3.2*i*   Schwarz reflection principle

Classically, the Schwarz reflection principle is a theorem for holomorphic functions, but it is also a theorem for harmonic functions. We will prove the corresponding holomorphic version (Theorem 10.1.1) later separately.

Basically the reflection principle says that if a harmonic function vanishes on a nice enough curve—such as the real line—then it extends (reflects) across.

Theorem 7.3.3 (Schwarz reflection principle for harmonic functions). *Suppose $U \subset \mathbb{C}$ is a domain symmetric across the real axis, that is, $z \in U$ if and only if $\bar{z} \in U$. Let $U_+ = \{z \in U : \operatorname{Im} z > 0\}$ and $L = U \cap \mathbb{R}$. Suppose $f : U_+ \cup L \to \mathbb{R}$ is a continuous function that is harmonic on U_+ and $f(z) = 0$ for all $z \in L$.*

Then there exists a harmonic $F : U \to \mathbb{R}$ such that $F|_{U_+ \cup L} = f$.

See Figure 7.6 for a diagram.

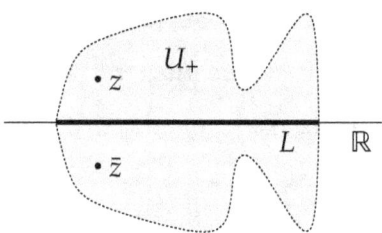

Figure 7.6: Schwarz reflection principle.

Proof. The trick is to define what we want F to be and then check that it is harmonic. For $z \in U$, define

$$F(z) = \begin{cases} f(z) & \text{if } \operatorname{Im} z \geq 0, \\ -f(\bar{z}) & \text{else.} \end{cases}$$

If $z \in U$ and $\operatorname{Im} z > 0$, then F is harmonic at z by hypothesis. Suppose $z \in U$ and $\operatorname{Im} z < 0$. Write $F(z) = F(x, y) = -f(x, -y)$, then

$$\nabla^2|_{(x,y)} F = \nabla^2|_{(x,y)} \left(-f(x, -y) \right) = -\frac{\partial^2 f}{\partial x^2}\bigg|_{(x,-y)} - \frac{\partial^2 f}{\partial y^2}\bigg|_{(x,-y)} = -\nabla^2|_{(x,-y)} f = 0.$$

Suppose that $z \in L$, that is, $z \in \mathbb{R}$. Compute the mean value at z around any $\partial \Delta_r(z)$ where $\overline{\Delta_r(z)} \subset U$:

$$\frac{1}{2\pi} \int_{-\pi}^{\pi} F(z + re^{i\theta}) \, d\theta = \frac{1}{2\pi} \int_{-\pi}^{0} -f(z + re^{-i\theta}) \, d\theta + \frac{1}{2\pi} \int_{0}^{\pi} f(z + re^{i\theta}) \, d\theta = 0.$$

That is, the mean value equals $F(z) = 0$, and the mean-value property is satisfied for all small enough r at every $z \in L$. We proved above that F is harmonic on $U \setminus L$ and so the mean-value property is also satisfied for all small enough circles around any $z \in U \setminus L$. By the mean-value property (Theorem 7.2.5), F is harmonic in U. $\qquad \square$

Exercise 7.3.5: Suppose $f : \mathbb{C} \to \mathbb{C}$ is an entire holomorphic function. Suppose $f(x)$ is real for all $x \in \mathbb{R}$. Prove:

a) If $f(iy)$ is purely imaginary for all $y \in \mathbb{R}$, then $f(z) = -f(-z)$ for all $z \in \mathbb{C}$.

b) If $f(iy)$ is real for all $y \in \mathbb{R}$, then $f(z) = f(-z)$ for all $z \in \mathbb{C}$.

Exercise 7.3.6: *Prove that the Dirichlet problem has a unique bounded solution in* $\mathbb{H}$. *That is, suppose* $f, g : \overline{\mathbb{H}} \to \mathbb{R}$ *are continuous and harmonic in* $\mathbb{H}$ *such that* $f - g$ *is bounded and* $f = g$ *on* $\partial \mathbb{H} = \mathbb{R}$. *Prove that* $f = g$ *everywhere. Compare to Exercise 7.2.5.*

Exercise 7.3.7: *Prove a version of reflection across the circle: Let* $U \subset \mathbb{C}$ *be a domain symmetric with respect to the inversion* $z \mapsto z/|z|^2$. *Suppose* f *is a harmonic function defined on* $U \cap \overline{\mathbb{D}}$ *and zero on* $U \cap \partial \mathbb{D}$. *Prove that* f *extends to a harmonic function on* U.

Exercise 7.3.8: *Suppose* $f : \mathbb{D} \setminus \{0\} \to \mathbb{R}$ *is harmonic and* $f = 0$ *on* $\mathbb{R} \cap (\mathbb{D} \setminus \{0\})$.
  a) *Show that* f *has a harmonic conjugate in* $\mathbb{D} \setminus \{0\}$.
  b) *Find an example* f *that does not extend to be harmonic through the origin.*

Exercise 7.3.9: *Suppose* $f : \overline{\mathbb{H}} \to \mathbb{R}$ *is continuous, harmonic on* $\mathbb{H}$, *zero on* $\mathbb{R} = \partial \mathbb{H}$, *and positive on* $\mathbb{H}$ *(a Martin function). Prove that* $f(z) = c \operatorname{Im} z$ *for some* $c > 0$. *Hint: Find an entire function whose imaginary part is* f *and show that it has a pole at infinity.*

Exercise 7.3.10:
  a) *Allow singularities in Exercise 7.3.6. Suppose* $S \subset \mathbb{R}$ *is finite, and* $f, g : \overline{\mathbb{H}} \setminus S \to \mathbb{R}$ *are continuous and harmonic in* $\mathbb{H}$ *such that* $f - g$ *is bounded and* $f = g$ *on* $\mathbb{R} \setminus S$. *Prove that* $f = g$ *everywhere.*
  b) *Show uniqueness of the bounded Dirichlet problem in* $\mathbb{D}$ *with discontinuities: Suppose* $S \subset \partial \mathbb{D}$ *is finite and* $f : \partial \mathbb{D} \setminus S \to \mathbb{R}$ *is continuous and bounded. By Exercise 7.2.4, a continuous* $g : \overline{\mathbb{D}} \setminus S \to \mathbb{R}$ *harmonic in* $\mathbb{D}$ *exists such that* $g = f$ *on* $\partial \mathbb{D} \setminus S$. *Prove that there is a unique such bounded* g. *Hint: Part a) and Cayley.*

Exercise 7.3.11: *Suppose* f *is an entire holomorphic function and* $\operatorname{Re} f(iy) = \operatorname{Re} f(1 + iy) = 0$ *for all* $y \in \mathbb{R}$. *Prove that* f *is 2-periodic:* $f(z + 2) = f(z)$ *for all* $z \in \mathbb{C}$.

7.4*i* Subharmonic functions ⋆

Harmonic (and holomorphic) functions are very rigid. There is a less restrictive (and much larger) set of functions that allows us to study harmonic functions. In essence, we replace equalities, which are hard, by inequalities, which are easier to work with.

7.4.1*i* Basic properties

Recall that $f : U \to \mathbb{R} \cup \{-\infty\}$ is *upper-semicontinuous** if

$$\limsup_{\zeta \to z} f(\zeta) \le f(z) \quad \text{for all } z \in U.$$

Definition 7.4.1. Let $U \subset \mathbb{C}$ be open. A function $f : U \to \mathbb{R} \cup \{-\infty\}$ is *subharmonic* if it is upper-semicontinuous and for every closed disc $\overline{\Delta_r(p)} \subset U$, and every continuous

---

*We do not require U to be open for semicontinuity.

$g \colon \overline{\Delta_r(p)} \to \mathbb{R}$, harmonic on $\Delta_r(p)$, such that $f(z) \leq g(z)$ for $z \in \partial \Delta_r(p)$, we have

$$f(z) \leq g(z) \quad \text{for all } z \in \Delta_r(p).$$

In other words, a subharmonic function is a function that is less than every harmonic function on every disc whenever that holds on the boundary. The best way to think about subharmonic functions is an analogy to convex functions in $\mathbb{R}$. We saw that harmonic functions in $\mathbb{R}$ are the affine linear functions: A function $g(x)$ on $\mathbb{R}$ is harmonic if $g'' \equiv 0$, that is, $g(x) = Ax + B$. A function of one real variable is *convex* if for every interval it is less than the affine linear function with the same end points. That is, the function f is convex if for every $\alpha < \beta$, and every affine linear g such that $f(\alpha) \leq g(\alpha)$ and $f(\beta) \leq g(\beta)$, we have $f(x) \leq g(x)$ for all $x \in [\alpha, \beta]$. See Figure 7.7. In $\mathbb{R}$, an interval $[\alpha, \beta]$ plays the role of a closed disc. So in $\mathbb{R}$, *convex* is the same as *subharmonic*. Graphs of real-valued functions of one real variable are also much easier to draw than functions on $\mathbb{C}$.

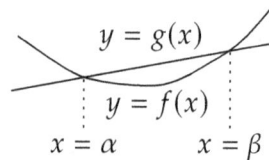

Figure 7.7: A convex function.

Exercise 7.4.1: Consider $f \colon U \to \mathbb{R} \cup \{-\infty\}$.
  a) *Prove that f is upper-semicontinuous if and only if for every $a \in \mathbb{R}$ the set $V = f^{-1}([-\infty, a)) = \{z \in U : f(z) < a\}$ is open (in the subspace topology of U).*
  b) *Prove that f is upper-semicontinuous if and only if for every $a \in \mathbb{R}$ the set $X = f^{-1}([a, +\infty)) = \{z \in U : f(z) \geq a\}$ is closed (in the subspace topology of U).*

Exercise 7.4.2: Prove that an upper-semicontinuous function defined on a compact set achieves a maximum.

Exercise 7.4.3: Prove that if $f \colon U \to \mathbb{R}$ is upper-semicontinuous and $-f$ is also upper-semicontinuous (that is, f is also lower-semicontinuous), then f is continuous.

Adding or subtracting harmonic functions does not kill subharmonicity. The proof is rather simple as a sum or difference of harmonic functions is harmonic and we leave it as an exercise.

Proposition 7.4.2. If $f \colon U \to \mathbb{R} \cup \{-\infty\}$ is subharmonic and $h \colon U \to \mathbb{R}$ is harmonic, then $f + h$ is subharmonic.

> *Exercise 7.4.4:* Prove the proposition.

Subharmonic functions are also classified by a mean-value-like property, although it is an inequality rather than an equality. There is a subtle issue of integrability. For an upper-semicontinuous function the integral in the theorem need not exist as a Riemann integral. We will only give a proof for the upper Darboux integral. The proof is similar for the Lebesgue integral if the reader knows that, although the statement with the Darboux integral is sufficient for us.

Theorem 7.4.3 (Sub-mean-value property). *Suppose $U \subset \mathbb{C}$ is open. An upper-semicontinuous $f \colon U \to \mathbb{R} \cup \{-\infty\}$ is subharmonic if and only if for every $p \in U$ there exists an $R_p > 0$ such that $\Delta_{R_p}(p) \subset U$ and*

$$f(p) \le \frac{1}{2\pi} \int_{-\pi}^{\pi} f(p + re^{i\theta}) \, d\theta \qquad \text{for all } r < R_p.$$

The integral is either the Lebesgue integral or the upper Darboux integral. Moreover, if f is subharmonic, then we may choose any $R_p > 0$ such that $\Delta_{R_p}(p) \subset U$.

As $\partial \Delta_r(p)$ is compact and f is upper-semicontinuous, then f is bounded from above and hence the upper Darboux integral is defined and finite. The upper Darboux integral of a function f on $[a, b]$ bounded above is normally defined as

$$\overline{\int_a^b} f(t) \, dt \overset{\text{def}}{=} \inf \left\{ \int_a^b s(t) \, dt : s \text{ is a step function and } f(t) \le s(t) \text{ for } t \in [a, b] \right\}.$$

A step function is a finite sum of characteristic functions of intervals and hence Riemann integrable. Since continuous functions are Riemann integrable, we can approximate from above by continuous functions g such that $\int_a^b g(t) \, dt$ approximates $\int_a^b s(t) \, dt$. In other words, in the definition we could replace step functions with continuous functions, and that is what we will do in the proof.

Proof. First suppose that f is subharmonic. Take $p \in U$ and any $R_p > 0$ such that $\Delta_{R_p}(p) \subset U$. Fix $r < R_p$ and $\epsilon > 0$. Find a continuous function $g \colon \partial \Delta_r(p) \to \mathbb{R}$ such that $f(p + re^{i\theta}) \le g(p + re^{i\theta})$ for all θ and such that

$$\frac{1}{2\pi} \int_{-\pi}^{\pi} g(p + re^{i\theta}) \, d\theta < \frac{1}{2\pi} \overline{\int_{-\pi}^{\pi}} f(p + re^{i\theta}) \, d\theta + \epsilon.$$

Solve the Dirichlet problem in the disc $\Delta_r(p)$ for g and with a slight abuse of notation call the solution on $\overline{\Delta_r(p)}$ also g. As g is harmonic and bigger than f, we have by definition of subharmonicity and the mean-value property of harmonic functions

$$f(p) \le g(p) = \frac{1}{2\pi} \int_{-\pi}^{\pi} g(p + re^{i\theta}) \, d\theta < \frac{1}{2\pi} \overline{\int_{-\pi}^{\pi}} f(p + re^{i\theta}) \, d\theta + \epsilon.$$

For the converse, suppose that f is upper-semicontinuous and the estimate holds for all p and all $r < R_p$. Suppose for contradiction that there exists a closed disc $\overline{\Delta_s(q)} \subset U$ and a continuous $h\colon \overline{\Delta_s(q)} \to \mathbb{R}$, harmonic in $\Delta_s(q)$, such that $f(z) \le h(z)$ on $\partial\Delta_s(q)$ and $f(p) > h(p)$ at some point $p \in \Delta_s(q)$. Consider $\varphi = f - h$, which is upper-semicontinuous on $\overline{\Delta_s(q)}$. Let p be the point where φ attains the maximum (see Exercise 7.4.2). Then $p \in \Delta_s(q)$, $\varphi(p) > 0$, and $\varphi(z) \le 0$ for all $z \in \partial\Delta_s(q)$. The set $X \subset \Delta_s(q)$ where $\varphi(z) = \varphi(p)$ is compact (as a subset of $\overline{\Delta_s(q)}$ it is closed via Exercise 7.4.1). Assume p is the point of X closest to $\partial\Delta_s(q)$. For some small $r < R_p$, the circle $\partial\Delta_r(p) \subset \Delta_s(q)$ and for a nonempty open subset of $\partial\Delta_r(p)$ the function φ must be less than some fixed constant less than $\varphi(p)$. This fact again follows via Exercise 7.4.1. The setup is the same as in Figure 7.4.

$$\frac{1}{2\pi}\overline{\int_{-\pi}^{\pi}}\varphi(p + re^{i\theta})\,d\theta < \varphi(p).$$

We obtain a contradiction,

$$f(p) - h(p) \le \frac{1}{2\pi}\overline{\int_{-\pi}^{\pi}}\bigl(\varphi(p + re^{i\theta}) + h(p + re^{i\theta})\bigr)\,d\theta - \frac{1}{2\pi}\int_{-\pi}^{\pi} h(p + re^{i\theta})\,d\theta$$

$$\le \frac{1}{2\pi}\overline{\int_{-\pi}^{\pi}}\varphi(p + re^{i\theta})\,d\theta + \frac{1}{2\pi}\overline{\int_{-\pi}^{\pi}}h(p + re^{i\theta})\,d\theta - \frac{1}{2\pi}\int_{-\pi}^{\pi} h(p + re^{i\theta})\,d\theta$$

$$= \frac{1}{2\pi}\overline{\int_{-\pi}^{\pi}}\varphi(p + re^{i\theta})\,d\theta < \varphi(p) = f(p) - h(p).$$

The first inequality follows by the hypothesis for $f = \varphi + h$ and the mean-value property for h. The second inequality follows because the Darboux upper integral is only subadditive ($\overline{\int}(a + b) \le \overline{\int}a + \overline{\int}b$). Then that h is Riemann integrable gives the next equality, and the final inequality is the inequality we just proved above.   □

From now on, we simply write the integral and the reader substitutes the upper Darboux integral or the Lebesgue integral according to the reader's taste.

An easy consequence of the sub-mean-value property is that subharmonic functions, while not necessarily continuous, are a bit better than just upper-semicontinuous.

Proposition 7.4.4. *Let $U \subset \mathbb{C}$ be open and $f\colon U \to \mathbb{R} \cup \{-\infty\}$ be subharmonic. Then*

$$\limsup_{\zeta \to z} f(\zeta) = f(z) \qquad \text{for all } z \in U.$$

Exercise 7.4.5: Prove the proposition.

When we said that the maximum principle was really something about harmonic functions, we lied. It is really there because harmonic functions are subharmonic. Be careful, however, there is no minimum principle for subharmonic functions. To get a minimum principle, you look at superharmonic functions.

Theorem 7.4.5 (Maximum principle). *Suppose $U \subset \mathbb{C}$ is a domain and $f : U \to \mathbb{R} \cup \{-\infty\}$ is subharmonic. If f attains a maximum* in U, then f is constant.*

Proof. Suppose f attains a maximum at $p \in U$. If $\overline{\Delta_r(p)} \subset U$, then

$$f(p) \leq \frac{1}{2\pi} \int_{-\pi}^{\pi} f(p + re^{i\theta}) \, d\theta \leq f(p).$$

Hence, on every subinterval of θ, $f(p + re^{i\theta}) = f(p)$ somewhere (using either the Darboux or Lebesgue integral). By upper-semicontinuity for any θ_0, $f(p + re^{i\theta_0}) = \limsup_{\theta \to \theta_0} f(p + re^{i\theta}) = f(p)$. So $f = f(p)$ everywhere on $\partial\Delta_r(p)$. This equality holds for all r with $\overline{\Delta_r(p)} \subset U$, so $f = f(p)$ on $\Delta_r(p)$, and the set where $f = f(p)$ is open. The set where an upper-semicontinuous function attains a maximum is closed. So $f = f(p)$ on U as U is connected. □

Exercise 7.4.6: *Prove that subharmonicity is a local property. That is, given an open set $U \subset \mathbb{C}$, a function $f : U \to \mathbb{R} \cup \{-\infty\}$ is subharmonic if and only if for every $p \in U$ there exists an open neighborhood W of p, $W \subset U$, such that $f|_W$ is subharmonic.*

Exercise 7.4.7: *Suppose $U \subset \mathbb{C}$ is bounded and open, $f : \overline{U} \to \mathbb{R} \cup \{-\infty\}$ is upper-semicontinuous such that $f|_U$ is subharmonic, and $g : \overline{U} \to \mathbb{R}$ is continuous such that $g|_U$ is harmonic and $f(z) \leq g(z)$ for all $z \in \partial U$. Prove that $f(z) \leq g(z)$ for all $z \in U$.*

Exercise 7.4.8: *Let g be a function harmonic on a disc $\Delta \subset \mathbb{C}$ and continuous on $\overline{\Delta}$. Prove that for every $\epsilon > 0$ there exists a function g_ϵ, harmonic in an open neighborhood of $\overline{\Delta}$, such that $g(z) \leq g_\epsilon(z) \leq g(z) + \epsilon$ for all $z \in \overline{\Delta}$. In particular, to test subharmonicity, we only need to consider those g that are harmonic a bit past the boundary of the disc.*

Exercise 7.4.9: *Prove the minimum principle for superharmonic functions (f is superharmonic if $-f$ is subharmonic). That is, if a superharmonic function defined on a domain U achieves a minimum inside U, then it is constant.*

To continue the analogy to convex functions, a C^2 function f of one real variable is convex if and only if $f''(x) \geq 0$ for all x. We obtain the same kind of result for subharmonic functions by replacing f'' by the Laplacian as before.

Proposition 7.4.6. *Suppose $U \subset \mathbb{C}$ is an open set and $f : U \to \mathbb{R}$ is a C^2 (twice continuously differentiable) function. The function f is subharmonic if and only if $\nabla^2 f \geq 0$.*

Proof. Suppose f is a C^2 function on a subset of $\mathbb{C}$ with $\nabla^2 f \geq 0$. We wish to show that f is subharmonic. Take a disc Δ such that $\overline{\Delta} \subset U$. Consider a function g continuous on $\overline{\Delta}$, harmonic on Δ, and such that $f \leq g$ on the boundary $\partial\Delta$. Because $\nabla^2(f - g) = \nabla^2 f \geq 0$, we assume $g = 0$ and $f \leq 0$ on the boundary $\partial\Delta$.

---

*We do mean the global maximum; if the maximum is local, one only obtains constancy nearby.

Suppose $\nabla^2 f > 0$ at all points on Δ. The Laplacian $\nabla^2 f$ is the trace of the Hessian matrix, that is, the sum of the eigenvalues. Thus f has no maximum in Δ, since at a maximum both eigenvalues of the Hessian matrix would be nonpositive. Therefore, $f \leq 0$ on all of $\overline{\Delta}$.

Next suppose only that $\nabla^2 f \geq 0$. Let M be the maximum of $x^2 + y^2$ on $\overline{\Delta}$. Take $f_n(x, y) = f(x, y) + \frac{1}{n}(x^2 + y^2) - \frac{1}{n}M$. Clearly $\nabla^2 f_n > 0$ everywhere on Δ and $f_n \leq 0$ on the boundary, so $f_n \leq 0$ on all of $\overline{\Delta}$. As $f_n \to f$, we obtain that $f \leq 0$ on all of $\overline{\Delta}$, that is, $f \leq g$ on $\overline{\Delta}$. So f is subharmonic.

The other direction is left as an exercise.   □

Exercise 7.4.10: Finish the proof of Proposition 7.4.6.

The supremum of convex functions is convex. Similarly, the supremum of subharmonic functions is subharmonic, as long as the supremum is upper-semicontinuous. We can therefore "piece together" many subharmonic functions by taking suprema.

Proposition 7.4.7. *Suppose $U \subset \mathbb{C}$ is an open set and $f_\alpha \colon U \to \mathbb{R} \cup \{-\infty\}$ is a family of subharmonic functions. Let*

$$\varphi(z) = \sup_{\alpha} f_\alpha(z).$$

If the family is finite, then φ is subharmonic. If the family is infinite, $\varphi(z) \neq +\infty$ for all z, and φ is upper-semicontinuous, then φ is subharmonic.

Proof. Suppose $\overline{\Delta_r(p)} \subset U$. For any α,

$$\frac{1}{2\pi} \int_{-\pi}^{\pi} \varphi(p + re^{i\theta})\, d\theta \geq \frac{1}{2\pi} \int_{-\pi}^{\pi} f_\alpha(p + re^{i\theta})\, d\theta \geq f_\alpha(p).$$

Taking the supremum on the right over α obtains the result.   □

Exercise 7.4.11: Prove that if $\varphi \colon \mathbb{R} \to \mathbb{R}$ is a monotonically increasing convex function, $U \subset \mathbb{C}$ is an open set, and $f \colon U \to \mathbb{R}$ is subharmonic, then $\varphi \circ f$ is subharmonic.

Exercise 7.4.12: Let $U \subset \mathbb{C}$ be open, $\{f_n\}$ a sequence of subharmonic functions uniformly bounded above on compact subsets, and $\{c_n\}$ a sequence of positive real numbers such that $\sum_{n=1}^{\infty} c_n < +\infty$. Prove that $f = \sum_{n=1}^{\infty} c_n f_n$ is subharmonic. Make sure to prove the function is upper-semicontinuous.

Exercise 7.4.13: Suppose $U \subset \mathbb{C}$ is a bounded open set, and $\{p_n\}$ a sequence of points in U. For $z \in U$, define $f(z) = \sum_{n=1}^{\infty} 2^{-n} \log|z - p_n|$, possibly taking on the value $-\infty$.
   a) Show that f is a subharmonic function in U.
   b) If $U = \mathbb{D}$ and $p_n = 1/n$, show that f is discontinuous at 0 (the natural topology on $\mathbb{R} \cup \{-\infty\}$).
   c) If $\{p_n\}$ is dense in U, show that f is nowhere continuous. Hint: Prove $f^{-1}(-\infty)$ is a small but dense set. Hint #2: Integrate the partial sums, and use polar coordinates.

7.4.2*i*   Applications, Radó's theorem

In complex analysis, we are really interested in proving results for harmonic functions, as we are interested in proving results for holomorphic functions. However, harmonic functions are rigid, they cannot be "put together" easily. Furthermore, there aren't that many of them. There are a lot of subharmonic functions. An example of the use of subharmonic functions to the theory of holomorphic functions is the theorem of Radó, which is a complementary result to the Riemann extension theorem. Here, on the one hand, the function is continuous and vanishes on the set you wish to extend across, but on the other hand, you know nothing about the size of this set.

Theorem 7.4.8 (Radó). *Let $U \subset \mathbb{C}$ be open and $f : U \to \mathbb{C}$ a continuous function that is holomorphic on the set where it is nonzero, that is, f is holomorphic on $\{z \in U : f(z) \neq 0\}$. Then f is holomorphic.*

Proof. Holomorphicity is local, so it is enough to prove the theorem for a small disc Δ assuming f is continuous on the closure $\overline{\Delta}$. Let $\Delta' \subset \Delta$ be the set where f is nonzero. If Δ' is empty, then we are done, as f is just identically zero and hence holomorphic.

Let u be the real part of f. On Δ', u is a harmonic function. Write $Pu = P[u|_{\partial\Delta}]$ for the Poisson integral of u on $\overline{\Delta}$. Hence Pu equals u on $\partial\Delta$, and Pu is harmonic in all of Δ. Consider the function $Pu(z) - u(z)$ on $\overline{\Delta}$. The function is zero on $\partial\Delta$ and it is harmonic on Δ'. By rescaling f, we assume that $|f(z)| < 1$ for all $z \in \overline{\Delta}$. The function $z \mapsto \log|f(z)|$ is harmonic on Δ', it is $-\infty$ when $f(z) = 0$, and hence it is upper-semicontinuous on $\overline{\Delta}$. Applying the sub-mean-value property near points where f vanishes, we find that $\log|f(z)|$ is subharmonic on Δ. As $|f(z)| < 1$, we find that $\log|f(z)|$ is negative on $\overline{\Delta}$. So for every $t > 0$, the function $z \mapsto t \log|f(z)|$ is subharmonic and negative and the function $z \mapsto -t \log|f(z)|$ is superharmonic (minus a subharmonic function) and positive. See Figure 7.8. It is immediate that for all $t > 0$ and $z \in \partial\Delta$, we have

$$t \log|f(z)| \leq Pu(z) - u(z) \leq -t \log|f(z)|. \tag{7.3}$$

The functions $z \mapsto t \log|f(z)| - (Pu(z) - u(z))$ and $z \mapsto t \log|f(z)| - (u(z) - Pu(z))$ are harmonic on Δ' and $-\infty$ whenever $f(z) = 0$. Thus both are upper-semicontinuous on $\overline{\Delta}$ and subharmonic on Δ. The maximum principle shows that (7.3) holds for all $z \in \overline{\Delta}$ and all $t > 0$.

Taking the limit $t \to 0$ shows that $Pu = u$ on Δ'. Let $W = \Delta \setminus \overline{\Delta'}$. On W, $u = 0$ and so $Pu - u$ is harmonic on W and continuous on $\overline{W}$. Furthermore, $Pu - u = 0$ on $\overline{\Delta'} \cup \partial\Delta$, and so $Pu - u = 0$ on ∂W. By the maximum principle, $Pu = u$ on W and therefore on all of $\overline{\Delta}$. All in all, u is harmonic on Δ. Repeating the whole procedure for v, the imaginary part of f, we find that v is harmonic as well. As Δ is simply connected, let $\tilde{v}$ be the harmonic conjugate of u that equals v at some point of Δ'. As f is holomorphic on Δ', the harmonic functions $\tilde{v}$ and v are equal on the nonempty open subset Δ' of Δ and so they are equal everywhere. Consequently, $f = u + iv$ is holomorphic on Δ. $\square$

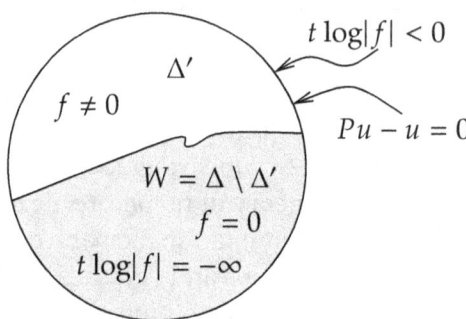

Figure 7.8: Proof of Radó's theorem.

Exercise 7.4.14: Let $U \subset \mathbb{C}$ be a domain and $p \in \partial U$ be a nonisolated point of ∂U. Suppose $f \colon \overline{U} \to \mathbb{C}$ is continuous, holomorphic in U, and such that for some small disc Δ centered at p, f is zero on $\Delta \cap \partial U$. Prove that $f \equiv 0$.

Another example of the use of subharmonic functions is another solution to the Dirichlet problem. A solution can be had by considering all the subharmonic functions that are less than the function given on the boundary. Then one takes a supremum to obtain a harmonic function. We will not go through this technique, which is called the *Perron method*. Clearly, that technique would work far better than the Poisson kernel for more complicated domains. For instance, the Poisson kernel can be computed for simply connected domains with nice boundary provided we know the Riemann map. However, the kernel is difficult to compute in general, and it requires a very nice boundary to be able to integrate. The Perron method works much more generally provided you can construct enough subharmonic functions (which can, after all, be pieced together unlike harmonic functions).

If a solution exists, it clearly must equal the Perron solution.

Exercise 7.4.15 (Easy): *Suppose $U \subset \mathbb{C}$ is a domain and $u \colon \overline{U} \to \mathbb{R}$ is continuous and harmonic on U. Prove that for all $p \in U$,*

$$u(p) = \sup_{v} v(p),$$

where v ranges over all upper-semicontinuous functions on $\overline{U}$ subharmonic on U, such that $v|_{\partial U} \leq u|_{\partial U}$.

8i Weierstrass Factorization

I became insane, with long intervals of horrible sanity.

—*Edgar Allan Poe*

8.1i Infinite products

If a function has zeros at 0, 1, and i, we can write f as $f(z) = g(z)z(z-1)(z-i)$. If those are the only zeros, then g is never zero. If there are infinitely many zeros, however, things become difficult. Can we factor the zeros out of $\sin z$? Can we write $\sin z$ as something times

$$\prod_{n \in \mathbb{Z}} (z - \pi n)?$$

Not quite, but sort of, if we take care of convergence.* Of course, we need to first figure out what we mean by convergence. Once we figure that out, we will show that convergence happens the way we want to if we use slightly more complicated factors.

Definition 8.1.1. The product

$$\prod_{n=1}^{\infty} (1 + a_n)$$

converges if the limit of the sequence of *partial products*

$$\lim_{k \to \infty} \prod_{n=1}^{k} (1 + a_n) = \lim_{k \to \infty} (1 + a_1)(1 + a_2) \cdots (1 + a_k)$$

exists. The product *converges absolutely* if $\prod_{n=1}^{\infty} (1 + |a_n|)$ converges.

A product could converge in two different ways. It either goes to 0 or it does not. One way to go to zero is if $1 + a_n = 0$ for some n, but there are other possibilities. For instance, if $|1 + a_n| \leq r < 1$ for all n, then the product also goes to zero, although in this case the convergence will not be absolute (exercise below).

---

*Beware of formal expressions bearing gifts. Especially ones with infinite sets in them. For example, $\prod_{n \in \mathbb{Z}} (z - \pi n)$ does not actually make any sense.

Suppose that the product converges to a nonzero number. In particular, $a_n \neq -1$ for any n. Then

$$\frac{\prod_{n=1}^{k}(1 + a_n)}{\prod_{n=1}^{k-1}(1 + a_n)} = 1 + a_k.$$

Taking the limit, we see that $1 + a_k$ must go to 1, and so a_k must go to 0.

A key idea about products is to relate infinite products to sums using the logarithm, which is something we can easily do if dealing with positive numbers. This way, absolutely convergent products can be understood if we understand absolutely convergent sums, which we do.

Proposition 8.1.2. *The product $\prod_{n=1}^{\infty}(1 + |a_n|)$ converges if and only if $\sum_{n=1}^{\infty}|a_n|$ converges.*

Proof. Suppose $\sum_{n=1}^{\infty}|a_n|$ converges. The sequence of partial products is increasing:

$$\prod_{n=1}^{k}(1 + |a_n|) \leq \prod_{n=1}^{k+1}(1 + |a_n|).$$

Thus it is sufficient to show that it is bounded. Consider

$$\log \prod_{n=1}^{k}(1 + |a_n|) = \sum_{n=1}^{k}\log(1 + |a_n|) \leq \sum_{n=1}^{k}|a_n| \leq \sum_{n=1}^{\infty}|a_n|.$$

For the other direction, suppose that the sequence of partial products converges. For all but finitely many n, $|a_n| < 1$. Otherwise the partial products would double infinitely often and go to infinity. So suppose that $|a_n| < 1$ for all $n \geq N$. If $|a_n| < 1$, then $|a_n| \leq 2\log(1 + |a_n|)$, and

$$\sum_{n=N}^{m}|a_n| \leq \sum_{n=N}^{m}2\log(1 + |a_n|) \leq 2\log \prod_{n=N}^{m}(1 + |a_n|).$$

The right-hand side is bounded as m goes to infinity, and so the tail of $\sum_{n=1}^{\infty}|a_n|$ converges. □

An immediate consequence is that if $\prod_{n=1}^{\infty}(1 + a_n)$ converges absolutely, then $\{a_n\}$ converges to 0. Not only that, the sequence goes to zero fast enough to be absolutely summable. As for series, we need to know that absolute convergence really is convergence.

Proposition 8.1.3. *Suppose $\prod_{n=1}^{\infty}(1 + a_n)$ converges absolutely. Then the product converges. Moreover, if $\mathrm{Re}\,a_n > -1$ for all n, then*

$$\prod_{n=1}^{\infty}(1 + a_n) = \exp\left(\sum_{n=1}^{\infty}\mathrm{Log}(1 + a_n)\right),$$

and the series converges absolutely. In particular, the product converges to a nonzero number.

Proof. If the product converges absolutely, then $|a_n|$ goes to 0, and we may, without loss of generality, assume that $\operatorname{Re} a_n > -1$ for all n, ignoring the first few terms if necessary. In particular, $\operatorname{Log}(1 + a_n)$ is well-defined for the principal branch of the logarithm. The derivative of $\operatorname{Log} z$ at 1 is 1, and a_n goes to zero. So

$$\lim_{n\to\infty} \frac{\left|\operatorname{Log}(1 + a_n)\right|}{|a_n|} = \lim_{n\to\infty}\left|\frac{\operatorname{Log}(1 + a_n) - \operatorname{Log} 1}{a_n}\right| = 1.$$

By the limit comparison test, $\sum_{n=1}^{\infty}\left|\operatorname{Log}(1 + a_n)\right|$ converges if and only if $\sum_{n=1}^{\infty}|a_n|$ converges. The series $\sum_{n=1}^{\infty}|a_n|$ converges via Proposition 8.1.2.

It is not necessarily true that $\sum_{n=1}^{k}\operatorname{Log}(1 + a_n) = \operatorname{Log}\prod_{n=1}^{k}(1 + a_n)$. However, $\sum_{n=1}^{k}\operatorname{Log}(1 + a_n) = \log\prod_{n=1}^{k}(1 + a_n)$ for some branch of the logarithm, because while the arguments might add up to something outside the range $(-\pi, \pi]$, the sum will be off by some multiple of $2\pi i$. Therefore, for this branch of the log

$$\exp\left(\sum_{n=1}^{k}\operatorname{Log}(1 + a_n)\right) = \exp\left(\log\prod_{n=1}^{k}(1 + a_n)\right) = \prod_{n=1}^{k}(1 + a_n).$$

Now taking the limit we find that the product converges. $\qquad\square$

It is important that the principal branch is used as otherwise the imaginary parts of the logs may not converge. Namely, it is important that $\operatorname{Log} 1 = 0$.

Proposition 8.1.4. *If $\prod_{n=1}^{\infty}(1 + a_n)$ converges absolutely to 0, then there exists a n such that $a_n = -1$.*

Proof. As before, because $|a_n|$ goes to zero, $\operatorname{Re} a_n > -1$ for all large enough n. The product of those terms converges to a nonzero number by the "Moreover" part of the proposition. The only way to get zero as a limit is for one of the initial terms to be zero, that is, if $a_n = -1$ for some n. $\qquad\square$

Just as with series, an absolutely convergent infinite product can be reordered.

Proposition 8.1.5. *Suppose $\prod_{n=1}^{\infty}(1 + a_n)$ converges absolutely to L. If $\varphi\colon \mathbb{N} \to \mathbb{N}$ is a bijection, then $\prod_{n=1}^{\infty}(1 + a_{\varphi(n)})$ also converges absolutely to L.*

Proof. That the reordered product converges absolutely follows from Proposition 8.1.2 and the corresponding result for series. If $L = 0$, then we just proved that one of the terms is zero, and the reordered product converges to 0. Suppose $L \neq 0$. By ignoring finitely many terms, we assume without loss of generality that $\operatorname{Re} a_n > -1$. Then $\sum \operatorname{Log}(1 + a_n)$ converges absolutely and

$$L = \exp\left(\sum_{n=1}^{\infty}\operatorname{Log}(1 + a_n)\right).$$

The sum converges absolutely, so it can be reordered and hence the product can be reordered and the limit remains the same. $\qquad\square$

Definition 8.1.6. Given functions $g_n \colon X \to \mathbb{C}$, the product $\prod_{n=1}^{\infty}(1 + g_n(x))$ *converges uniformly absolutely* if $\prod_{n=1}^{\infty}(1 + |g_n(x)|)$ converges uniformly in $x \in X$.

Uniformly absolute convergence of the product is the same as uniformly absolute convergence of the corresponding series.

Proposition 8.1.7. *For $g_n \colon X \to \mathbb{C}$, the product $\prod_{n=1}^{\infty}(1 + |g_n(x)|)$ converges uniformly if and only if $\sum_{n=1}^{\infty}|g_n(x)|$ converges uniformly.*

> *Exercise 8.1.1: Prove the proposition. Hint: Use log to convert the partial products to partial sums. Then apply the estimates in Proposition 8.1.2 to show that one sequence of partial sums is uniformly Cauchy if and only if the other one is.*

As for series, uniform absolute convergence means uniform convergence.

Proposition 8.1.8. *For $g_n \colon X \to \mathbb{C}$, if $\prod_{n=1}^{\infty}(1 + g_n(x))$ converges uniformly absolutely, then $\prod_{n=1}^{\infty}(1 + g_n(x))$ converges uniformly.*

> *Exercise 8.1.2: Prove the proposition.*

In the sequel, we will simply write products in the form $\prod_{n=1}^{\infty} f_n(z)$ for some functions f_n. In this case, of course, to apply the definitions one needs to think of the products as

$$\prod_{n=1}^{\infty}\left(1 + (f_n(z) - 1)\right).$$

In particular, $\prod_{n=1}^{\infty} f_n(z)$ converges (uniformly) absolutely if $\prod_{n=1}^{\infty}(1 + |f_n(z) - 1|)$ converges (uniformly), which we have seen happens if and only if $\sum_{n=1}^{\infty}|f_n(z) - 1|$ converges (uniformly).

Corollary 8.1.9. *Suppose $U \subset \mathbb{C}$ is open, $f_n \colon U \to \mathbb{C}$ are holomorphic, and $f(z) = \prod_{n=1}^{\infty} f_n(z)$ converges uniformly absolutely on compact subsets of U. Then f is holomorphic. Furthermore, $f(z) = 0$ for some $z \in U$ if and only if there exists an n such that $f_n(z) = 0$.*

Proof. As uniform absolute convergence implies uniform convergence, f is holomorphic. For absolutely convergent products, the only way to get zero is for one of the terms to be zero.   □

> *Exercise 8.1.3: Suppose $|1 + a_n| \leq r < 1$ for all n. Prove that $\prod_{n=1}^{\infty}(1 + a_n)$ converges to 0, but that the convergence is not absolute.*

> *Exercise 8.1.4: Suppose $\operatorname{Re} a_n > -1$ for all n. Prove $\prod_{n=1}^{\infty}(1 + a_n)$ converges to a nonzero number if and only if $\sum_{n=1}^{\infty} \operatorname{Log}(1 + a_n)$ converges. Note: We already proved one direction.*

Exercise 8.1.5: *Suppose $U \subset \mathbb{C}$ is a domain, $f_n : U \to \mathbb{C}$ are holomorphic, and*

$$F(z) = \prod_{n=1}^{\infty} f_n(z)$$

converges uniformly absolutely on compact subsets of U. Prove that

$$F'(z) = \sum_{n=1}^{\infty} f_n'(z) \prod_{k=1, k \neq n}^{\infty} f_k(z),$$

converging uniformly absolutely on compact subsets of U.

8.2i Weierstrass factorization and product theorems

8.2.1i In the plane

To factor an arbitrary holomorphic function such as the sine, we have to be a smidge trickier than just trying to factor out $(z - z_n)$ for all the zeros. To get the product to converge, we multiply the factor by something to make the factor closer to 1. We start with functions that have a zero at $z = 1$.

Definition 8.2.1. Define the *elementary factors*

$$E_0(z) = 1 - z, \qquad E_m(z) = (1 - z) \exp\left(z + \frac{z^2}{2} + \cdots + \frac{z^m}{m}\right).$$

The function $E_m(z/a)$ has a zero of order 1 at a. As we are relating the absolute convergence of $\prod(1 + a_n)$ to $\sum |a_n|$, let us consider what happens to $\left|E_m(z) - 1\right| = \left|1 - E_m(z)\right|$. Before we do so, we prove a useful estimate for holomorphic functions on the disc. By the maximum modulus principle, the maximum is attained at the boundary. If the derivatives at the origin are all positive, then it is attained at a very specific point.

Lemma 8.2.2. *Suppose f is a holomorphic function on a neighborhood of the closed disc $\overline{\mathbb{D}}$ and suppose $f^{(n)}(0) \geq 0$ for all n. Then $f(1)$ is real and for $z \in \overline{\mathbb{D}}$,*

$$\left|f(z)\right| \leq f(1).$$

Proof. Expand $f(z) = \sum c_n z^n$. As $f^{(n)}(0) \geq 0$, then $c_n \geq 0$. Thus for $z \in \overline{\mathbb{D}}$,

$$\left|f(z)\right| = \left|\sum_{n=0}^{\infty} c_n z^n\right| \leq \sum_{n=0}^{\infty} c_n |z|^n \leq \sum_{n=0}^{\infty} c_n = f(1). \qquad \square$$

Lemma 8.2.3. $|1 - E_m(z)| \le |z|^{m+1}$ for all $m = 0, 1, 2, \dots$ and all $z \in \overline{\mathbb{D}}$.

Proof. The lemma clearly holds for $m = 0$. Differentiating $1 - E_m(z)$ and using the finite geometric sum, we find

$$-E'_m(z) = \exp\left(z + \frac{z^2}{2} + \cdots + \frac{z^m}{m}\right) - (1-z)(1+z+\cdots+z^{m-1})\exp\left(z + \frac{z^2}{2} + \cdots + \frac{z^m}{m}\right)$$

$$= z^m \exp\left(z + \frac{z^2}{2} + \cdots + \frac{z^m}{m}\right).$$

Notably, $1 - E_m(z)$ has a zero of order $m + 1$ at $z = 0$. Thus $f(z) = \frac{1-E_m(z)}{z^{m+1}}$ has a removable singularity. From the formula for the derivative $-E'_m(z)$, it is clear the coefficients of the power series for f at 0 are nonnegative. By Lemma 8.2.2,

$$\left|\frac{1 - E_m(z)}{z^{m+1}}\right| \le \frac{1 - E_m(1)}{1^{m+1}} = 1 \qquad \text{for } z \in \overline{\mathbb{D}}. \qquad \square$$

The Weierstrass product theorem says that we can prescribe the zeros of an entire function arbitrarily. The only requirement is the obvious one from the identity theorem: The zeros have no limit point in $\mathbb{C}$.

Theorem 8.2.4 (Weierstrass product theorem in $\mathbb{C}$). *Suppose $\{c_k\}$ is a sequence of distinct points in $\mathbb{C}$ with no limit points in $\mathbb{C}$ and $\{m_k\}$ is a sequence of natural numbers. Then there exists an entire holomorphic $f : \mathbb{C} \to \mathbb{C}$ that has zeros exactly at c_k, with orders given by m_k.*
*   More precisely, suppose $\{a_n\}$ is the sequence of nonzero $\{c_k\}$ with points repeated according to the multiplicities $\{m_k\}$ and m is the order of the zero at the origin. Then there exists a sequence $\{\ell_n\}$ such that one such f is given by*

$$f(z) = z^m \prod_{n=1}^{\infty} E_{\ell_n}\left(\frac{z}{a_n}\right),$$

converging uniformly absolutely on compact subsets. In fact, any sequence $\{\ell_n\}$ such that

$$\sum_{n=1}^{\infty} \left|\frac{r}{a_n}\right|^{\ell_n+1} \tag{8.1}$$

converges for all $r > 0$ can be used.

Proof. Ignore the zero at the origin, and just consider the nonzero zeros $\{a_n\}$. We claim that at least one sequence $\{\ell_n\}$ such that (8.1) converges for all $r > 0$ exists. Indeed, choosing $\ell_n = n - 1$ would suffice: As $\{a_n\}$ has no limit points in $\mathbb{C}$ it must "escape to infinity." For any $r > 0$, we get $|a_n| \ge 2r$ or $r/|a_n| < 1/2$ for all large enough n. If $\ell_n = n - 1$, a tail of the series is bounded by the geometric series $\sum (1/2)^n$.

Consider a compact set K in the plane. It is contained in some closed disc $\overline{\Delta_r(0)}$. We want to get uniformly absolute convergence of the product, and so we need

$$\sum_{n=1}^{\infty} \left| E_{\ell_n}\left(\frac{z}{a_n}\right) - 1 \right|$$

to converge uniformly on K. If $z \in K$, then $|z| \leq r$. As a_n goes to infinity, $|a_n| \geq r$ and so $\left|\frac{z}{a_n}\right| \leq 1$ for all n large enough. By Lemma 8.2.3,

$$\left| E_{\ell_n}\left(\frac{z}{a_n}\right) - 1 \right| \leq \left|\frac{z}{a_n}\right|^{\ell_n+1} \leq \left|\frac{r}{a_n}\right|^{\ell_n+1}.$$

Thus the series converges as its tail converges. The convergence is uniform in K, as the far right-hand side above does not depend on z. $\qquad\square$

There are many choices for the sequence $\{\ell_n\}$. The proof says that the convergence of (8.1) for every $r > 0$ guarantees convergence of the product, but we may try to make a convenient choice of $\{\ell_n\}$, and often there is a more convenient choice than $\ell_n = n - 1$.

Now that we can prescribe zeros of an entire function, we use it to divide out all the zeros of any other entire function, and obtain a factorization.

Corollary 8.2.5 (Weierstrass factorization theorem). *Let f be an entire holomorphic function, not identically zero, with zeros (repeated according to multiplicity) at points of the sequence $\{a_n\}$ except the zero at the origin, whose order is m (possibly $m = 0$). Then there exists an entire holomorphic function g and a sequence $\{\ell_n\}$ such that*

$$f(z) = z^m e^{g(z)} \prod_{n=1}^{\infty} E_{\ell_n}\left(\frac{z}{a_n}\right),$$

converges uniformly absolutely on compact subsets.

Proof. Let $h \colon \mathbb{C} \to \mathbb{C}$ be the entire function

$$h(z) = \prod_{n=1}^{\infty} E_{\ell_n}\left(\frac{z}{a_n}\right),$$

where the $\{\ell_n\}$ comes from the product theorem. The function $\varphi(z) = \frac{f(z)}{h(z)z^m}$ has only removable singularities, so φ can be made entire. As φ has no zeros, $\varphi(z) = e^{g(z)}$ for an entire function g because $\mathbb{C}$ is simply connected. $\qquad\square$

Exercise 8.2.1 (Easy): *Suppose $\{a_n\}$ is a sequence such that there is some $\epsilon > 0$ and $|a_n| \geq n^{1+\epsilon}$ for all n. Prove that*

$$\prod_{n=1}^{\infty} \left(1 - \frac{z}{a_n}\right)$$

converges uniformly absolutely on compact subsets of $\mathbb{C}$.

Exercise 8.2.2 (Easy): *Explicitly find an infinite product for an entire holomorphic function with simple zeros precisely at $\mathbb{Z} \times \mathbb{Z}$ (that is, all $x + iy$ where $x, y \in \mathbb{Z}$).*

Exercise 8.2.3: Suppose that $\{a_n\}$ is a sequence converging to 0. Show that there exists a holomorphic function $f \colon \mathbb{C} \setminus \{0\} \to \mathbb{C}$ with zeros (counting multiplicity) at a_n.

Exercise 8.2.4: Suppose $\{a_k\}$, $\{b_k\}$ are sequences of distinct points in $\mathbb{C}$ with no limit points, and $\{n_k\}$, $\{m_k\}$ are sequences of natural numbers. Prove that there exists a meromorphic function $f \colon \mathbb{C} \to \mathbb{C}_{\infty}$ whose zeros are exactly at a_k, with orders given by n_k, and poles are exactly at b_k, with orders given by m_k.

Exercise 8.2.5: Suppose $\{a_n\}$ is a sequence of distinct points in $\mathbb{C}$ with no limit points, and $\{c_n\}$ an arbitrary sequence of complex numbers. Construct an entire function f such that $f(a_n) = c_n$. Hint: For any radius $r > 0$, a point $p \in \mathbb{C}$, $|p| > r$, and an $\epsilon > 0$, try to find a function with some prescribed zeros such that $f(p) = 1$ and $|f(z)| < \epsilon$ for all $z \in \Delta_r(0)$. Another hint: If $|w| < 1$, then $|w|^n$ goes to zero.

8.2.2*i* Factorization of sine

Let us use the Weierstrass factorization theorem to factor the sine function as promised. The function $\sin(\pi z)$ has zeros at the integers $\mathbb{Z}$. We start with the positive integers. Note that

$$\sum_{n=1}^{\infty} \left|\frac{r}{n}\right|^2$$

converges for every $r > 0$. Similarly for the negative integers. Thus we may choose $\ell_n = 1$ in the product theorem. Write

$$f(z) = \pi z \prod_{n \in \mathbb{Z} \setminus \{0\}} E_1\left(\frac{z}{n}\right) = \pi z \prod_{n \in \mathbb{Z} \setminus \{0\}} \left(1 - \frac{z}{n}\right) e^{z/n}.$$

We write the product in no particular order as the product converges absolutely. The π out front can be guessed by thinking what would we get if we differentiate $\sin(\pi z)$ and evaluate at 0. Differentiating the product (using product rule) would give you $\pi \cdot 1 + 0 = \pi$, as any time the derivative falls on some factor other than z, when you evaluate at 0, you get 0. We can do this formal computation on the finite products as the product converges uniformly on compact subsets and therefore so does the derivative. See also Exercise 8.1.5.

We group some terms together for convenience

$$f(z) = \pi z \prod_{n=1}^{\infty} \left(1 - \frac{z}{n}\right) e^{z/n} \left(1 - \frac{z}{-n}\right) e^{-z/n} = \pi z \prod_{n=1}^{\infty} \left(1 - \frac{z^2}{n^2}\right).$$

That's a rather nice factorization. We still do not know if $f(z)$ is $\sin(\pi z)$. All we know is that the two have the same zeros, and the derivative at 0 is π as it should be. Because f captures the zeros of $\sin(\pi z)$, we write (as in the factorization theorem)

$$\sin(\pi z) = e^{g(z)} f(z) = \pi z e^{g(z)} \prod_{n=1}^{\infty} \left(1 - \frac{z^2}{n^2}\right),$$

for some entire function g. We need to show that $g \equiv 0$. By the computation of the derivative above, $g(0) = 0$.

We wish to convert the product to a series using the logarithm, as series are simpler to handle*. Unfortunately, if $\varphi(z) = \sin(\pi z)$, the function $\log \varphi(z)$ is not well-defined. But as we saw a couple of times before, while $\log \varphi(z)$ is not well-defined, the logarithmic derivative

$$\frac{d}{dz} \left[\log \varphi(z)\right] = \frac{\varphi'(z)}{\varphi(z)}$$

is well-defined. We can find the logarithmic derivative by differentiating the series for $\log \varphi(z)$ since for the derivative we only need to work locally and we can use any branch of the logarithm. Locally, using any branch of the logarithm, pick k large enough so that $\operatorname{Re}\left(1 - \frac{z^2}{n^2}\right) > 0$ for all $n \geq k$, and Proposition 8.1.3 applies:

$$\pi \cot(\pi z) = \frac{\pi \cos(\pi z)}{\sin(\pi z)} = \frac{d}{dz} \left[\log \sin(\pi z)\right] = \frac{d}{dz} \left[\log \left(\pi z e^{g(z)} \prod_{n=1}^{\infty} \left(1 - \frac{z^2}{n^2}\right)\right)\right]$$

$$= \frac{d}{dz} \left[\log(\pi z) + g(z) + \sum_{n=1}^{k} \log\left(1 - \frac{z^2}{n^2}\right) + \sum_{n=k+1}^{\infty} \operatorname{Log}\left(1 - \frac{z^2}{n^2}\right)\right]$$

$$= \frac{1}{z} + g'(z) + \sum_{n=1}^{\infty} \frac{2z}{z^2 - n^2}.$$

The penultimate equality would not be true without the derivative as what's inside the square brackets may differ by a constant. Since the far left-hand side and the far right-hand side do not have any logarithms in them they are clearly well-defined. The equality holds with $g' = 0$, which is a nice exercise in applying the residue theorem, and we leave it to the reader.

---

*Any calculus student will tell you so when they try to differentiate a product with many factors.

Exercise 8.2.6: *Let γ_n be the rectangular path with vertices $n + 1/2 - in$, $n + 1/2 + in$, $-(n + 1/2) + in$, $-(n + 1/2) - in$.*

a) Using the residue theorem, for $z \notin \mathbb{Z}$, evaluate

$$\int_{\gamma_n} \frac{\pi \cot(\pi\xi)}{\xi^2 - z^2} \, d\xi.$$

b) Show that

$$\lim_{n \to \infty} \int_{\gamma_n} \frac{\pi \cot(\pi\xi)}{\xi^2 - z^2} \, d\xi = 0.$$

c) Prove that

$$\pi \cot(\pi z) = \frac{1}{z} + \sum_{n=1}^{\infty} \frac{2z}{z^2 - n^2},$$

with uniform convergence on compact subsets of $\mathbb{C} \setminus \mathbb{Z}$.

In particular, the exercise says that $g'(z) = 0$, and as $g(0) = 0$, we find $g \equiv 0$. We have thus proved the following proposition.

Proposition 8.2.6. *For all $z \in \mathbb{C}$,*

$$\sin(\pi z) = \pi z \prod_{n=1}^{\infty} \left(1 - \frac{z^2}{n^2}\right),$$

with uniform absolute convergence on compact subsets.

Exercise 8.2.7: *Find a factorization for cosine. Hint: There is a hard way and an easy way (now that we know the factorization of sine).*

Exercise 8.2.8: *Find a factorization for $\sinh(z) = \frac{e^z - e^{-z}}{2}$.*

Exercise 8.2.9: *Suppose $U \subset \mathbb{C}$ is a domain and*

$$F(z) = \prod_{n=1}^{\infty} f_n(z)$$

for holomorphic functions $f_n \colon U \to \mathbb{C}$ converging uniformly on compact subsets of U. Prove that

$$\frac{F'(z)}{F(z)} = \sum_{n=1}^{\infty} \frac{f_n'(z)}{f_n(z)},$$

converging uniformly on compact subsets of U.

8.2.3*i*   The product theorem in any open set

The product theorem holds in any open set, with a similar proof. What changes is that the zeros can now congregate on points of the boundary as well as at infinity, and we must handle these differently. There are a couple of ways to handle this complication. One way is to move infinity to ensure all the zeros are bounded. The other way is to not move anything and split the sequence of zeros in a smart way into a sequence that goes to infinity with no finite limit points and a sequence that has limit points on the boundary. The tricky business is that the part of the sequence that has limit points at the boundary could also go to infinity, but we can ensure that it at least gets closer and closer to the boundary as it marches off to infinity.

Theorem 8.2.7 (Weierstrass product theorem). *Suppose $U \subset \mathbb{C}$ is open, $\{c_k\}$ is a sequence of distinct points in U with no limit points in U, and $\{m_k\}$ is a sequence of natural numbers. Then there exists a holomorphic $f : U \to \mathbb{C}$ with zeros exactly at c_k, with orders given by m_k.*

Proof. Suppose $\{a_n\}$ is the sequence of points $\{c_k\}$ repeated according to the multiplicities $\{m_k\}$. As the case $U = \mathbb{C}$ has already been proved, we assume $\mathbb{C} \setminus U$ is nonempty. Let

$$D = \left\{ z \in U : d(z, \mathbb{C} \setminus U) < \frac{1}{|z| + 1} \right\},$$

where $d(z, \mathbb{C} \setminus U) = \inf_{\zeta \in \mathbb{C} \setminus U} |z - \zeta|$. For $z \in U$ it is the distance to the boundary. Divide $\{a_n\}$ into two parts: Let $\{a_n^1\}$ be those a_n that lie in D and $\{a_n^2\}$ be those that do not. We will generate a function f_1 with zeros at $\{a_n^1\}$ and a function f_2 with zeros at $\{a_n^2\}$. Let $f = f_1 f_2$ to finish the proof. Without loss of generality assume both sequences are infinite.

First consider $\{a_n^1\}$. Let $\{p_n\}$ be a sequence of points in $\mathbb{C} \setminus U$ such that

$$|a_n^1 - p_n| = d(a_n^1, \mathbb{C} \setminus U).$$

See Figure 8.1. We claim that $|a_n^1 - p_n|$ goes to zero as n goes to infinity. Indeed, suppose not, then there is an $\epsilon > 0$ and a subsequence such that $|a_{n_k}^1 - p_{n_k}| > \epsilon$ for all k. As these are the distances to the boundary and $a_{n_k}^1 \in D$, we must have that $|a_{n_k}^1| < 1/\epsilon - 1$ for all k. In particular, these $\{a_{n_k}^1\}$ must have a limit point and since they are all at least ϵ away from the boundary this limit point would be in U. That is a contradiction. Hence $|a_n^1 - p_n|$ goes to 0.

Let $K \subset U$ be compact, then for $z \in K$

$$\left| \frac{a_n^1 - p_n}{z - p_n} \right| \leq \frac{|a_n^1 - p_n|}{d(p_n, K)} \leq \frac{|a_n^1 - p_n|}{d(\mathbb{C} \setminus U, K)}.$$

So for all n large enough $\left| \frac{a_n^1 - p_n}{z - p_n} \right| < 1/2$. Then by Lemma 8.2.3,

$$\left| E_n \left(\frac{a_n^1 - p_n}{z - p_n} \right) - 1 \right| \leq \frac{1}{2^{n+1}},$$

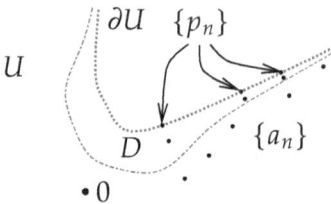

Figure 8.1: The sequence of $\{a_n\}$ with respect to D. The points of $\{a_n\}$ that lie in D are the $\{a_n^1\}$ and for those we pick the $\{p_n\}$, which will lie at the closest place on the boundary of U (the thick dotted line).

and hence

$$f_1(z) = \prod_{n=1}^{\infty} E_n \left(\frac{a_n^1 - p_n}{z - p_n} \right)$$

converges uniformly on K. As K is arbitrary, f_1 converges uniformly on compact subsets of U. We are thus done with f_1.

For f_2, note that $\{a_n^2\}$ must go to infinity and has no finite limit points. Indeed, if $|a_n^2| \le r$, then $d(a_n^2, \mathbb{C} \setminus U) \ge \frac{1}{r+1}$, and $\{a_n^2\}$ has no limit points in U. We can, therefore, construct an entire function with these zeros.   □

If $f = g/h$ for two holomorphic functions g and h, then f is meromorphic with poles at the zeros of h. Conversely, if f has a pole of order m at p, then $f(z) = \frac{g(z)}{(z-p)^m}$ for some holomorphic g, so f is a quotient of holomorphic functions (near p). Using the Weierstrass product theorem, we get this converse globally: A meromorphic function on a domain is a quotient of holomorphic functions on that same domain. In more fancy language, the set of meromorphic functions on a domain in $\mathbb{C}$ is the field of fractions of the ring of holomorphic functions.*

Corollary 8.2.8. *Suppose $U \subset \mathbb{C}$ is open and $f \colon U \to \mathbb{C}_\infty$ is meromorphic. Then there exist holomorphic $g, h \colon U \to \mathbb{C}$ such that $f = g/h$.*

Proof. By the product theorem, there exists a holomorphic function h that has zeros exactly where f has poles, and of the same order. The function fh therefore has removable singularities at all the poles of f. In other words, there is a holomorphic g such that $g = fh$.   □

Remark 8.2.9. When we introduced the corollary, we mentioned "domain," but then we proved it for an open set. The issue is a bit of algebra. If U is not connected, then the set of holomorphic functions is not an integral domain, it has zero divisors, and a field of fractions is only defined for integral domains.

---

*It is important that it is a domain in $\mathbb{C}$. We saw meromorphic functions on $\mathbb{C}_\infty$, and those are not quotients of two holomorphic functions, since there are no nonconstant holomorphic functions on $\mathbb{C}_\infty$.

Exercise 8.2.10: *Given a domain $U \subset \mathbb{C}$ and $f, g \colon U \to \mathbb{C}$ holomorphic, neither of which is identically zero. Prove that there exist functions h, F, and G holomorphic on U, such that F and G have no common zeros and $f = hF$ and $g = hG$.*

Exercise 8.2.11: *Let $S = \left\{ e^n \in \mathbb{C} : n \in \mathbb{Z} \right\}$. Explicitly construct a holomorphic function on $\mathbb{C} \setminus \{0\}$ that has simple zeros at points of S.*

Exercise 8.2.12: *Given a domain $U \subset \mathbb{C}$ and a sequence $\{a_n\}$ of distinct points in U with no limit points in U, find a holomorphic function on $U \setminus \{a_n : n \in \mathbb{N}\}$ that has essential singularities at $\{a_n\}$.*

Exercise 8.2.13: *Show that there exists a holomorphic function $f \colon \mathbb{D} \to \mathbb{C}$ that does not extend holomorphically to any domain U where $\mathbb{D} \subsetneq U$. Hint: Consider a sequence of points in $\mathbb{D}$ whose limit set is the circle.*

Exercise 8.2.14: *Suppose $U \subset \mathbb{C}$ is a simply connected domain and $f \colon U \to \mathbb{C}$ a nonconstant function. Show that there exists a holomorphic $g \colon U \to \mathbb{C}$ such that $g^2 = f$ if and only if every zero of f has even order.*

9*i* | Rational Approximation

It has been said that man is a rational animal. All my life I have been searching for evidence which could support this.

—Bertrand Russell

9.1*i* | Polynomial approximation

In real analysis, you may have seen the very useful Weierstrass approximation theorem (or the more general Stone–Weierstrass approximation theorem), which says that a continuous function on a compact interval $[a, b]$ can be uniformly approximated by polynomials of a real variable. Utilizing holomorphic polynomials—polynomials in z—we have a similar theorem, but for holomorphic functions not continuous functions. That makes sense. After all, uniform limits of holomorphic functions are holomorphic, so we can't expect to approximate all continuous functions.

Example 9.1.1: Let $z = x + iy$. By Stone–Weierstrass, every continuous function on the closed unit disc $\overline{\mathbb{D}}$ is a uniform limit of a sequence of polynomials $Q_n(x, y)$ (polynomials of both x and y), but only a function that is holomorphic on $\mathbb{D}$ can be the uniform limit of polynomials $P_n(z)$ (polynomials in z). More explicitly, the function $z \mapsto \bar{z} = x - iy$ is not only a limit of polynomials in x and y, it is a polynomial in x and y. However, as the function is not holomorphic on $\mathbb{D}$, it cannot be a uniform limit of holomorphic polynomials on $\overline{\mathbb{D}}$.

From now on, all polynomials will again be polynomials in z, as they were up until the example above. So when we say polynomial, it will be a polynomial in z, in other words, a holomorphic polynomial.

A function holomorphic on a disc $\Delta_r(p)$ is a limit of the partial sums $\sum_{n=0}^{m} c_n (z - p)^n$ of its power series. See also Exercise 3.4.9. On the other hand, consider a function holomorphic on an open neighborhood of the square $[-1, 1] \times [-1, 1]$, say $f(z) = \frac{1}{(z-1.1)(z+1.1)}$. If we expand f around any point, the series will never converge on all of $[-1, 1] \times [-1, 1]$, as the domain of convergence is a disc which cannot include ± 1.1. However, we can (exercise below) still find a sequence of polynomials that converge to f on $[-1, 1] \times [-1, 1]$.

Exercise 9.1.1: *Let $f(z) = \frac{1}{(z-1.1)(z+1.1)}$. Find an explicit (that is, find a formula for it, do not just prove that it exists) sequence of polynomials $P_n(z)$ that converges uniformly to f on the square $[-1,1] \times [-1,1]$.*

We may be thwarted in polynomial approximation by topology. For example, no sequence of polynomials $P_n(z)$ converges to $1/z$ uniformly on $\partial \mathbb{D}$. This fact follows by Rouché (or Exercise 3.4.6 and Cauchy's theorem), see the following exercise.

Exercise 9.1.2: *For any polynomial $P(z)$, there exists a $z_0 \in \partial \mathbb{D}$ such that $\left| P(z_0) - \frac{1}{z_0} \right| \geq 1$.*

The problem with $1/z$ and the unit circle is that the circle goes around a hole. Any function holomorphic on a neighborhood of $\partial \mathbb{D}$ can be approximated on $\partial \mathbb{D}$ by rational functions if we allow a pole at 0. The function $1/z$ is itself a rational function with a pole at 0. A polynomial is really just a rational function that has a pole at infinity. Once we prove Runge's theorem, we will have proved that it really is the hole enclosed by $\partial \mathbb{D}$ that is the problem for $1/z$. We can approximate a holomorphic function by polynomials on any set whose complement is connected.

Exercise 9.1.3: *Suppose a sequence of polynomials $\{P_n\}$ converges uniformly on $\partial \mathbb{D}$. Show that $\{P_n\}$ converges uniformly on $\overline{\mathbb{D}}$.*

Exercise 9.1.4: *Prove that if a sequence of polynomials $\{P_n\}$ converges uniformly on $\mathbb{C}$, then there is an N such that $P_n - P_m$ is a constant for all $n, m \geq N$ (so the limit is P_N plus a constant).*

9.2i Runge's theorem

We first prove rational approximation without any control of the poles. The key is to find a cycle around a compact set K (see § 6.3.3) and then to apply Cauchy's integral formula for points of K. The Riemann sums of the integral are the rational functions.

Lemma 9.2.1. *Let $U \subset \mathbb{C}$ be open, $K \subset U$ compact, $f : U \to \mathbb{C}$ holomorphic, and Γ a cycle in U homologous to zero in U such that $\Gamma \cap K = \emptyset$ and $n(\Gamma; z) = 1$ for all $z \in K$. Then for every $\epsilon > 0$, there exists a rational function $R(z)$ with poles on Γ such that $|f(z) - R(z)| < \epsilon$ for all $z \in K$.*

Proof. The hypotheses mean that Cauchy's integral formula applies for any $z \in K$:

$$f(z) = \frac{1}{2\pi i} \int_\Gamma \frac{f(\zeta)}{\zeta - z} \, d\zeta.$$

The cycle Γ is a finite sum of closed piecewise-C^1 paths. If we prove that the Riemann sums corresponding to each path converge uniformly on K, then their sum also converges uniformly and it converges to f. Thus consider just one path $\gamma \colon [0,1] \to U$, and let $\epsilon > 0$ be given. The function

$$(z,t) \mapsto \frac{f(\gamma(t))}{\gamma(t) - z}$$

is uniformly continuous on the compact set $K \times [0,1]$. As γ' is bounded, there is a $\delta > 0$ such that

$$\left| \frac{f(\gamma(t))}{\gamma(t) - z} \gamma'(t) - \frac{f(\gamma(\tau))}{\gamma(\tau) - z} \gamma'(t) \right| < \epsilon$$

for all $z \in K$ and all $t, \tau \in [0,1]$ such that $|t - \tau| < \delta$. Partition $[0,1]$ into $0 = t_0 < t_1 < \cdots < t_k = 1$, where $t_j - t_{j-1} < \delta$. Write

$$\int_\gamma \frac{f(\zeta)}{\zeta - z} \, d\zeta = \sum_{j=1}^k \int_{t_{j-1}}^{t_j} \frac{f(\gamma(t))}{\gamma(t) - z} \gamma'(t) \, dt.$$

We estimate each bit of this integral by a rational function of z (note the use of the fundamental theorem of calculus):

$$\left| \int_{t_{j-1}}^{t_j} \frac{f(\gamma(t))}{\gamma(t) - z} \gamma'(t) \, dt \; - \; \frac{f(\gamma(t_j))}{\gamma(t_j) - z} \big(\gamma(t_j) - \gamma(t_{j-1})\big) \right|$$

$$= \left| \int_{t_{j-1}}^{t_j} \left(\frac{f(\gamma(t))}{\gamma(t) - z} \gamma'(t) - \frac{f(\gamma(t_j))}{\gamma(t_j) - z} \gamma'(t) \right) dt \right|$$

$$\leq \int_{t_{j-1}}^{t_j} \left| \frac{f(\gamma(t))}{\gamma(t) - z} \gamma'(t) - \frac{f(\gamma(t_j))}{\gamma(t_j) - z} \gamma'(t) \right| dt \leq \epsilon(t_j - t_{j-1}).$$

Summing these bits together and using the triangle inequality, we find

$$\left| \int_\gamma \frac{f(\zeta)}{\zeta - z} \, d\zeta - \sum_{j=1}^k \frac{f(\gamma(t_j))}{\gamma(t_j) - z} \big(\gamma(t_j) - \gamma(t_{j-1})\big) \right| \leq \sum_{j=1}^k \epsilon(t_j - t_{j-1}) = \epsilon.$$

Thus the integral over γ converges uniformly for $z \in K$, and we are done by summing up over all the paths in Γ. $\square$

In some parts of the proof below, to simplify the verbiage, we will say that functions of a certain type (uniformly) approximate f on K if for every $\epsilon > 0$ there exists a function g of the given type such that $|f(z) - g(z)| < \epsilon$ on K. We leave the following statement as a simple exercise in chasing those epsilons.

> ***Exercise* 9.2.1:** *Let K be a set and $\mathcal{F}$ an algebra (a vector space that is closed under multiplication) of functions on K. Rigorously prove:*
>   a) *If functions from $\mathcal{F}$ uniformly approximate a function f on K, then functions from $\mathcal{F}$ uniformly approximate f^n for any $n \in \mathbb{N}$.*
>   b) *Suppose $f_1, \ldots, f_n$ are functions on K that can be individually uniformly approximated by functions from $\mathcal{F}$. Then for any numbers $c_1, \ldots, c_n$, the function $c_1 f_1 + \cdots + c_n f_n$ can be uniformly approximated on K by functions in $\mathcal{F}$.*
>   c) *If $\mathcal{G}$ is another set of functions on K, f can be uniformly approximated by functions in $\mathcal{G}$, and every function in $\mathcal{G}$ can be uniformly approximated by functions in $\mathcal{F}$, then f can be uniformly approximated by functions in $\mathcal{F}$.*

We now approximate simple poles of the form $\frac{1}{z-p}$ by rational functions with poles in a given set. This procedure is called *pole pushing* as we are going to "push" the poles along a path to where we need them to be.

Lemma 9.2.2. *Suppose $K \subset \mathbb{C}$ is compact, $p \in \mathbb{C} \setminus K$, and $q \in \mathbb{C}_\infty \setminus K$ is in the same component of $\mathbb{C}_\infty \setminus K$ as p. Then for every $\epsilon > 0$, there exists a rational function R with pole at q such that*

$$\left| \frac{1}{z-p} - R(z) \right| < \epsilon \qquad \text{for all } z \in K.$$

Proof. Suppose first that $q \neq \infty$. As the component of $\mathbb{C}_\infty \setminus K$ is open and connected, it is path connected, so we connect p and q by a path. And as a path is compact, we cover the path by finitely many discs as follows: There exist points $p = z_1, z_2, \ldots, z_n = q$ and finitely many discs $\Delta_r(z_1), \ldots, \Delta_r(z_n)$ of some radius $r > 0$ such that $z_{j+1} \in \Delta_r(z_j)$ and $\Delta_{2r}(z_j) \cap K = \emptyset$ for all j. See Figure 9.1.

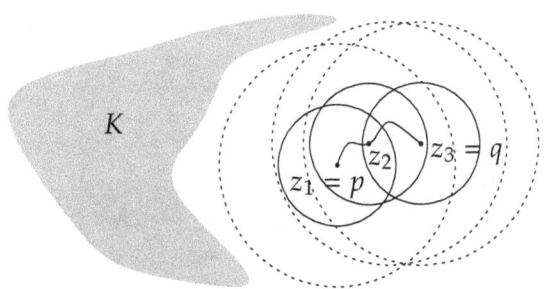

Figure 9.1: Pole pushing from p to q.

As there are finitely many discs, if we show that we can approximate $\frac{1}{z-p}$ on K by a rational function with a pole at $z_2 \in \Delta_r(p)$, then we claim we are done. A rational function R with a pole at z_2 may be written as a finite linear combination of terms of the form $\frac{1}{(z-z_2)^k}$ or $(z - z_2)^k$. If we can uniformly approximate $\frac{1}{z-z_2}$ on K by a rational function with a pole at z_3, we can approximate R by a rational function with a pole at z_3, see Exercise 9.2.1. And so we can approximate $\frac{1}{z-p}$ by a rational function with a pole at z_3. Rinse and repeat.

So without loss of generality suppose $z_1 = p$ and $z_2 = q$. Then

$$\frac{1}{z-p} = \frac{1}{z-q} \frac{1}{1 - \frac{p-q}{z-q}} = \frac{1}{z-q} \sum_{k=0}^{\infty} \left(\frac{p-q}{z-q}\right)^k,$$

which converges uniformly on K as for $z \in K$, we have

$$\left|\frac{p-q}{z-q}\right| \le \frac{r}{|z-q|} \le \frac{r}{2r} = \frac{1}{2}.$$

A partial sum of the series is the rational function we seek.

Now suppose $q = \infty$. Find a rational function that approximates $\frac{1}{z-p}$ uniformly on K and has a pole at some q_1, where $K \subset \Delta_M(0)$ and $|q_1| > M$. As above, without loss of generality, suppose that this function is $\frac{1}{z-q_1}$. Next

$$\frac{1}{z-q_1} = \frac{-1}{q_1} \frac{1}{1 - z/q_1} = \frac{-1}{q_1} \sum_{k=0}^{\infty} \left(\frac{z}{q_1}\right)^k,$$

which converges uniformly on $\overline{\Delta_M(0)}$ as $|q_1| > M$. Taking partial sums, $\frac{1}{z-q_1}$ can be approximated uniformly on K by polynomials (rational functions with a pole at ∞), and hence $\frac{1}{z-p}$ can also be uniformly approximated on K by polynomials.   □

We can now prove Runge on compact sets.

Theorem 9.2.3 (Runge on a compact set). *Suppose $U \subset \mathbb{C}$ is open, $K \subset U$ is compact, $S \subset \mathbb{C}_\infty \setminus K$ intersects every component of $\mathbb{C}_\infty \setminus K$, and $f : U \to \mathbb{C}$ is holomorphic. Then for every $\epsilon > 0$, there exists a rational function R with poles in S such that*

$$\left|f(z) - R(z)\right| < \epsilon \qquad \text{for all } z \in K.$$

Proof. Lemma 6.3.7 gives a cycle Γ around K homotopic to zero in U. Use Lemma 9.2.1 to approximate f uniformly on K by a rational function with poles on Γ. That rational function is a linear combination of terms such as $\frac{1}{z-p_j}$ for $p_j \in \Gamma$. Using Lemma 9.2.2, approximate each of these terms by a rational function with poles in S.   □

Example 9.2.4: There exists a sequence of polynomials $\{P_n\}$ such that

$$\lim_{n \to \infty} P_n(z) = \begin{cases} 1 & \text{if } z \in \mathbb{R}, \\ 0 & \text{else.} \end{cases}$$

Proof: Let $K_n = \{z \in \mathbb{C} : |z| \le n, \operatorname{Im} z \in (-\infty, -1/n] \cup \{0\} \cup [1/n, \infty)\}$. The set $\mathbb{C} \setminus K_n$ is connected and K_n is compact and has three components. There is a function f

that is holomorphic on a neighborhood of K_n, and it equals 0 on $K_n \setminus \mathbb{R}$ and 1 on $K_n \cap \mathbb{R}$. Indeed, pick three disconnected neighborhoods of the three components of K_n, and set the function to be identically 0 or 1 on the corresponding component. Constants are holomorphic. Runge says that there is a polynomial $P_n(z)$ such that $|P_n(z) - f(z)| \leq 1/n$ for $z \in K_n$. The sequence $\{P_n\}$ then does the job. Since $K_n \subset K_{n+1}$ for all n and $\bigcup K_n = \mathbb{C}$, eventually any z is in some K_n (and all the later sets). It follows that $\lim P_n(z)$ is 0 or 1 as required.

But be careful with what we proved. We have a pointwise limit only. The sequence does *not* converge uniformly on compact subsets of $\mathbb{C}$. That is easy to see, otherwise the limit would be continuous.

To prove a version of Runge for open sets, we need a slightly stronger version of Lemma 6.1.7. We need to add an extra property about the complements.

Lemma 9.2.5. *Let $U \subset \mathbb{C}$ be open. Then there exists a sequence K_n of compact subsets of U such that $K_n \subset K_{n+1}^\circ$, $\bigcup K_n = U$, for any compact $K \subset U$, there is an n such that $K \subset K_n$, and each component of $\mathbb{C}_\infty \setminus K_n$ contains a point of $\mathbb{C}_\infty \setminus U$.*

Proof. Let $\{K_n'\}$ be the sequence of sets from Lemma 6.1.7. For each n, let K_n be the set K_n' together with any component of $\mathbb{C}_\infty \setminus K_n'$ that is completely contained in U. In particular, we added only bounded components. The set K_n is still closed (in $\mathbb{C}$) and bounded and hence compact. If X is a component of $\mathbb{C}_\infty \setminus K_n'$ that we added into K_n, then $X \subset K_m$ for all $m > n$, and as X is open, X is in the interior of K_m. So all the conditions are satisfied. $\square$

Corollary 9.2.6 (Runge). *Suppose $U \subset \mathbb{C}$ is open, $S \subset \mathbb{C}_\infty \setminus U$ intersects every component of $\mathbb{C}_\infty \setminus U$, and $f \colon U \to \mathbb{C}$ is holomorphic. Then there exists a sequence $\{R_n\}$ of rational functions with poles in S that converges to f uniformly on compact subsets of U.*

Proof. Let $\{K_n\}$ be the sequence of sets from the lemma. Each component of $\mathbb{C}_\infty \setminus K_n$ intersects S. Let R_n be a rational function such that

$$\left| f(z) - R_n(z) \right| < 1/n \quad \text{for all } z \in K_n.$$

For a compact $K \subset U$, find an N such that $K \subset K_n$ for all $n \geq N$. Then $\left| f(z) - R_n(z) \right| < 1/n$ for all $n \geq N$. So $\{R_n\}$ converges to f uniformly on compact subsets. $\square$

Exercise 9.2.2: Prove a version of Corollary 9.2.6, where we only require that the closure $\bar{S}$ intersects every component of $\mathbb{C}_\infty \setminus U$.

Exercise 9.2.3: Prove that if $U \subset V \subset \mathbb{C}$ are open sets such that $\mathbb{C}_\infty \setminus V$ intersects every component of $\mathbb{C}_\infty \setminus U$, then every holomorphic $f \colon U \to \mathbb{C}$ can be written as a limit of a sequence of functions holomorphic in V converging uniformly on compact subsets of U.

Exercise 9.2.4: *Suppose $U \subset \mathbb{C}$ is open, $K \subset U$ is compact, $S \subset \mathbb{C}_\infty \setminus K$ intersects every component of $\mathbb{C}_\infty \setminus K$, and $f : U \to \mathbb{C}$ is holomorphic.*

  a) *If S is open, prove that f can be uniformly approximated on K by a rational function with only simple poles in S.*

  b) *Find an example U, K, S, and f, where S is not open and f cannot be approximated by a rational functions with only simple poles in S.*

Exercise 9.2.5: *Prove that there exists a sequence of polynomials $\{P_n\}$ such that pointwise for all $z \in \mathbb{C}$,*

$$\lim_{n \to \infty} P_n(z) = \begin{cases} 1 & \text{if } z = 0, \\ 0 & \text{else.} \end{cases}$$

Note that the convergence will not be uniform in any neighborhood of the origin.

Exercise 9.2.6: *Prove that there exists a sequence of polynomials $\{P_n\}$ such that pointwise for all $z \in \mathbb{C}$,*

$$\lim_{n \to \infty} P_n(z) = \lfloor \operatorname{Re} z \rfloor,$$

where $\lfloor x \rfloor$ means the largest integer less than or equal to x.

Exercise 9.2.7: *Let $U \subset \mathbb{C}$ be a domain and $f : U \to \mathbb{C}$ be holomorphic.*

  a) *Suppose that for every $p \in \partial U$ there exists a disc $\Delta_\epsilon(p)$ such that $\Delta_\epsilon(p) \setminus \overline{U}$ is nonempty and connected. Prove that there exists an open $W \subset \mathbb{C}$ such that $\overline{U} \subset W$ such that for every holomorphic $f : U \to \mathbb{C}$ there is a sequence of holomorphic $f_n : W \to \mathbb{C}$ such that converge uniformly on compact subsets of U to f.*

  b) *Find a counterexample U and f to the conclusion of part a) where there is at least one $p \in \partial U$ such that $\Delta_\epsilon(p) \setminus \overline{U}$ is empty for some $\epsilon > 0$.*

  c) *Find a counterexample U and f to the conclusion of part a) where there is at least one $p \in \partial U$ such that $\Delta_\epsilon(p) \setminus \overline{U}$ is nonempty but disconnected for every $\epsilon > 0$.*

9.3*i*   Polynomial hull and simply-connectedness

Definition 9.3.1. Let $K \subset \mathbb{C}$ be a compact set. The *polynomial hull* of K is the set

$$\widehat{K} \stackrel{\text{def}}{=} \left\{ z \in \mathbb{C} : |P(z)| \leq \sup_{\zeta \in K} |P(\zeta)| \text{ for every polynomial } P \right\}.$$

A set $U \subset \mathbb{C}$ is *polynomially convex* if for every compact $K \subset U$, we get $\widehat{K} \subset U$.

Hulls can (and often are) defined for other classes of functions than polynomials, and they are a generalization of the idea of convexity. Classical convexity is convexity with respect to affine linear functions (see the exercises).

Proposition 9.3.2. *If $K \subset \mathbb{C}$ is compact, then $K \subset \widehat{K}$ and $\widehat{K}$ is compact.*

Proof. Clearly $K \subset \widehat{K}$. As $\widehat{K}$ is the intersection of closed sets, that is, sets where $|P(z)| \leq \sup_{\zeta \in K} |P(\zeta)|$ for some specific P, it is closed. It must be bounded since K is bounded: If $|z| \leq M$ for some M on K, then all points of $\widehat{K}$ also satisfy $|z| \leq M$ (use $P(z) = z$). Thus $\widehat{K}$ is compact. $\qquad\square$

Example 9.3.3: Let $K = \partial \mathbb{D}$. By the maximum principle, $|P(z)| \leq \sup_{\zeta \in K} |P(\zeta)|$ for any $z \in \mathbb{D}$ and any polynomial P. Furthermore, as above, $|z| \leq 1$ for all $z \in \widehat{K}$. So, $\widehat{K} = \overline{\mathbb{D}}$.

Proposition 9.3.4. *Suppose $K \subset \mathbb{C}$ is compact, $\widehat{K}$ is its polynomial hull, and $\{P_n\}$ is a sequence of polynomials converging uniformly on K. Then $\{P_n\}$ converges uniformly on $\widehat{K}$.*

Exercise **9.3.1:** *Prove the proposition. Hint: Show that $\{P_n\}$ is uniformly Cauchy on $\widehat{K}$.*

In one complex variable, polynomial hulls are easy to describe.* They are given simply by a topological property: The polynomial hull just "fills in the holes."

Theorem 9.3.5. *Suppose $K \subset \mathbb{C}$ is compact and X is the unbounded component of $\mathbb{C} \setminus K$. Then $\widehat{K} = \mathbb{C} \setminus X$.*

Proof. Suppose $p \in X$. Then $Q = \{p\} \cup (\mathbb{C} \setminus X)$ is a compact set whose complement (in either $\mathbb{C}$ or $\mathbb{C}_\infty$) is connected. The function that is 1 at p and 0 on $\mathbb{C} \setminus X$ is holomorphic in a neighborhood of Q. Use Runge to approximate this function on Q by a polynomial $P(z)$ to within $1/2$. In other words, $|P(p) - 1| < 1/2$ so $|P(p)| > 1/2$, and for all $z \in \mathbb{C} \setminus X$ (and therefore all $z \in K$), we have $|P(z)| < 1/2$. Thus, $p \notin \widehat{K}$.

Conversely, suppose $p \notin X$. If $p \in K$, then $p \in \widehat{K}$, so suppose $p \notin K$. Then $p \in B$ where B is one of the bounded components of $\mathbb{C} \setminus K$. The boundary $\partial B \subset K$, and so the second version of the maximum principle says that

$$|P(p)| \leq \sup_{z \in \partial B} |P(z)| \leq \sup_{z \in K} |P(z)|.$$

So $p \in \widehat{K}$. $\qquad\square$

Note that the components of $\mathbb{C} \setminus K$ are open sets. The polynomial hull of K is K together with all the bounded components of $\mathbb{C} \setminus K$. See Figure 9.2. As a corollary we get the following equivalence.

Corollary 9.3.6. *Let $U \subset \mathbb{C}$ be a domain. The following are equivalent.*

(i) *The domain U is simply connected.*

(ii) *Every holomorphic $f: U \to \mathbb{C}$ is a limit of polynomials converging uniformly on compact subsets of U.*

(iii) *The domain U is polynomially convex.*

---

*In the analysis of several complex variables, polynomial hulls are far more difficult to handle, and they are the subject of ongoing research.

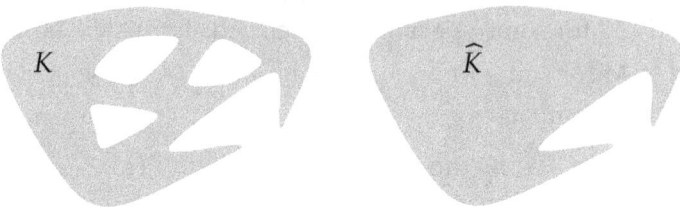

Figure 9.2: Polynomial hull is "filling holes."

Proof. Suppose (i) is true. Then $\mathbb{C}_\infty \setminus U$ has exactly one component. Runge says that we can approximate any function holomorphic on U by polynomials (rational functions with a pole at ∞) uniformly on compact subsets. Namely (ii) holds.

Suppose (ii) is true. Consider any $K \subset U$. Suppose $p \in \widehat{K} \setminus U$ exists (for contradiction). Then p is not in K and so p is in one of the bounded components of $\mathbb{C} \setminus K$. Consider $f(z) = \frac{1}{z-p}$, which is holomorphic on U, so find a sequence of polynomials $\{P_n\}$ converging uniformly to f on compact subsets. By Proposition 9.3.4, the sequence converges uniformly on $\widehat{K}$. Thus it converges to a holomorphic function g defined on $U \cup \widehat{K}$, which is an open set by Theorem 9.3.5. The set $U \cup \widehat{K}$ is connected as each component of $\mathbb{C} \setminus K$ we added has boundary in U, which is connected. The function f is defined on $\mathbb{C} \setminus \{p\}$, and the boundary of the component of $\mathbb{C} \setminus K$ that includes p must have a limit point (exercise below). Thus g equals f on $(U \cup \widehat{K}) \setminus \{p\}$, which is impossible as g is bounded near p while f has a pole there. So no such p exists and (iii) is true.

Suppose (iii) is true. Suppose $\mathbb{C}_\infty \setminus U$ has a component K that does not include ∞. The set K is compact. By Lemma 6.3.7, there exists a cycle Γ in U such that $n(\Gamma; z) = 1$ on K. Proposition 4.1.3 says that K is not in the unbounded component of $\mathbb{C} \setminus \Gamma$. By Theorem 9.3.5, $K \subset \widehat{\Gamma}$, which contradicts U being polynomially convex. Thus the only component of $\mathbb{C}_\infty \setminus U$ is the one that contains ∞ and so U is simply connected. $\square$

Exercise **9.3.2:** *Suppose $U \subset \mathbb{C}$ is a polynomially convex domain such that for each $p \in \mathbb{C}$ there exists an M such that $\partial \Delta_M(p) \subset U$. Prove that $U = \mathbb{C}$.*

Exercise **9.3.3:** *Prove that if $W \subset \mathbb{C}$ is a nonempty bounded open set, then ∂W has a limit point.*

Exercise **9.3.4:** *Suppose $K_1 \subset K_2 \subset \mathbb{C}$ are compact. Prove that $\widehat{K_1} \subset \widehat{K_2}$.*

Exercise **9.3.5:** *Let $U \subset \mathbb{C}$ be a domain. Prove that U is simply connected if and only if there exists a sequence K_n of compact subsets of U such that $K_n \subset K_{n+1}^\circ$, $\bigcup K_n = U$, and such that $\widehat{K_n} = K_n$.*

Exercise **9.3.6:** *Suppose $U_1 \subset U_2 \subset \cdots$ is a sequence of nested polynomially convex domains in $\mathbb{C}$. Let $U = \bigcup U_n$. Prove that U is polynomially convex.*

Exercise **9.3.7:** *Instead of using polynomials, define*

$$\widehat{K} = \left\{ z \in \mathbb{C} : f(z) \le \sup_{\zeta \in K} f(\zeta) \text{ for every } f(x + iy) = ax + by + c, a, b, c \in \mathbb{R} \right\}.$$

Prove that a set U is convex (in the usual sense) if and only if for every compact $K \subset U$, we have $\widehat{K} \subset U$.

Exercise **9.3.8:** *Prove that if the hull is defined in terms of continuous functions on $\mathbb{C}$ rather than polynomials, then $\widehat{K} = K$ for every compact K.*

9.4i Mittag-Leffler

Given a principal part of a function with a pole at p,

$$P(z) = \sum_{m=1}^{k} \frac{c_m}{(z - p)^m},$$

we ask for a meromorphic function with such a pole. That is not hard: $P(z)$. How about two principal parts, $P_1(z)$ and $P_2(z)$, for two poles? Well, $P_1(z) + P_2(z)$ works wonderfully, no? Given a sequence of poles and principal parts $P_1(z), P_2(z), \dots$, we wish to take

$$\sum_{\ell=1}^{\infty} P_\ell(z).$$

But can we? The sum may not converge. We must be a tad trickier, and here is where Runge's theorem is useful: Adding a holomorphic function doesn't change the principal part, but it may make the terms in the sum smaller and make things converge. We obtain the Mittag-Leffler theorem.[*]

Theorem 9.4.1 (Mittag-Leffler). *Suppose $U \subset \mathbb{C}$ is open, $S \subset U$ is a countable set with no limit point in U, and for every $p \in S$ there is a principal part*

$$P_p(z) = \sum_{m=1}^{k_p} \frac{c_{p,m}}{(z - p)^m}$$

of a pole of order k_p. Then there exists a meromorphic function f in U with poles precisely at points of S, and for each $p \in S$, the principal part of f at p is P_p.

Proof. Apply Lemma 9.2.5 to get an exhaustion of U by compact sets $\{K_n\}$, where every component of $\mathbb{C}_\infty \setminus K_n$ contains a point of $\mathbb{C}_\infty \setminus U$. Let $S_1 = K_1 \cap S$ and

---

[*]The downside of a hyphenated name: Mittag always has to share fame with that Leffler guy.

$S_n = K_n \cap S \setminus K_{n-1}$. Each S_n is finite as S has no limit point in U, and S is the disjoint union of $S_1, S_2, \ldots$. Let

$$f_n(z) = \sum_{p \in S_n} P_p(z).$$

Let $R_1 = 0$. Suppose $n \geq 2$. Every component of $\mathbb{C}_\infty \setminus K_{n-1}$ contains a point of $\mathbb{C}_\infty \setminus U$, and f_n is holomorphic on a neighborhood of K_{n-1} (it has finitely many poles on $K_n \setminus K_{n-1}$). Runge's theorem gives a rational function R_n with poles in $\mathbb{C}_\infty \setminus U$ such that $|f_n(z) - R_n(z)| < 2^{-n}$ for all $z \in K_{n-1}$. We only care that R_n is holomorphic in U; we do not use its poles or that it is rational.

We wish to define

$$f(z) = \sum_{n=1}^{\infty} \big(f_n(z) - R_n(z)\big).$$

We claim that this series converges uniformly on compact subsets of $U \setminus S$, and the resulting function has poles on S with principal parts P_p at $p \in S$.

First suppose that $K \subset U \setminus S$ is compact. Then $K \subset K_\ell$ for some ℓ. Write

$$\sum_{n=1}^{\infty} \big(f_n(z) - R_n(z)\big) = \sum_{n=1}^{\ell} \big(f_n(z) - R_n(z)\big) + \sum_{n=\ell+1}^{\infty} \big(f_n(z) - R_n(z)\big).$$

The first term has poles on K_ℓ (but not on K), but the second term does not. Furthermore, for all $z \in K_\ell$, and hence in K, we have

$$\sum_{n=\ell+1}^{\infty} \big|f_n(z) - R_n(z)\big| \leq \sum_{n=\ell+1}^{\infty} 2^{-n}.$$

The sum converges uniformly absolutely on K (Weierstrass M-test, Exercise 2.3.8). So the series for f converges uniformly absolutely on compact subsets of $U \setminus S$ and f is holomorphic on $U \setminus S$. Let us see that it has the right sort of singularities. Suppose $p \in S_\ell$. A neighborhood of p contains no other singularities. Then

$$f(z) = \sum_{n=1}^{\ell-1} \big(f_n(z) - R_n(z)\big) + \big(f_\ell(z) - R_\ell(z)\big) + \sum_{n=\ell+1}^{\infty} \big(f_n(z) - R_n(z)\big).$$

The first sum and the last sum are holomorphic on a neighborhood of p: The first sum has all its poles in $K_{\ell-1}$ and the last sum has no singularities on K_ℓ. The function R_ℓ is holomorphic in all of U. So f has the same singularity at p as f_ℓ and the same principal part, and f_ℓ was set up precisely so that it has a principal part P_p at p. $\square$

Example 9.4.2: Sometimes, though not always, convergence happens simply by grouping the terms correctly. In other words, grouping the principal parts correctly (a choice of $\{K_n\}$ in the proof), convergence happens even with $R_n = 0$. Suppose we want a function with a pole at every $p \in \mathbb{Z}$ with singular part $\frac{1}{z-p}$. Regrettably,

$$\sum_{p \in \mathbb{Z}} \frac{1}{z - p}$$

does not converge, but

$$\frac{1}{z} + \sum_{n=1}^{\infty} \left(\frac{1}{z+n} + \frac{1}{z-n} \right) = \frac{1}{z} + \sum_{n=1}^{\infty} \frac{2z}{z^2 - n^2}$$

does converge uniformly on compact subsets of $\mathbb{C} \setminus \mathbb{Z}$. In terms of the proof above, this grouping corresponds to, say, $K_n = \overline{\Delta_n(0)}$. To be clear, we're not saying the proof as is goes through; in particular, it is not true that $\left| \frac{2z}{z^2 - n^2} \right| < 2^{-n}$ on K_{n-1}. All we are saying is that the series converges to the desired function. See Exercise 9.4.1.

Example 9.4.3: In general, the R_n are necessary and grouping is not enough. Let $U = \mathbb{C}$, $S = \mathbb{N}$, and $P_p(z) = \frac{-p}{z-p}$ for $p \in \mathbb{N}$. No matter how we group the terms $P_p(z)$, the series will not converge. Fix a positive $z \notin \mathbb{N}$. For p large enough, $P_p(z)$ is close to 1, say $P_p(z) > 1/2$. Suppose you group terms as $f_n(z)$ in some way. The terms of the series $\sum f_n(z)$ are eventually (for large n) finite sums of only such $P_p(z)$, and so each such $f_n(z)$ is larger than $1/2$. The series cannot converge. In Exercise 9.4.2, you will find the R_n that work.

The Mittag-Leffler theorem is a sister theorem to the Weierstrass product theorem. In Weierstrass's theorem, we prescribe zeros rather than poles. In fact, with just a little bit of trickery, one could use the Mittag-Leffler theorem to prove the Weierstrass theorem. Again, see the exercises.

Exercise 9.4.1: Prove the statements Example 9.4.2: $\sum_{p \in \mathbb{Z}} \frac{1}{z-p}$ converges for no $z \in \mathbb{C} \setminus \mathbb{Z}$, but $\sum_{n=1}^{\infty} \frac{2z}{z^2 - n^2}$ converges uniformly on compact subsets of $\mathbb{C} \setminus \mathbb{Z}$.

Exercise 9.4.2: Suppose $U = \mathbb{C}$, $S = \mathbb{N}$, $P_n(z) = \frac{-n}{z-n}$, $K_n = \overline{\Delta_n(0)}$. Use the geometric sum $1 + x + \cdots + x^n = \frac{1 - x^{n+1}}{1-x}$ to find an explicit R_n that will make the proof of the theorem work. Note that one part of the formula may be somewhat ugly, c'est la vie.

Exercise 9.4.3: Suppose $U \subset \mathbb{C}$ is open, $f : U \to \mathbb{C}$ holomorphic, $p \in U$, and $P(z) = \sum_{m=1}^{k} \frac{c_m}{(z-p)^m}$. Suppose $K \subset U \setminus \{p\}$ is compact such that p is in the unbounded component of $\mathbb{C} \setminus K$. Show that for every $\epsilon > 0$, there exists a function g holomorphic in $U \setminus \{p\}$ with a pole at p and principal part P at p and such that $|f(z) - g(z)| < \epsilon$ for all $z \in K$.

Exercise 9.4.4: Prove that for every open $U \subset \mathbb{C}$ there exists a meromorphic function $f : U \to \mathbb{C}_\infty$ such that for every $p \in \partial U$ and every $\epsilon > 0$, there are infinitely many poles of f in $\Delta_\epsilon(p) \cap U$. Hint: The trick is constructing S.

Exercise 9.4.5: Suppose that instead of principal parts of poles, $P_p(z)$ are principal parts of essential singularities that converge in $\mathbb{C} \setminus \{p\}$. Prove the Mittag-Leffler with this setup.

Exercise 9.4.6:
   a) Prove that in the Mittag-Leffler theorem, the function f can be chosen so that f has no zeros in $U \setminus S$.
   b) Use part a) to prove the Weierstrass product theorem.

10i   Analytic Continuation

May the forces of evil become confused on the way to your house.

—George Carlin

10.1i   Schwarz reflection principle

One of the consequences of the identity theorem is that once we know a function in a neighborhood we know it in the whole domain. If a holomorphic function f is defined in a domain U and $U \subset W$ for some other domain W, then we may want to find a holomorphic function in W that agrees with f on U. By the identity theorem, the extension is unique, but it may not always exist. If $f(z) = 1/z$ in $U = \mathbb{C} \setminus \{0\}$, there is no way of extending it to $W = \mathbb{C}$.

One type of continuation is reflection, namely the Schwarz reflection principle*, which says that we can, under some conditions, reflect values across some boundary. We have seen the harmonic version (see Theorem 7.3.3) of this theorem. The proof of the principle works the same for both harmonic and holomorphic functions. We simply write down the candidate function by using the right reflection and then show that the reflection is harmonic or holomorphic. Then over the line where they meet, we use either the mean-value property or Morera's theorem.

Theorem 10.1.1 (Schwarz reflection principle). *Suppose $U \subset \mathbb{C}$ is a domain symmetric across the real axis, that is, $z \in U$ if and only if $\bar{z} \in U$. Let $U_+ = \{z \in U : \operatorname{Im} z > 0\}$ and $L = U \cap \mathbb{R}$. Suppose $f \colon U_+ \cup L \to \mathbb{C}$ is a continuous function that is holomorphic on U_+ and real-valued on L, that is, $\operatorname{Im} f(z) = 0$ for all $z \in L$.*

Then there exists a holomorphic function $F \colon U \to \mathbb{C}$ such that $F|_{U_+ \cup L} = f$.

The setup is the same as for the harmonic version of the theorem. See Figure 7.6 for a diagram of the setup.

Proof. For $z \in U$, define

$$F(z) = f(z) \quad \text{if } \operatorname{Im} z \geq 0, \qquad F(z) = \overline{f(\bar{z})} \quad \text{else.}$$

---

*Sometimes it is called *Riemann–Schwarz principle* as Riemann saw it first, but he didn't properly justify it, so Schwarz has dibs on it.

If $z \in U$ and $\mathrm{Im}\, z > 0$, then F is holomorphic near z by hypothesis. Suppose $z \in U$ and $\mathrm{Im}\, z < 0$. The easiest way to see that F is holomorphic is by using the Wirtinger derivative and the identities proved in Exercise 2.2.8 and the Wirtinger chain rule Exercise 2.2.10:

$$\frac{\partial}{\partial \bar{z}} F(z) = \frac{\partial}{\partial \bar{z}} \overline{f(\bar{z})} = \overline{\frac{\partial}{\partial z} f(\bar{z})} = 0.$$

The chain rule came up because we are taking the z derivative of f composed with a conjugation map, and the z derivative of the conjugation map is zero. Another way to see it is to write down the power series representation.

Finally, we need to prove that F is holomorphic near L, which we can do by applying Morera's theorem. It is not hard to see that F is continuous. The only tricky part is to check integrals over triangles that intersect the real line. Such a triangle can be split into several triangles, each of which lies on one side of the line L and intersects L either at a vertex or along a side. See Figure 10.1.

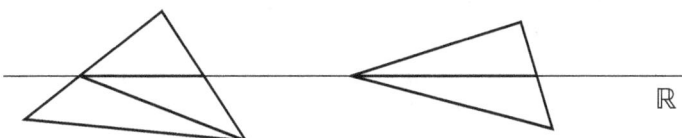

Figure 10.1: Splitting triangles to use Morera.

Call such a triangle T, and suppose it is above the axis. We can translate the triangle by ϵ, that is, consider $T_\epsilon = \{z \in \mathbb{C} : z - i\epsilon \in T\}$. For small $\epsilon > 0$, we have $T_\epsilon \subset U_+$ and so the integral of F over ∂T_ϵ is zero. We can write the integral over ∂T_ϵ as

$$0 = \int_{\partial T_\epsilon} F(z)\, dz = \int_{\partial T} F(z + i\epsilon)\, dz.$$

The function F is uniformly continuous on some neighborhood of T, thus $F(z + i\epsilon)$ converges uniformly to $F(z)$ as $\epsilon \to 0$. Consequently, $\int_{\partial T} F(z)\, dz = 0$. Morera then implies that F is holomorphic. □

The theorem is often applied by first mapping some curve to the real line. This curve must be nice enough, that is, real-analytic. Similarly the fact that f is real-valued can be replaced by being valued in some real-analytic curve. We will define a real-analytic curve as the image of an interval under a real-analytic map $\gamma \colon (a, b) \to \mathbb{C}$. By γ being real-analytic we mean that at each point it has a power series expansion. Equivalently, a real-analytic function is a restriction of a holomorphic function to the real line by just plugging complex numbers into the power series. Therefore, the actual definition we will use is the following.

Definition 10.1.2. A set $C \subset \mathbb{C}$ is a *real-analytic curve* if there exist a domain $V \subset \mathbb{C}$ such that $V \cap \mathbb{R} = (a, b)$ and an injective holomorphic $\varphi \colon V \to \mathbb{C}$ such that $\varphi\big((a, b)\big) = C$.

The images $\varphi(V_+)$ and $\varphi(V_-)$ of the sets $V_+ = \{z \in V : \mathrm{Im}\, z > 0\}$ and $V_- = \{z \in V : \mathrm{Im}\, z < 0\}$ are the two "sides" of C.

Corollary 10.1.3 (Schwarz reflection principle for curves). *Let $U \subset \mathbb{C}$ be open, $C \subset \partial U$ a real-analytic curve with one side not in U and one side in U, that is, there is domain V, $V \cap \mathbb{R} = (a, b)$ and an injective holomorphic $\varphi\colon V \to \mathbb{C}$ such that $\varphi((a, b)) = C \subset \partial U$, and such that $\varphi(V_+) \subset U$ and $\varphi(V_-) \cap U = \emptyset$. Suppose that $f\colon U \cup C \to \mathbb{C}$ is a continuous function, holomorphic on U, such that $f(C) \subset D$ for some other real-analytic curve D.*

Then there exists a neighborhood W of C and a holomorphic function $F\colon U \cup W \to \mathbb{C}$ such that $F|_{U \cup C} = f$.

Proof. First start with D. As it is defined by some invertible holomorphic function, there must be a neighborhood H of D, and an injective holomorphic $\psi\colon H \to \mathbb{C}$ such that $\psi(D) \subset \mathbb{R}$. We can pick V small enough so that it is symmetric across the real-axis and such that $f\big(\varphi(V_+)\big) \subset H$. Write $L = (a, b) = V \cap \mathbb{R}$ as before. Consider the composition $\psi \circ f \circ \varphi$, which is holomorphic in V_+, continuous on $V_+ \cup L$, and real-valued on L. So the Schwarz reflection principle holds, and we get a holomorphic $G\colon V \to \mathbb{C}$ extending $\psi \circ f \circ \varphi$.

We possibly make V smaller yet to ensure that $G(V) \subset \psi(H)$ so that we may invert ψ on the image. The set W is $\varphi(V)$ and on W we write $F = \psi^{-1} \circ G \circ \varphi^{-1}$. This definition agrees with f on $W \cap U = \varphi(V_+)$, and so we can define F on U as f. $\square$

The corollary says that we can extend holomorphic functions across real-analytic boundaries as long as the function is continuous and valued in a real-analytic curve on the boundary. However, the downside of this more general statement is that it does not say how to figure out exactly the size of the neighborhood W. One could note the sizes of the neighborhoods V and H and then follow "reductions of V" in the proof to work out how far the extension goes, but that does require understanding how far the functions defining the curves extend as invertible functions.

In practice, the extension is often applied to rather special curves such as a straight line or a circle and so we can explicitly figure out the form of the extension and where it is defined. The original version is for a straight line, so let us give the circle version. The reflection across the unit circle, $1/\bar{z}$, is the inversion from euclidean geometry. If $z = re^{i\theta}$, then the reflection $1/\bar{z} = (1/r)e^{i\theta}$ is on the same ray from the origin but the modulus is the reciprocal.

Corollary 10.1.4 (Schwarz reflection principle for a circle). *Suppose $U \subset \mathbb{C} \setminus \{0\}$ is a domain, symmetric with respect to reflection across $\partial\mathbb{D}$, that is, $z \in U$ implies $1/\bar{z} \in U$. Let $U_{in} = U \cap \mathbb{D}$ and $U_{out} = U \setminus \overline{\mathbb{D}}$. If $f\colon U \cap \overline{\mathbb{D}} \to \mathbb{C}$ is continuous, holomorphic on U_{in}, and $f(U \cap \partial\mathbb{D}) \subset \partial\mathbb{D}$, then there exists a holomorphic $F\colon U \to \mathbb{C}$ such that $F|_{U \cap \overline{\mathbb{D}}} = f$.*

The proof is left as an exercise: The idea is to define $F(z) = f(z)$ on $U \cap \overline{\mathbb{D}}$, and for $z \in U_{out}$, define

$$F(z) = \frac{1}{\overline{f(1/\bar{z})}}.$$

Exercise **10.1.1:** *Prove Corollary 10.1.4. Hint: Cayley.*

Exercise **10.1.2:** *Suppose $U = \{z \in \mathbb{C} : \operatorname{Im} z > 0 \text{ and } \operatorname{Re} z > 0\}$ and $f \colon \overline{U} \to \mathbb{C}$ is continuous, holomorphic on U, real-valued when $\operatorname{Im} z = 0$, and imaginary-valued when $\operatorname{Re} z = 0$. Prove that f extends to an entire holomorphic function.*

Exercise **10.1.3:** *Suppose $U \subset \mathbb{H}$ is a domain in the upper half-plane and Δ is a disc centered at some real number with $\Delta \cap \partial U = \Delta \cap \mathbb{R} = (a, b)$. Suppose $f \colon U \cup (a, b) \to \mathbb{C}$ is continuous, holomorphic on U, and f is zero on (a, b). Prove that f is identically zero.*

Exercise **10.1.4:** *Let $T = \{e^{it} : \alpha < t < \beta\}$ be a small arc of the unit circle. Suppose $f \colon \mathbb{D} \cup T \to \mathbb{C}$ and $g \colon \mathbb{D} \cup T \to \mathbb{C}$ are continuous, holomorphic in $\mathbb{D}$, and $g = f$ on T. Prove that $f = g$ on $\mathbb{D}$.*

Exercise **10.1.5:** *Suppose $f \colon \mathbb{C} \to \mathbb{C}$ is holomorphic and f is real-valued on the line $\operatorname{Re} z = 0$ and the line $\operatorname{Re} z = 1$. Prove that f is 2-periodic, that is, $f(z + 2) = f(z)$ for all z.*

Exercise **10.1.6:** *Let $U = \{z \in \mathbb{C} : \operatorname{Re} z > 0\}$ and $g \colon U \to \mathbb{C}$ be holomorphic such that $\big(g(z)\big)^2 = z$ for all $z \in U$ and $g(1) = 1$ (g is one of the square roots).*
  a) Show that $|\operatorname{Im} g(z)| \leq \operatorname{Re} g(z)$ for all $z \in U$.
  b) Show that $f(z) = e^{-1/g(z)}$ is holomorphic in U and extends to a continuous function on the closure $\overline{U}$. (The hard part is continuity at $z = 0$.)
  c) Show that f does not extend holomorphically through the origin. That is, for no open neighborhood V of 0 does there exist a holomorphic $\varphi \colon V \to \mathbb{C}$ such that $\varphi = f$ on $U \cap V$.
  d) Show that f extends as a C^∞ (infinitely real differentiable) function on $\overline{U}$. Hint: It is enough to show that all real partial derivatives of f of all orders on U are locally bounded on $\overline{U}$: In particular, for any $z \in \partial U$ and any real partial derivative, there is a neighborhood V of z such that the derivative is bounded on $U \cap V$, and that is only tricky at $z = 0$.
This exercise shows that a holomorphic function can be smooth up to the boundary but still not extend past the boundary holomorphically.

10.2i Analytic continuation along paths

10.2.1i Definition

More generally than a reflection, we define a continuation along a path. We have seen such continuation with the logarithm. We define a function locally and "continue" it uniquely to another point of the domain along some path. We have also seen some problems with this approach: When dealing with the log in the punctured plane, we could go around the origin and not end up where we started.

Geometry and topology may get in the way, and so it is easiest to work with discs. We will cover a path with discs and try to extend from one disc to another. The advantage of discs is that if two discs intersect, then their intersection is connected, and a union of two intersecting discs is always simply connected, in fact, it is star-like. Since any path can be reparametrized to $[0, 1]$, we will, for simplicity, consider all paths to have the parameter in $[0, 1]$ in this section.

Definition 10.2.1. Suppose $\gamma \colon [0, 1] \to \mathbb{C}$ is continuous with $\gamma(0) = p$, and $U \subset \mathbb{C}$ an open connected neighborhood of p. A holomorphic $f \colon U \to \mathbb{C}$ can be *analytically continued* along γ if for every $t \in [0, 1]$ there exists a disc D_t centered at $\gamma(t)$ and a holomorphic function $f_t \colon D_t \to \mathbb{C}$, such that:

(i) $D_0 \subset U$ and $f_0 = f|_{D_0}$.

(ii) For each $s \in [0, 1]$, there is an $\epsilon > 0$ such that if $|t - s| < \epsilon$, then $f_t = f_s$ in $D_t \cap D_s$.

We refer to the continuation as f_t or perhaps (f_t, D_t). The function together with the domain such as (f, U) or (f_t, D_t) is called a *function element*.

The value $f_t(\gamma(t))$ is really what we mean by the value of the continuation at $\gamma(t)$, and what we would really want "$f(\gamma(t))$" to be. Though of course, $f(\gamma(t))$ is not defined for $\gamma(t)$ outside of U. The identity theorem implies that the definition of $f_t(\gamma(t))$ is unique.

Proposition 10.2.2. *Suppose that f and γ are as in Definition 10.2.1, and f can be continued along γ. The value $f_t(\gamma(t))$ is uniquely defined for every $t \in [0, 1]$.*

Exercise 10.2.1: Prove the proposition.

Exercise 10.2.2: Prove that in the definition we could take D_t to not be centered at $\gamma(t)$. We simply need to require $\gamma(t) \in D_t$, and we would get the same values of $f_t(\gamma(t))$.

Since γ (as a set) is compact, we can use only finitely many discs. See Figure 10.2. We were continuing the logarithm in this way back in chapter 4.

Proposition 10.2.3. *Suppose that f, U, and γ are as in Definition 10.2.1. Then f can be analytically continued along γ if and only if there exist numbers $0 = t_0 < t_1 < \cdots < t_n = 1$, open discs $\Delta_1, \ldots, \Delta_n$ that cover γ (as a set), and holomorphic functions $\varphi_j \colon \Delta_j \to \mathbb{C}$, such that for all j, $\gamma([t_{j-1}, t_j]) \subset \Delta_j$, $\varphi_{j-1} = \varphi_j$ on $\Delta_{j-1} \cap \Delta_j$, and such that $\varphi_1 = f$ on $\Delta_1 \cap U$. Furthermore, if (f_t, D_t) is a continuation and $t \in [t_{j-1}, t_j]$, then $\varphi_j(\gamma(t)) = f_t(\gamma(t))$.*

Exercise 10.2.3: Prove the proposition.

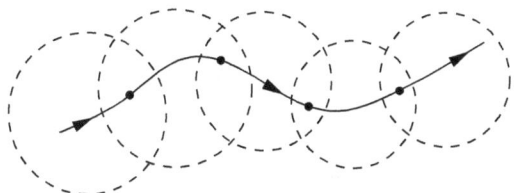

Figure 10.2: Analytic continuation with finitely many discs. The endpoints of the subintervals are marked.

Exercise 10.2.4: *There's good reason to use discs in analytic continuation. Find two domains $U_1, U_2 \subset \mathbb{C}$, $U_1 \cap U_2 \neq \emptyset$, $p \in U_1$, $q \in U_2$, holomorphic functions $f_1 \colon U_1 \to \mathbb{C}$ and $f_2 \colon U_2 \to \mathbb{C}$, a path γ in $U_1 \cup U_2$ from p to q, and two discs Δ_1, Δ_2 as in the proposition such that $\Delta_1 \subset U_1$ and $\Delta_2 \subset U_2$, such that $\gamma \subset \Delta_1 \cup \Delta_2$, and these are the discs giving the analytic continuation of f_1 from p to q, and $f_1 = f_2$ on $\Delta_1 \cap \Delta_2$, but such that f_1 is not equal to f_2 on $U_1 \cap U_2$.*

Proposition 10.2.4. *Let f and γ be as in Definition 10.2.1. Suppose f continues analytically along γ, and let f_t be the continuation. Then there exists an $\epsilon > 0$ such that if a continuous $\sigma \colon [0,1] \to \mathbb{C}$ is such that $p = \sigma(0) = \gamma(0)$, $q = \sigma(1) = \gamma(1)$, and $|\sigma(t) - \gamma(t)| < \epsilon$ for all t, then f can be analytically continued as g_t along σ and $g_1(q) = f_1(q)$.*

Proof. We use Proposition 10.2.3 and its notation. The image of each subinterval $\gamma\big([t_{j-1}, t_j]\big)$ is compact in Δ_j and so it is a positive distance away from the boundary $\partial\Delta_j$. Let $\epsilon > 0$ be smaller than this distance for all $j = 1, \ldots, n$. Suppose σ is as in the statement. Then $\sigma\big([t_{j-1}, t_j]\big)$ is still in Δ_j. We can thus use the same φ_j and Δ_j to get a continuation using Proposition 10.2.3 again. □

We defined continuation for continuous γ, but we could have used piecewise-C^1 paths since we can approximate γ by a piecewise-C^1 path, or even a polygonal path. See the following exercise.

Exercise 10.2.5: *Suppose U is open, $p \in U$, and a holomorphic $f \colon U \to \mathbb{C}$ continues analytically along a continuous $\gamma \colon [0,1] \to \mathbb{C}$ with $\gamma(0) = p$, $\gamma(1) = q$, then for every $\epsilon > 0$, there exists a polygonal path $\sigma \colon [0,1] \to \mathbb{C}$ such that $|\sigma(t) - \gamma(t)| < \epsilon$ for all t, $\sigma(0) = \gamma(0)$, $\sigma(1) = \gamma(1)$, the function f continues analytically along σ, and the value of the continuation at $\gamma(1) = \sigma(1)$ is the same for σ or γ.*

10.2.2*i* Unrestricted continuation

The general problem we are interested in is to start with a domain U and a holomorphic f defined in a small subset of U. Then we want to extend f to all of U.

Definition 10.2.5. Let $U \subset \mathbb{C}$ be a domain, $p \in U$, and $W \subset U$ an open connected neighborhood of p. A holomorphic $f: W \to \mathbb{C}$ *admits unrestricted continuation* to U if for every $q \in U$ and *every* continuous $\gamma: [0, 1] \to U$, $\gamma(0) = p$, $\gamma(1) = q$, f can be analytically continued along γ.

It may appear that the definition depends on p, but any other $p \in W$ could be used as W is connected, see Exercise 10.2.7.

The logarithm, and in general every primitive (exercise below), admits unrestricted continuation in the domain where the derivative is defined. We do not necessarily get a unique value. For instance for the logarithm, if $U = \mathbb{C} \setminus \{0\}$, $W = \Delta_r(p) \subset U$, then we can define a branch of the logarithm in $\Delta_r(p)$, and it admits unrestricted continuation to all of $\mathbb{C} \setminus \{0\}$, but the value of the continuation is not well-defined, it depends on the path taken. Moreover, not every function allows unrestricted continuation even if it allows continuation along some path to every point.

Example 10.2.6: Consider a branch of the square root $\sqrt{z}$ such that $\sqrt{1} = 1$ defined in some neighborhood of 1. The logarithm allows unrestricted continuation in $\mathbb{C} \setminus \{0\}$, and so does the square root. If we continue from $z = 1$ along a closed path that does not go around the origin and comes back to $z = 1$, the continuation of the root also has the value 1 there. However, if we go around a path that goes once around the origin, the continuation will have a value of -1 at $z = 1$. OK, nothing new so far.

Now consider the function $f(z) = \frac{1}{1+\sqrt{z}}$, where the square root is as before. As we can continue the square root, we can try to continue f. If we take a path starting at $z = 1$ that does not go around the origin (such as a small loop near 1), then we obtain the square root being 1 and so f also continues along this loop. However, if we take the unit circle and go once around the origin, the square root becomes -1 once we get back to $z = 1$, and so f cannot be analytically continued along this path. That is, f does not allow unrestricted continuation to $\mathbb{C} \setminus \{0\}$ even though it allows continuation to every point of $\mathbb{C} \setminus \{0\}$ along some path. Just not every path.

Inverses can often be continued for some paths, but as we saw in the example above, where we tried to continue the inverse of $w \mapsto (1/w - 1)^2$, inverses may not admit unrestricted continuation. However, if the mapping is a so-called covering map (for example, a k-to-1 onto holomorphic map), then its inverse does admit such continuation.

Definition 10.2.7. Suppose $U, V \subset \mathbb{C}$ are open, $f: U \to V$ is holomorphic and onto, and for every $p \in V$, there exists a neighborhood $W \subset V$ such that $f^{-1}(W)$ is a disjoint union of open connected sets $\Omega_1, \Omega_2, \ldots$ such that $f|_{\Omega_j}$ is a biholomorphism of Ω_j and W. Then we call f a *holomorphic covering map*.

Example 10.2.8: The map z^2 is a covering map of $\mathbb{C} \setminus \{0\}$ onto $\mathbb{C} \setminus \{0\}$.

Example 10.2.9: The exponential e^z is a covering map of $\mathbb{C}$ onto $\mathbb{C} \setminus \{0\}$.

Example 10.2.10: Suppose $U, V \subset \mathbb{C}$ are open, $f: U \to V$ holomorphic, k-to-1 ($k \in \mathbb{N}$), and onto. Using Exercise 5.6.4, f' never vanishes and f is locally invertible. It is not hard to prove (exercise below) that f is a covering map.

Proposition 10.2.11. *Suppose $U \subset \mathbb{C}$ is open, $V \subset \mathbb{C}$ is a domain, $f: U \to V$ is a holomorphic covering map, and $p \in V$. Then starting with any value of $f^{-1}(p)$, a branch of f^{-1} can be defined in some neighborhood of p (a holomorphic g defined near p such that $f \circ g$ is the identity) and admits unrestricted analytic continuation to V.*

Proof. Consider a continuous $\gamma: [0,1] \to V$, where $\gamma(0) = p$. At each point w of γ using the definition of covering map, we can find a disc $\Delta \subset V$ centered at w such that $f^{-1}(\Delta)$ has disjoint components and for each component, a branch of f^{-1} can be defined as a holomorphic map onto that component. As γ is compact, finitely many such discs $\Delta_1, \ldots, \Delta_n$ cover γ such that $\Delta_{j-1} \cap \Delta_j \neq \emptyset$ and $p \in \Delta_1$. Given any choice of f^{-1} at p, we can define a branch of f^{-1} in Δ_j that agrees with our choice on Δ_{j-1} for each j. In other words, we can continue our branch of f^{-1} along γ. $\square$

Exercise 10.2.6: Let $U \subset \mathbb{C}$ be a domain, $W \subset U$ a connected open subset, $p \in W$, and $f: U \to \mathbb{C}$ holomorphic. Prove that the restriction $f|_W$ allows unrestricted continuation to U with p as a starting point.

Exercise 10.2.7: Let $U \subset \mathbb{C}$ be a domain, $W \subset U$ a nonempty connected open subset, and $f: W \to \mathbb{C}$ holomorphic. Suppose f admits unrestricted continuation to U with $p_1 \in W$ as a starting point. Prove that for any other $p_2 \in W$, f admits unrestricted continuation to U with p_2 as a starting point.

Exercise 10.2.8: Let $U \subset \mathbb{C}$ be a domain and let $f: U \to \mathbb{C}$ be holomorphic. Locally near some $p \in U$, suppose F is an antiderivative of f. Prove that F admits unrestricted continuation to U.

Exercise 10.2.9: Let $U \subset \mathbb{C}$ be a domain and $f: U \to \mathbb{C}$ be holomorphic and not identically zero. Let Z_f be the set of zeros of f. Given any $p \in U \setminus Z_f$, we can locally (in some neighborhood) define some branch of $\log f(z)$. Show that this branch allows unrestricted continuation to $U \setminus Z_f$.

Exercise 10.2.10: Given two domains $U, V \subset \mathbb{C}$, prove that a k-to-1 onto holomorphic mapping $f: U \to V$ is a covering map.

10.2.3*i* Monodromy theorem

We can continue a function along many different paths and the continuation will be the same if the paths do not change much. We therefore introduce a topological equivalence that tells us how we navigate the domain around the various holes, an equivalence that doesn't care exactly what path we take as long as we can deform one path to the other. We have seen that for closed paths, homotopy is such an equivalence, although here we look at paths from one fixed point to another. The definition is almost the same (compare Definition 4.5.1), except instead of "closed" we require that the paths are "from p to q."

Definition 10.2.12. Let $U \subset \mathbb{C}$ be open and $p, q \in U$. Two continuous functions $\gamma_0 \colon [0,1] \to U$ and $\gamma_1 \colon [0,1] \to U$ where $\gamma_0(0) = \gamma_1(0) = p$ and $\gamma_0(1) = \gamma_1(1) = q$ are *fixed-endpoint homotopic* in U (or relative to U) if there exists a continuous function $H \colon [0,1] \times [0,1] \to U$ such that for all s and t in $[0,1]$

$$H(t,0) = \gamma_0(t), \qquad H(t,1) = \gamma_1(t), \qquad H(0,s) = p, \qquad \text{and} \qquad H(1,s) = q.$$

See Figure 10.3. We also write γ_s, where $\gamma_s(t) = H(t,s)$, for the paths in the homotopy.

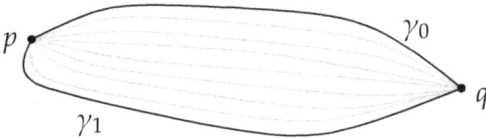

Figure 10.3: Fixed endpoint homotopy of two paths γ_0 and γ_1 with intermediate paths marked in gray.

The key property of fixed-endpoint homotopy in the context of continuation is that the value of the continuation at q is the same for homotopic paths.

Proposition 10.2.13. *Suppose $U \subset \mathbb{C}$ is a domain, $p \in U$, $W \subset U$ is an open connected neighborhood of p, and $f \colon W \to \mathbb{C}$ is holomorphic and admits unrestricted continuation to U. Suppose further that $\gamma_0 \colon [0,1] \to U$ and $\gamma_1 \colon [0,1] \to U$ are continuous, $\gamma_0(0) = \gamma_1(0) = p$ and $\gamma_0(1) = \gamma_1(1) = q$, and γ_0 and γ_1 are fixed-endpoint homotopic in U. Then the value at q of the continuation of f along γ_0 is equal to the value at q of the continuation of f along γ_1.*

Proof. Let $H(t,s)$ be the homotopy. Let $\varphi(s)$ be the value of the continuation at q for the path γ_s. By Proposition 10.2.4, $\varphi(s)$ is locally constant (that is, each s has a neighborhood in which φ is constant). As $[0,1]$ is connected, φ is constant. $\qquad \square$

The monodromy theorem says that as long as there are no holes, analytic continuation defines a function uniquely.

Theorem 10.2.14 (Monodromy theorem). *Suppose $U \subset \mathbb{C}$ is a simply connected domain, $W \subset U$ is a nonempty connected open subset, and $f \colon W \to \mathbb{C}$ is holomorphic and admits unrestricted continuation to U. Then there exists a unique holomorphic function $F \colon U \to \mathbb{C}$ such that $F|_W = f$.*

Proof. As U is simply connected, apply the Riemann mapping theorem and assume that $U = \mathbb{D}$ or $U = \mathbb{C}$. We must show that if we continue f to any $q \in U$, we always get the same value, no matter what path we continue along. Suppose $\gamma_0 \colon [0,1] \to U$ and $\gamma_1 \colon [0,1] \to U$ are continuous, $\gamma_0(0) = \gamma_1(0) \in W$, and $\gamma_0(1) = \gamma_1(1) = q$. Let

$$H(t,s) = (1-s)\gamma_0(t) + s\gamma_1(t).$$

As U is convex, $H(t,s) \in U$ for all $(t,s) \in [0,1] \times [0,1]$. Also, $H(0,s) = \gamma_0(0) = \gamma_1(0)$ and $H(1,s) = \gamma_0(1) = \gamma_1(1)$. So H is a fixed-endpoint homotopy in U. By the proposition, the value of the continuation at q is the same whether we continued f along γ_0 or γ_1. We define $F(q)$ to be this value. $\qquad\square$

We remark that the proof above contains a proof of the useful topological fact that in a simply connected domain $U \subset \mathbb{C}$, any two paths with the same endpoints are fixed-endpoint homotopic. Converse is also true, see Exercise 10.2.17.

Corollary 10.2.15. *Suppose that $U, V \subset \mathbb{C}$ are domains, V is simply connected, and $f \colon U \to V$ is a holomorphic covering map. Then f is a biholomorphism.*

Proof. We have seen that any local inverse of a covering map admits unrestricted continuation. Given a choice of $f^{-1}(p)$ for some $p \in V$, a local inverse extends to a global one defined on all of V by the monodromy theorem. In other words, f is one-to-one and hence a biholomorphism. $\qquad\square$

If U is a simply connected, then a covering map $f \colon U \to V$ is called a *universal covering map* (or *universal cover* for short) of V, and U is called the *universal covering space* of V. We will not prove so, but every domain in $\mathbb{C}$ has a universal cover.[*] What we proved above is that the only universal covering of a simply connected domain is essentially just a biholomorphism. By Riemann mapping theorem, every domain has a universal cover that is either $\mathbb{C}$ or $\mathbb{D}$. In fact, the little Picard theorem, which we do not prove, says that any nonconstant entire function misses at most one value. So the only domains with $\mathbb{C}$ as a universal cover are $\mathbb{C}$ itself (the identity) and $\mathbb{C} \setminus \{p\}$ (the exponential, $e^z + p$). Every other domain in $\mathbb{C}$ has $\mathbb{D}$ as the universal cover.

Why is it called a universal cover? Because it covers any cover. We leave the proof as an exercise.

Corollary 10.2.16. *Suppose $U, V, W \subset \mathbb{C}$ are domains, $h \colon U \to V$ is a holomorphic covering map and $f \colon W \to V$ is a holomorphic universal cover (W is simply connected). Then there exists a holomorphic covering map $g \colon W \to U$ such that $f = h \circ g$.*

In other words, you have the following commutative diagram:

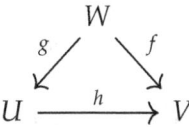

---

[*]Actually every reasonably nice topological space has a topological universal cover, where the "local biholomorphism" is replaced by "local homeomorphism."

After you prove this corollary, via the monodromy theorem, you can quickly prove that the universal cover is unique up to biholomorphism.

Corollary 10.2.17. *Suppose $U_1, U_2, V \subset \mathbb{C}$ are domains and $f\colon U_1 \to V$ and $g\colon U_2 \to V$ are universal holomorphic covering maps (U_1 and U_2 are simply connected). Then there exists a biholomorphism $\varphi\colon U_1 \to U_2$ such that $f = g \circ \varphi$.*

In other words, you have the following commutative diagram:

$$
\begin{array}{ccc}
U_1 & \xrightarrow{\ \varphi\ } & U_2 \\
& f\searrow \quad \swarrow g & \\
& V &
\end{array}
$$

As an example for the setup of both corollaries, notice that e^z is a universal covering map taking $\mathbb{C}$ to $\mathbb{C} \setminus \{0\}$. The map z^2 is a covering map from $\mathbb{C} \setminus \{0\}$ to itself. We might think we can compose the two and obtain a new universal covering map:

$$
\left(e^z\right)^2 = e^{2z},
$$

and darnit, $2z$ is an automorphism of $\mathbb{C}$. Since we're already drawing diagrams:

$$
\begin{array}{ccc}
\mathbb{C} & \xrightarrow{\ 2z\ } & \mathbb{C} \\
e^z\nearrow \quad \searrow e^{2z} & & \swarrow e^z \\
\mathbb{C} \setminus \{0\} & \xrightarrow[z^2]{} & \mathbb{C} \setminus \{0\}
\end{array}
$$

Exercise 10.2.11: Explicitly find the universal cover of $\mathbb{D} \setminus \{0\}$.

Exercise 10.2.12: Prove Corollary 10.2.16. Hint: A holomorphic covering map admits unrestricted analytic continuation.

Exercise 10.2.13: Prove Corollary 10.2.17.

Exercise 10.2.14: Explicitly find the universal cover of $\mathbb{C} \setminus [-2, 2]$. See Exercise 2.2.17.

Exercise 10.2.15: Suppose $K \subset \mathbb{C}$ is compact, connected, contains more than one point, and $\mathbb{C} \setminus K$ is connected. Show that there exists a universal holomorphic cover $f\colon \mathbb{D} \to \mathbb{C} \setminus K$.

Exercise 10.2.16: Suppose $U \subset \mathbb{C}$ is a domain and there exists a holomorphic covering map $f\colon U \to U$ that is not injective. Prove that the universal covering map of U is infinite-to-one.

Exercise 10.2.17: Suppose $U \subset \mathbb{C}$ is a domain such that every pair of continuous $\gamma_0\colon [0,1] \to U$ and $\gamma_1\colon [0,1] \to U$ with $\gamma_0(0) = \gamma_1(0)$ and $\gamma_0(1) = \gamma_1(1)$ are fixed-endpoint homotopic in U. Prove that every holomorphic $f\colon U \to \mathbb{C}$ has a primitive and hence U is simply connected (in the homology sense).

A*i*　Metric Spaces

Except in mathematics, the shortest distance between point A and point B is seldom a straight line. I don't believe in mathematics.

—Albert Einstein

Let us give an introduction to metric spaces for the student that may not have seen metric spaces in full generality. This appendix is an adapted and shortened version of chapter 7 from [L1].

A.1*i*　Metric spaces

The main idea in analysis is to take limits and talk about continuity. We wish to abstract what it means to be able to take limits in various contexts. The most basic such abstraction is a *metric space*. While it is not sufficient to describe every type of limit we find in modern analysis, it gets us very far indeed.

Definition A.1.1. Let X be a set, and let $d \colon X \times X \to \mathbb{R}$ be a function such that for all $x, y, z \in X$,

(i)　$d(x, y) \geq 0$　　　　　　　　　　　(nonnegativity),

(ii)　$d(x, y) = 0$ if and only if $x = y$　　　(identity of indiscernibles),

(iii)　$d(x, y) = d(y, x)$　　　　　　　　(symmetry),

(iv)　$d(x, z) \leq d(x, y) + d(y, z)$　　　　(*triangle inequality*).

The pair (X, d) is called a *metric space*. The function d is called the *metric* or the *distance function*. If the metric is clear from context, we may write simply X instead of (X, d).

The geometric idea is that d is the distance between two points. Items (i)–(iii) have obvious geometric interpretation: Distance is always nonnegative, the only point that is distance 0 away from x is x itself, and that the distance from x to y is the same as the distance from y to x. The triangle inequality (iv) has the interpretation given in Figure A.1.

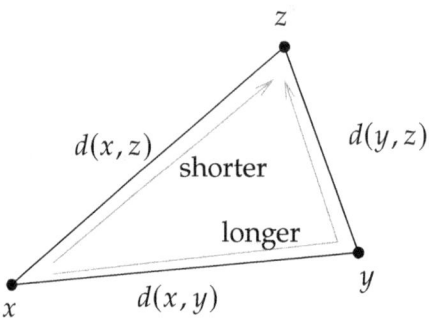

Figure A.1: Diagram of the triangle inequality in metric spaces.

For the purposes of drawing, it is convenient to draw figures and diagrams in the plane with the metric being the euclidean distance. However, that is only one particular metric space. Just because a certain fact seems to be clear from drawing a picture does not mean it is true in every metric space. You might be getting sidetracked by intuition from euclidean geometry, whereas the concept of a metric space is a lot more general.

Example A.1.2: The set of real numbers $\mathbb{R}$ is a metric space with the metric

$$d(x, y) = |x - y|.$$

Items (i)–(iii) of the definition are easy to verify. The triangle inequality (iv) follows immediately from the standard triangle inequality for real numbers:

$$d(x, z) = |x - z| = |x - y + y - z| \le |x - y| + |y - z| = d(x, y) + d(y, z).$$

This metric is the *standard metric on* $\mathbb{R}$. If we talk about $\mathbb{R}$ as a metric space without mentioning a specific metric, we mean this particular metric.

The n-dimensional *euclidean space* $\mathbb{R}^n = \mathbb{R} \times \mathbb{R} \times \cdots \times \mathbb{R}$ is also a metric space. In this book we mostly see $\mathbb{R}^2$, but let us give the example in more generality. We use the following notation for points: $x = (x_1, x_2, \ldots, x_n) \in \mathbb{R}^n$. Before making $\mathbb{R}^n$ a metric space, let us prove an important inequality, the so-called Cauchy–Schwarz inequality.

Lemma A.1.3 (Cauchy–Schwarz inequality*). *If* $x = (x_1, x_2, \ldots, x_n) \in \mathbb{R}^n$, $y = (y_1, y_2, \ldots, y_n) \in \mathbb{R}^n$, *then*

$$\left(\sum_{j=1}^{n} x_j y_j \right)^2 \le \left(\sum_{j=1}^{n} x_j^2 \right) \left(\sum_{j=1}^{n} y_j^2 \right).$$

*Sometimes it is called the Cauchy–Bunyakovsky–Schwarz inequality. What we stated should really be called the Cauchy inequality, as Bunyakovsky and Schwarz provided proofs for infinite-dimensional versions.

Proof. A square of a real number is nonnegative and so a sum of squares is nonnegative:

$$0 \le \sum_{j=1}^{n} \sum_{k=1}^{n} (x_j y_k - x_k y_j)^2$$

$$= \sum_{j=1}^{n} \sum_{k=1}^{n} (x_j^2 y_k^2 + x_k^2 y_j^2 - 2 x_j x_k y_j y_k)$$

$$= \left(\sum_{j=1}^{n} x_j^2\right)\left(\sum_{k=1}^{n} y_k^2\right) + \left(\sum_{j=1}^{n} y_j^2\right)\left(\sum_{k=1}^{n} x_k^2\right) - 2\left(\sum_{j=1}^{n} x_j y_j\right)\left(\sum_{k=1}^{n} x_k y_k\right).$$

We relabel and divide by 2 to obtain the needed inequality:

$$0 \le \left(\sum_{j=1}^{n} x_j^2\right)\left(\sum_{j=1}^{n} y_j^2\right) - \left(\sum_{j=1}^{n} x_j y_j\right)^2. \qquad \square$$

Example A.1.4: Let us construct the standard metric for $\mathbb{R}^n$. Define

$$d(x, y) = \sqrt{\sum_{j=1}^{n} (x_j - y_j)^2}.$$

For $n = 1$, the real line, this metric agrees with what we did above. Again, the only tricky part of the definition to check is the triangle inequality. The trick is to work with the square of the metric and apply the Cauchy–Schwarz inequality.

$$\left(d(x, z)\right)^2 = \sum_{j=1}^{n} (x_j - z_j)^2$$

$$= \sum_{j=1}^{n} (x_j - y_j + y_j - z_j)^2$$

$$= \sum_{j=1}^{n} (x_j - y_j)^2 + \sum_{j=1}^{n} (y_j - z_j)^2 + 2 \sum_{j=1}^{n} (x_j - y_j)(y_j - z_j)$$

$$\le \sum_{j=1}^{n} (x_j - y_j)^2 + \sum_{j=1}^{n} (y_j - z_j)^2 + 2\sqrt{\sum_{j=1}^{n} (x_j - y_j)^2 \sum_{j=1}^{n} (y_j - z_j)^2}$$

$$= \left(\sqrt{\sum_{j=1}^{n} (x_j - y_j)^2} + \sqrt{\sum_{j=1}^{n} (y_j - z_j)^2}\right)^2 = \left(d(x, y) + d(y, z)\right)^2.$$

Taking the square root of both sides we obtain the correct inequality.

Example A.1.5: The set of complex numbers $\mathbb{C}$ is a metric space using the standard euclidean metric on $\mathbb{R}^2$ by identifying $x + iy \in \mathbb{C}$ with $(x, y) \in \mathbb{R}^2$.

Example A.1.6: Let $C([a,b], \mathbb{R})$ be the set of continuous real-valued functions on the interval $[a,b]$. Define the metric on $C([a,b], \mathbb{R})$ as

$$d(f,g) = \sup_{x\in[a,b]} \left| f(x) - g(x) \right|.$$

Let us check the properties. First, $d(f,g)$ is finite as $\left| f(x) - g(x) \right|$ is a continuous function on a closed bounded interval $[a,b]$, and so is bounded. Clearly $d(f,g) \geq 0$. If $f = g$, then $\left| f(x) - g(x) \right| = 0$ for all x and hence $d(f,g) = 0$. Conversely, if $d(f,g) = 0$, then for any x, we have $\left| f(x) - g(x) \right| \leq d(f,g) = 0$, and hence $f = g$. That $d(f,g) = d(g,f)$ is equally trivial. The triangle inequality follows from the triangle inequality on $\mathbb{R}$.

$$d(f,g) = \sup_{x\in[a,b]} \left| f(x) - g(x) \right| = \sup_{x\in[a,b]} \left| f(x) - h(x) + h(x) - g(x) \right|$$

$$\leq \sup_{x\in[a,b]} \left(\left| f(x) - h(x) \right| + \left| h(x) - g(x) \right| \right)$$

$$\leq \sup_{x\in[a,b]} \left| f(x) - h(x) \right| + \sup_{x\in[a,b]} \left| h(x) - g(x) \right| = d(f,h) + d(h,g).$$

When treating $C([a,b], \mathbb{R})$ as a metric space without mentioning a metric, we mean this particular metric.

Example A.1.7: The sphere with the so-called *great circle distance* is also a metric space. Let S^2 be the unit sphere in $\mathbb{R}^3$, that is, $S^2 = \left\{ x \in \mathbb{R}^3 : x_1^2 + x_2^2 + x_3^2 = 1 \right\}$. Take x and y in S^2, draw a line through the origin and x, and another line through the origin and y, and let θ be the angle that the two lines make. Then define $d(x,y) = \theta$, see Figure A.2. The law of cosines from vector calculus says $d(x,y) = \arccos(x_1 y_1 + x_2 y_2 + x_3 y_3)$. It is relatively easy to see that this function satisfies the first three properties of a metric. Triangle inequality is harder to prove, and requires a bit more trigonometry and linear algebra than we wish to indulge in right now, so let us leave it without proof.

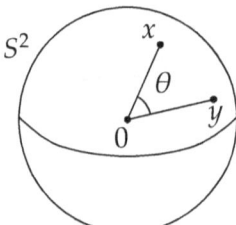

Figure A.2: The great circle distance on the unit sphere.

Oftentimes it is useful to consider a subset of a larger metric space as a metric space itself. We obtain the following proposition, which has a trivial proof.

Proposition A.1.8. *Let (X, d) be a metric space and $Y \subset X$. Then the restriction $d|_{Y \times Y}$ is a metric on Y.*

Definition A.1.9. *If (X, d) is a metric space, $Y \subset X$, and $d' = d|_{Y \times Y}$, then (Y, d') is said to be a subspace of (X, d).*

It is common to simply write d for the metric on Y, as it is the restriction of the metric on X. We say d' is the *subspace metric* and Y has the *subspace topology*.

Definition A.1.10. *Let (X, d) be a metric space. A subset $S \subset X$ is said to be bounded if there exists a $p \in X$ and a $B \in \mathbb{R}$ such that*

$$d(p, x) \leq B \quad \text{for all } x \in S.$$

We say (X, d) is bounded if X itself is a bounded subset.

For instance, the set of real numbers with the standard metric is not a bounded metric space. On the other hand, the real numbers with the *discrete metric*, $d(x, y) = 1$ if $x \neq y$, and $d(x, x) = 0$, is a bounded metric space. Any set with the discrete metric is bounded.

Suppose X is nonempty. Then $S \subset X$ is bounded if and only if

(i) For every $p \in X$, there exists a $B > 0$ such that $d(p, x) \leq B$ for all $x \in S$.

(ii) $\text{diam}(S) = \sup\{d(x, y) : x, y \in S\} < \infty$.

See the exercises. The quantity $\text{diam}(S)$ is called the *diameter* of a set and is usually only defined for a nonempty set.

Exercise A.1.1: *Show that for any given set X, the discrete metric ($d(x, y) = 1$ if $x \neq y$ and $d(x, x) = 0$) does give a metric space (X, d).*

Exercise A.1.2: *Suppose (X, d) is a metric space and $\varphi \colon [0, \infty) \to \mathbb{R}$ is an increasing function such that $\varphi(t) \geq 0$ for all t and $\varphi(t) = 0$ if and only if $t = 0$. Also suppose φ is subadditive, that is, $\varphi(s + t) \leq \varphi(s) + \varphi(t)$. Show that with $d'(x, y) = \varphi\big(d(x, y)\big)$, we obtain a new metric space (X, d').*

Exercise A.1.3: *Let (X, d_X) and (Y, d_Y) be metric spaces.*
  a) *Show that $(X \times Y, d)$ with $d\big((x_1, y_1), (x_2, y_2)\big) = d_X(x_1, x_2) + d_Y(y_1, y_2)$ is a metric space.*
  b) *Show that $(X \times Y, d)$ with $d\big((x_1, y_1), (x_2, y_2)\big) = \max\{d_X(x_1, x_2), d_Y(y_1, y_2)\}$ is a metric space.*

Exercise A.1.4: *Let X be the set of continuous functions on $[0, 1]$. Let $\varphi \colon [0, 1] \to (0, \infty)$ be continuous. Define*

$$d(f, g) = \int_0^1 |f(x) - g(x)| \varphi(x) \, dx.$$

Show that (X, d) is a metric space.

Exercise A.1.5: *Let (X, d) be a metric space. For nonempty bounded subsets A and B let*

$$d(x, B) = \inf\{d(x, b) : b \in B\} \qquad and \qquad d(A, B) = \sup\{d(a, B) : a \in A\}.$$

Now define the **Hausdorff** **metric** *as*

$$d_H(A, B) = \max\{d(A, B), d(B, A)\}.$$

Note: d_H can be defined for arbitrary nonempty subsets if we allow the extended reals.

a) *Let Y be the set of bounded nonempty subsets of X. Prove that (Y, d_H) is a so-called* **pseudometric space**: *d_H satisfies the metric properties (i), (iii), (iv), and further $d_H(A, A) = 0$ for all $A \in Y$.*

b) *Show by example that d itself is not symmetric, that is, $d(A, B) \neq d(B, A)$.*

c) *Find a metric space X and two different nonempty bounded subsets A and B such that $d_H(A, B) = 0$.*

Exercise A.1.6: *Let (X, d) be a nonempty metric space and $S \subset X$ a subset. Prove:*

a) *S is bounded if and only if for every $p \in X$, there exists a $B > 0$ such that $d(p, x) \leq B$ for all $x \in S$.*

b) *A nonempty S is bounded if and only if $\operatorname{diam}(S) = \sup\{d(x, y) : x, y \in S\} < \infty$.*

Exercise A.1.7:

a) *Find a metric d on $\mathbb{N}$, such that $\mathbb{N}$ is an unbounded set in $(\mathbb{N}, d)$.*

b) *Find a metric d on $\mathbb{N}$, such that $\mathbb{N}$ is a bounded set in $(\mathbb{N}, d)$.*

c) *Find a metric d on $\mathbb{N}$ such that for every $n \in \mathbb{N}$ and every $\epsilon > 0$, there exists an $m \in \mathbb{N}$ such that $d(n, m) < \epsilon$.*

Exercise A.1.8: *Let $C^1([a, b], \mathbb{R})$ be the set of once continuously differentiable functions on $[a, b]$. Define*

$$d(f, g) = \|f - g\|_{[a,b]} + \|f' - g'\|_{[a,b]},$$

where $\|f\|_X = \sup_{x \in X} |f(x)|$ is the uniform norm. Prove that d is a metric.

Exercise A.1.9: *The set of sequences $\{x_n\}$ of real numbers such that $\sum_{n=1}^{\infty} x_n^2 < \infty$ is called ℓ^2.*

a) *Prove the Cauchy–Schwarz inequality for two sequences $\{x_n\}$ and $\{y_n\}$ in ℓ^2: Prove that $\sum_{n=1}^{\infty} x_n y_n$ converges (absolutely) and*

$$\left(\sum_{n=1}^{\infty} x_n y_n \right)^2 \leq \left(\sum_{n=1}^{\infty} x_n^2 \right) \left(\sum_{n=1}^{\infty} y_n^2 \right).$$

b) *Prove that ℓ^2 is a metric space with the metric $d(x, y) = \sqrt{\sum_{n=1}^{\infty} (x_n - y_n)^2}$. Hint: Don't forget to show that the series for $d(x, y)$ always converges to some finite number.*

A.2i Open and closed sets

A.2.1i Topology

Definition A.2.1. Let (X, d) be a metric space, $x \in X$, and $\delta > 0$. The *open ball* or simply *ball* of radius δ around x is

$$B(x, \delta) \stackrel{\text{def}}{=} \{y \in X : d(x, y) < \delta\}.$$

Similarly the *closed ball* is

$$C(x, \delta) \stackrel{\text{def}}{=} \{y \in X : d(x, y) \leq \delta\}.$$

When we are dealing with different metric spaces, we may emphasize which metric space the ball is in by writing $B_X(x, \delta) = B(x, \delta)$ or $C_X(x, \delta) = C(x, \delta)$.

Example A.2.2: Consider $\mathbb{R}$ with the standard metric. For $x \in \mathbb{R}$ and $\delta > 0$,

$$B(x, \delta) = (x - \delta, x + \delta) \qquad \text{and} \qquad C(x, \delta) = [x - \delta, x + \delta].$$

Example A.2.3: Consider the metric space $[0, 1]$ as a subspace of $\mathbb{R}$. Then

$$B(0, 1/2) = B_{[0,1]}(0, 1/2) = \{y \in [0, 1] : |0 - y| < 1/2\} = [0, 1/2).$$

This is different from $B_{\mathbb{R}}(0, 1/2) = (-1/2, 1/2)$. The important thing to keep in mind is which metric space we are working in.

Definition A.2.4. Let (X, d) be a metric space. A subset $V \subset X$ is *open* if for every $x \in V$, there exists a $\delta > 0$ such that $B(x, \delta) \subset V$. See Figure A.3. A subset $E \subset X$ is *closed* if the complement $E^c = X \setminus E$ is open. If the ambient space X is not clear from context, we say V *is open in* X and E *is closed in* X. The set of open sets is called the *topology* on X.

If $x \in V$ and V is open, then V is an *open neighborhood* of x (or simply *neighborhood*). More generally a *neighborhood* of x is a set that contains an open neighborhood of x, but unless otherwise specified we usually mean open neighborhood.

Intuitively, an open set V is a set that does not include its "boundary." Wherever we are in V, we are allowed to "wiggle" a little bit and stay in V. Similarly, a set E is closed if everything not in E is some distance away from E. The open and closed balls are examples of open and closed sets (which must still be proved). Not every set is either open or closed, most subsets are neither.

Example A.2.5: The set $(0, \infty) \subset \mathbb{R}$ is open: Given any $x \in (0, \infty)$, let $\delta = x$.

The set $[0, \infty) \subset \mathbb{R}$ is closed: Given $x \in (-\infty, 0) = [0, \infty)^c$, let $\delta = -x$.

The set $[0, 1) \subset \mathbb{R}$ is neither open nor closed. Every $B(0, \delta) = (-\delta, \delta)$, contains negative numbers and hence is not contained in $[0, 1)$. So $[0, 1)$ is not open. Every $B(1, \delta) = (1 - \delta, 1 + \delta)$, contains numbers in $[0, 1)$. Thus $[0, 1)^c = \mathbb{R} \setminus [0, 1)$ is not open, and $[0, 1)$ is not closed.

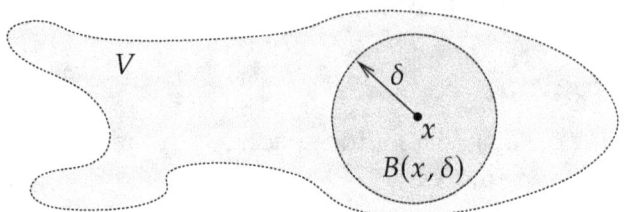

Figure A.3: Open set in a metric space. Note that δ depends on x.

Proposition A.2.6. *Let (X, d) be a metric space.*

(i) *$\emptyset$ and X are open.*

(ii) *If $V_1, V_2, \dots, V_k$ are open subsets of X, then*

$$V_1 \cap V_2 \cap \cdots \cap V_k$$

is also open. That is, a finite intersection of open sets is open.

(iii) *If $\{V_\lambda\}_{\lambda \in I}$ is an arbitrary collection of open subsets of X, then*

$$\bigcup_{\lambda \in I} V_\lambda$$

is also open. That is, a union of open sets is open.

Proof. Item (i) is obvious. Let us prove (ii). If $x \in \bigcap_{\ell=1}^{k} V_\ell$, then $x \in V_\ell$ for all ℓ. As V_ℓ are all open, for every ℓ there exists a $\delta_\ell > 0$ such that $B(x, \delta_\ell) \subset V_\ell$. Take $\delta = \min\{\delta_1, \delta_2, \dots, \delta_k\}$ and notice $\delta > 0$. Then $B(x, \delta) \subset B(x, \delta_\ell) \subset V_\ell$ for every ℓ and so $B(x, \delta) \subset \bigcap_{\ell=1}^{k} V_\ell$. Let us prove (iii). If $x \in \bigcup_{\lambda \in I} V_\lambda$, then $x \in V_\lambda$ for some $\lambda \in I$. As V_λ is open, $B(x, \delta) \subset V_\lambda$ for some $\delta > 0$. But then $B(x, \delta) \subset \bigcup_{\lambda \in I} V_\lambda$. $\square$

Item (ii) is not true for an arbitrary intersection: $\bigcap_{n \in \mathbb{N}} (-1/n, 1/n) = \{0\}$ is not open.

Proposition A.2.7. *Let (X, d) be a metric space.*

(i) *$\emptyset$ and X are closed.*

(ii) *If $\{E_\lambda\}_{\lambda \in I}$ is an arbitrary collection of closed subsets of X, then*

$$\bigcap_{\lambda \in I} E_\lambda$$

is also closed. That is, an intersection of closed sets is closed.

(iii) *If $E_1, E_2, \dots, E_k$ are closed subsets of X, then*

$$E_1 \cup E_2 \cup \cdots \cup E_k$$

is also closed. That is, a finite union of closed sets is closed.

Exercise **A.2.1:** *Prove Proposition A.2.7.*

We have not yet shown that the open ball is open and the closed ball is closed. Let us show this fact now to justify the terminology.

Proposition A.2.8. *Let (X, d) be a metric space, $x \in X$, and $\delta > 0$. Then $B(x, \delta)$ is open and $C(x, \delta)$ is closed.*

Proof. Let $y \in B(x, \delta)$. Let $\alpha = \delta - d(x, y)$. As $\alpha > 0$, consider $z \in B(y, \alpha)$. Then

$$d(x, z) \le d(x, y) + d(y, z) < d(x, y) + \alpha = d(x, y) + \delta - d(x, y) = \delta.$$

Thus, $z \in B(x, \delta)$ for every $z \in B(y, \alpha)$. So $B(y, \alpha) \subset B(x, \delta)$, and so $B(x, \delta)$ is open.

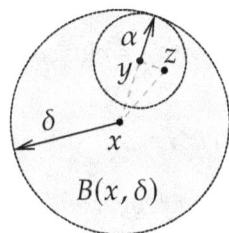

Figure A.4: Proof that $B(x, \delta)$ is open: $B(y, \alpha) \subset B(x, \delta)$ with the triangle inequality illustrated.

The proof that $C(x, \delta)$ is closed is left as an exercise. □

Exercise **A.2.2:** *Finish the proof of Proposition A.2.8 by proving that $C(x, \delta)$ is closed.*

Be careful about what metric space you find yourself in. As $[0, 1/2)$ is an open ball in $[0, 1]$, the set $[0, 1/2)$ is an open set in $[0, 1]$. On the other hand, $[0, 1/2)$ is neither open nor closed in $\mathbb{R}$.

Proposition A.2.9. *Let a, b be two real numbers, $a < b$. Then (a, b), (a, ∞), and $(-\infty, b)$ are open in $\mathbb{R}$. Also $[a, b]$, $[a, \infty)$, and $(-\infty, b]$ are closed in $\mathbb{R}$.*

Exercise **A.2.3:** *Prove Proposition A.2.9.*

Proposition A.2.10. *Suppose (X, d) is a metric space, and $Y \subset X$. Then $U \subset Y$ is open in Y (in the subspace topology) if and only if there exists an open set $V \subset X$ (so open in X), such that $V \cap Y = U$.*

For example, let $X = \mathbb{R}$, $Y = [0, 1]$, $U = [0, 1/2)$. We saw that U is an open set in Y. We may take $V = (-1/2, 1/2)$.

Proof. Suppose $V \subset X$ is open and $V \cap Y = U$. Let $x \in U$. As V is open and $x \in V$, there exists a $\delta > 0$ such that $B_X(x, \delta) \subset V$. Then

$$B_Y(x, \delta) = B_X(x, \delta) \cap Y \subset V \cap Y = U.$$

The proof of the opposite direction, that is, that if $U \subset Y$ is open in the subspace topology there exists a V is left as an exercise. □

Exercise A.2.4: *Finish the proof of Proposition A.2.10. Suppose (X, d) is a metric space and $Y \subset X$. Show that with the subspace metric on Y, if a set $U \subset Y$ is open (in Y), then there exists an open set $V \subset X$ such that $U = V \cap Y$.*

For an open subset of an open set or a closed subset of a closed set, matters are simpler.

Proposition A.2.11. *Suppose (X, d) is a metric space, $V \subset X$ is open, and $E \subset X$ is closed.*

(i) $U \subset V$ *is open in the subspace topology if and only if U is open in X.*

(ii) $F \subset E$ *is closed in the subspace topology if and only if F is closed in X.*

Proof. Let us prove (i) and leave (ii) to an exercise.

If $U \subset V$ is open in the subspace topology, by Proposition A.2.10, there exists a set $W \subset X$ open in X, such that $U = W \cap V$. Intersection of two open sets is open so U is open in X.

Now suppose U is open in X, then $U = U \cap V$. So U is open in V again by Proposition A.2.10. □

Exercise A.2.5: *Finish the proof of Proposition A.2.11.*

Exercise A.2.6: *Show that in any metric space, every open set can be written as a union of closed sets.*

Exercise A.2.7: *Let X be a set and d, d' be two metrics on X. Suppose there exists an $\alpha > 0$ and $\beta > 0$ such that $\alpha d(x, y) \leq d'(x, y) \leq \beta d(x, y)$ for all $x, y \in X$. Show that $U \subset X$ is open in (X, d) if and only if U is open in (X, d'). That is, the topologies of (X, d) and (X, d') are the same.*

Exercise A.2.8: *Let (X, d) be a metric space.*
   a) For every $x \in X$ and $\delta > 0$, show $\overline{B(x, \delta)} \subset C(x, \delta)$.
   b) Is it always true that $\overline{B(x, \delta)} = C(x, \delta)$? Prove or find a counterexample.

> **Exercise A.2.9:** *Let (X, d) be a metric space. Show that there exists a bounded metric d' such that (X, d') has the same open sets, that is, the topology is the same.*

> **Exercise A.2.10:** *For every $x \in \mathbb{R}^n$ and every $\delta > 0$ define the "rectangle" $R(x, \delta) = (x_1 - \delta, x_1 + \delta) \times (x_2 - \delta, x_2 + \delta) \times \cdots \times (x_n - \delta, x_n + \delta)$. Show that these sets generate the same open sets as the balls in standard metric. That is, show that a set $U \subset \mathbb{R}^n$ is open in the sense of the standard metric if and only if for every point $x \in U$, there exists a $\delta > 0$ such that $R(x, \delta) \subset U$.*

A.2.2*i*   Connected sets

A set is connected if we can continuously move from one point of it to another point without jumping. For example, an interval in $\mathbb{R}$. We usually study functions on connected sets.

Definition A.2.12. A nonempty* metric space (X, d) is *connected* if the only subsets of X that are both open and closed (so-called *clopen* subsets) are $\emptyset$ and X itself. If a nonempty (X, d) is not connected we say it is *disconnected*.

When we apply the term *connected* to a nonempty subset $A \subset X$, we mean that A with the subspace topology is connected.

In other words, a nonempty X is connected if whenever we write $X = X_1 \cup X_2$ where $X_1 \cap X_2 = \emptyset$ and X_1 and X_2 are open, then either $X_1 = \emptyset$ or $X_2 = \emptyset$. So to show X is disconnected, we find nonempty disjoint open sets X_1 and X_2 whose union is X. We state this idea as a proposition for subsets.

Proposition A.2.13. *Let (X, d) be a metric space. A nonempty set $S \subset X$ is disconnected if and only if there exist open sets U_1 and U_2 in X, such that $U_1 \cap U_2 \cap S = \emptyset$, $U_1 \cap S \neq \emptyset$, $U_2 \cap S \neq \emptyset$, and*

$$S = (U_1 \cap S) \cup (U_2 \cap S).$$

The proposition is illustrated in Figure A.5.

Proof. First suppose S is disconnected, that is, there are nonempty disjoint S_1 and S_2 that are open in S and $S = S_1 \cup S_2$. Proposition A.2.10 says there exist U_1 and U_2 that are open in X such that $U_1 \cap S = S_1$ and $U_2 \cap S = S_2$.

For the other direction start with the U_1 and U_2. Then $U_1 \cap S$ and $U_2 \cap S$ are open in S by Proposition A.2.10. Via the discussion before the proposition, S is disconnected. $\qquad\square$

---

*Some authors do not exclude the empty set from the definition, and the empty set would then be connected. We avoid the empty set for essentially the same reason why 1 is neither a prime nor a composite number: Our connected sets have exactly two clopen subsets and disconnected sets have more than two. The empty set has exactly one. We will not dwell on this technicality.

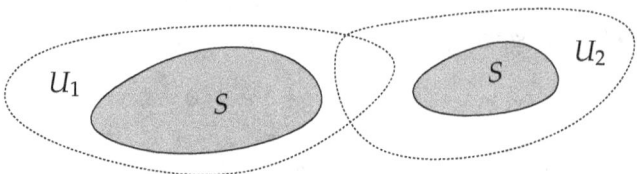

Figure A.5: Disconnected subset. Note that $U_1 \cap U_2$ need not be empty, but $U_1 \cap U_2 \cap S = \emptyset$.

Example A.2.14: Let $S \subset \mathbb{R}$ be such that $x < z < y$ with $x, y \in S$ and $z \notin S$. Claim: S is disconnected. Proof:

$$\big((-\infty, z) \cap S\big) \cup \big((z, \infty) \cap S\big) = S.$$

Proposition A.2.15. *A nonempty set $S \subset \mathbb{R}$ is connected if and only if it is an interval or a single point.*

Proof. Suppose S is connected. If S is a single point, then we are done. So suppose $x < y$ and $x, y \in S$. If $z \in \mathbb{R}$ is such that $x < z < y$, then by same logic as in Example A.2.14, $z \in S$. So S is an interval.

If S is a single point, it is connected. Therefore, suppose S is an interval. Consider open subsets U_1 and U_2 of $\mathbb{R}$, such that $U_1 \cap S$ and $U_2 \cap S$ are nonempty, and $S = (U_1 \cap S) \cup (U_2 \cap S)$. We will show that $U_1 \cap S$ and $U_2 \cap S$ contain a common point, so they are not disjoint, proving that S is connected. Suppose $x \in U_1 \cap S$ and $y \in U_2 \cap S$. Without loss of generality, assume $x < y$. As S is an interval, $[x, y] \subset S$. Note that $U_2 \cap [x, y] \neq \emptyset$, and let $z = \inf(U_2 \cap [x, y])$. If $z = x$, then $z \in U_1$. If $z > x$, then for every $\epsilon > 0$ the ball $B(z, \epsilon) = (z - \epsilon, z + \epsilon)$ contains points of $[x, y]$ not in U_2, as z is the infimum of such points. So $z \notin U_2$ as U_2 is open. Therefore, $z \in U_1$. As U_1 is open, $B(z, \delta) \subset U_1$ for a small enough $\delta > 0$. As z is the infimum of the nonempty set $U_2 \cap [x, y]$, there must exist some $w \in U_2 \cap [x, y]$ such that $w \in [z, z + \delta) \subset B(z, \delta) \subset U_1$. So $U_1 \cap S$ and $U_2 \cap S$ are not disjoint, and S is connected. $\qquad\square$

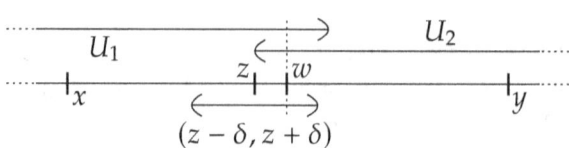

Figure A.6: Proof that an interval is connected.

Example A.2.16: The ball $B(x, \delta)$ may or may not be connected, depending on the metric space. Take the space $\{a, b\}$ with the discrete metric. The ball $B(a, 2) = \{a, b\}$ is not connected as $B(a, 1) = \{a\}$ and $B(b, 1) = \{b\}$ are open and disjoint.

Exercise A.2.11: *Suppose (X, d) is a nonempty metric space with the discrete topology. Show that X is connected if and only if it contains exactly one element.*

Exercise A.2.12: *Take $\mathbb{Q}$ with the standard metric, $d(x, y) = |x - y|$, as our metric space. Prove that $\mathbb{Q}$ is totally disconnected, that is, show that for every $x, y \in \mathbb{Q}$ with $x \neq y$, there exist two open sets U and V, such that $x \in U$, $y \in V$, $U \cap V = \emptyset$, and $U \cup V = \mathbb{Q}$.*

Exercise A.2.13: *Suppose $\{S_k\}$, $k \in \mathbb{N}$, is a collection of connected subsets of a metric space (X, d), and there exists an $x \in X$ such that $x \in S_k$ for all $k \in \mathbb{N}$. Show that $\bigcup_{k=1}^{\infty} S_k$ is connected.*

A.2.3i   Closure and boundary

Sometimes we wish to take a set and throw in everything that we can approach from within the set. This concept is called the closure. More precisely the closure of A is the intersection of all closed sets that contain A.

Definition A.2.17. Let (X, d) be a metric space and $A \subset X$. The *closure* of A is the set

$$\overline{A} \stackrel{\text{def}}{=} \bigcap \{E \subset X : E \text{ is closed and } A \subset E\}.$$

We say A is *dense* in X if $\overline{A} = X$.

Proposition A.2.18. *Let (X, d) be a metric space and $A \subset X$. Then, $A \subset \overline{A}$ and $\overline{A}$ is closed. Furthermore, if A is closed, then $\overline{A} = A$.*

Proof. There is at least one closed set containing A, the set X itself, so $A \subset \overline{A}$. The closure is an intersection of closed sets, so $\overline{A}$ is closed. If A is closed, then A is a closed set that contains A and $\overline{A} \subset A$. So $A = \overline{A}$. $\square$

Example A.2.19: The closure of $(0, 1)$ in $\mathbb{R}$ is $[0, 1]$. Proof: If E is closed and contains $(0, 1)$, then $0, 1 \in E$. Thus $[0, 1] \subset E$. But $[0, 1]$ is also closed. Thus, $\overline{(0, 1)} = [0, 1]$.

Example A.2.20: Always notice what ambient metric space you are working with. If $X = (0, \infty)$, then the closure of $(0, 1)$ in $(0, \infty)$ is $(0, 1]$. Proof: Similarly as above $(0, 1]$ is closed in $(0, \infty)$ (why?). Any closed set E that contains $(0, 1)$ must contain 1 (why?). Therefore, $(0, 1] \subset E$, and hence $\overline{(0, 1)} = (0, 1]$ when working in $(0, \infty)$.

Let us justify the statement that the closure is everything that we can "approach" from the set.

Proposition A.2.21. *Let (X, d) be a metric space and $A \subset X$. Then $x \in \overline{A}$ if and only if for every $\delta > 0$, $B(x, \delta) \cap A \neq \emptyset$.*

Proof. We will prove the two contrapositives. First suppose $x \notin \bar{A}$. As $\bar{A}$ is closed, $B(x, \delta) \subset \bar{A}^c$ for some $\delta > 0$. Furthermore, $\bar{A}^c \subset A^c$, and hence $B(x, \delta) \cap A = \emptyset$.

On the other hand, suppose $B(x, \delta) \cap A = \emptyset$ for some $\delta > 0$. In other words, $A \subset B(x, \delta)^c$. As $B(x, \delta)^c$ is a closed set, as $x \notin B(x, \delta)^c$, and as $\bar{A}$ is the intersection of closed sets containing A, we have $x \notin \bar{A}$. $\qquad\square$

We also talk about the *interior* of a set (points we cannot approach from the complement) and the *boundary* of a set (points we can approach both from the set and its complement).

Definition A.2.22. Let (X, d) be a metric space and $A \subset X$. The *interior* of A is the set

$$A^\circ \overset{\text{def}}{=} \{x \in A : \text{there exists a } \delta > 0 \text{ such that } B(x, \delta) \subset A\}.$$

The *boundary* of A is the set

$$\partial A \overset{\text{def}}{=} \bar{A} \setminus A^\circ.$$

Example A.2.23: Consider $X = \mathbb{R}$ and $A = (0, 1]$. Then $\bar{A} = [0, 1]$, $A^\circ = (0, 1)$, and $\partial A = \{0, 1\}$.

Example A.2.24: Suppose $X = \{a, b\}$ with the discrete metric is the metric space and $A = \{a\}$. Then $\bar{A} = A^\circ = A$ and $\partial A = \emptyset$.

Proposition A.2.25. *Let (X, d) be a metric space and $A \subset X$. Then A° is open and ∂A is closed.*

Proof. Given $x \in A^\circ$, there is a $\delta > 0$ such that $B(x, \delta) \subset A$. If $z \in B(x, \delta)$, then as open balls are open, there is an $\epsilon > 0$ such that $B(z, \epsilon) \subset B(x, \delta) \subset A$. So $z \in A^\circ$. Therefore, $B(x, \delta) \subset A^\circ$, and A° is open.

As A° is open, then $\partial A = \bar{A} \setminus A^\circ = \bar{A} \cap (A^\circ)^c$ is closed. $\qquad\square$

The boundary is the set of points that are close to both the set and its complement. See Figure A.7 for a diagram of the next proposition.

Proposition A.2.26. *Let (X, d) be a metric space and $A \subset X$. Then $x \in \partial A$ if and only if for every $\delta > 0$, $B(x, \delta) \cap A$ and $B(x, \delta) \cap A^c$ are both nonempty.*

Proof. Suppose $x \in \partial A = \bar{A} \setminus A^\circ$ and let $\delta > 0$ be arbitrary. By Proposition A.2.21, $B(x, \delta)$ contains a point of A. If $B(x, \delta)$ contained no points of A^c, then x would be in A°. Hence $B(x, \delta)$ contains a point of A^c as well.

Suppose $x \notin \partial A$, so $x \notin \bar{A}$ or $x \in A^\circ$. If $x \notin \bar{A}$, then $B(x, \delta) \subset \bar{A}^c$ for some $\delta > 0$ as $\bar{A}$ is closed. So $B(x, \delta) \cap A$ is empty, because $\bar{A}^c \subset A^c$. If $x \in A^\circ$, then $B(x, \delta) \subset A$ for some $\delta > 0$, so $B(x, \delta) \cap A^c$ is empty. $\qquad\square$

The proposition above and Proposition A.2.21 give the following corollary.

Corollary A.2.27. *Let (X, d) be a metric space and $A \subset X$. Then $\partial A = \bar{A} \cap \overline{A^c}$.*

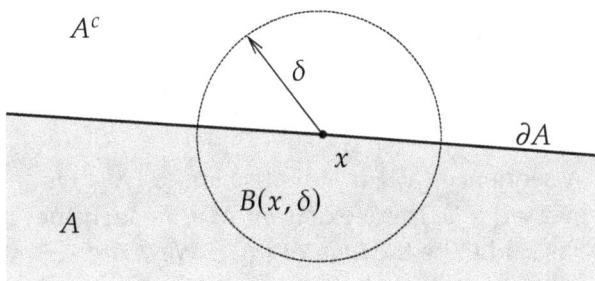

Figure A.7: Boundary is the set where every ball contains points in the set and also its complement.

Exercise **A.2.14:** *In any metric space, prove:*
    a) E *is closed if and only if* $\partial E \subset E$.
    b) U *is open if and only if* $\partial U \cap U = \emptyset$.

Exercise **A.2.15:** *In any metric space, prove:*
    a) Show that A *is open if and only if* $A^\circ = A$.
    b) Suppose that U *is an open set and* $U \subset A$. *Show that* $U \subset A^\circ$.

Exercise **A.2.16:** *Let* A *be a connected set in a metric space.*
    a) Is $\overline{A}$ *connected? Prove or find a counterexample.*
    b) Is A° *connected? Prove or find a counterexample.*
Hint: Think of sets in $\mathbb{R}^2$.

Exercise **A.2.17:** *Prove that* $A^\circ = \bigcup \{V : V \text{ is open and } V \subset A\}$.

A.3*i*   Sequences and convergence

A.3.1*i*   Sequences

Definition A.3.1. A *sequence* in a metric space (X, d) is a function $x \colon \mathbb{N} \to X$. We write x_n for the nth element in the sequence and

$$\{x_n\} \quad \text{or} \quad \{x_n\}_{n=1}^{\infty}$$

for the entire sequence.

    A sequence $\{x_n\}$ is *bounded* if there exists a $p \in X$ and $B \in \mathbb{R}$ such that

$$d(p, x_n) \leq B \qquad \text{for all } n \in \mathbb{N}.$$

That is, the sequence $\{x_n\}$ is bounded whenever the set $\{x_n : n \in \mathbb{N}\}$ is bounded.

If $\{n_k\}_{k=1}^{\infty}$ is a sequence of natural numbers such that $n_{k+1} > n_k$ for all k, then the sequence $\{x_{n_k}\}_{k=1}^{\infty}$ is said to be a *subsequence* of $\{x_n\}$.

In what follows, we cheat a little and use the definite article in front of the word *limit* before we prove that the limit is unique.

Definition A.3.2. A sequence $\{x_n\}$ in a metric space (X, d) is said to *converge* to a point $p \in X$ if for every $\epsilon > 0$, there exists an $M \in \mathbb{N}$ such that $d(x_n, p) < \epsilon$ for all $n \geq M$. The point p is said to be the *limit* of $\{x_n\}$. We write

$$\lim_{n \to \infty} x_n \overset{\text{def}}{=} p.$$

A sequence that converges is *convergent*. Otherwise, the sequence is *divergent*. See Figure A.8 for an idea of the definition.

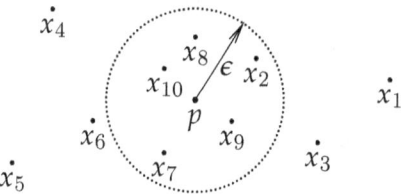

Figure A.8: Sequence converging to p. The first 10 points are shown and $M = 7$ for this ϵ.

Proposition A.3.3. *A convergent sequence in a metric space has a unique limit.*

Proof. Suppose the sequence $\{x_n\}$ has limits x and y. Take an arbitrary $\epsilon > 0$. From the definition find an n such that $d(x_n, x) < \epsilon/2$ and $d(x_n, y) < \epsilon/2$. Then

$$d(y, x) \leq d(y, x_n) + d(x_n, x) < \frac{\epsilon}{2} + \frac{\epsilon}{2} = \epsilon.$$

So $x = y$, and the limit (if it exists) is unique.   □

The proofs of the following propositions are left as exercises.

Proposition A.3.4. *A convergent sequence in a metric space is bounded.*

Proposition A.3.5. *A sequence $\{x_n\}$ in a metric space (X, d) converges to $p \in X$ if and only if there exists a sequence $\{a_n\}$ of real numbers such that*

$$d(x_n, p) \leq a_n \quad \text{for all } n \in \mathbb{N}, \qquad \text{and} \qquad \lim_{n \to \infty} a_n = 0.$$

Proposition A.3.6. *Let $\{x_n\}$ be a sequence in a metric space (X, d).*

(i) *If $\{x_n\}$ converges to $p \in X$, then every subsequence $\{x_{n_k}\}$ converges to p.*

(ii) *If for some $K \in \mathbb{N}$ the K-tail $\{x_n\}_{n=K+1}^{\infty}$ converges to $p \in X$, then $\{x_n\}$ converges to p.*

Exercise A.3.1: *Prove Proposition A.3.4.*

Exercise A.3.2: *Prove Proposition A.3.5.*

Exercise A.3.3: *Prove Proposition A.3.6.*

Example A.3.7: The set of continuous functions $C([a, b], \mathbb{R})$, see Example A.1.6, is a metric space. Convergence of a sequence of functions in this metric space is the same as uniform convergence. See also section B.1 in the next appendix.

Exercise A.3.4:
a) *Show that $d(x, y) = \min\{1, |x - y|\}$ defines a metric on $\mathbb{R}$.*
b) *Show that a sequence converges in $(\mathbb{R}, d)$ if and only if it converges in the standard metric.*
c) *Find a bounded sequence in $(\mathbb{R}, d)$ that contains no convergent subsequence.*

Exercise A.3.5: *Suppose $\{x_n\}_{n=1}^{\infty}$ converges to x. Suppose $f : \mathbb{N} \to \mathbb{N}$ is a one-to-one function. Show that $\{x_{f(n)}\}_{n=1}^{\infty}$ converges to x.*

Exercise A.3.6: *Let (X, d) be a metric space where d is the discrete metric. Suppose $\{x_n\}$ is a convergent sequence in X. Show that there exists a $K \in \mathbb{N}$ such that for all $n \geq K$, we have $x_n = x_K$.*

Exercise A.3.7: *A set $S \subset X$ is said to be dense in X if $X \subset \bar{S}$ or, in other words, if for every $x \in X$, there exists a sequence $\{x_n\}$ in S that converges to x. Prove that $\mathbb{R}^n$ contains a countable dense subset.*

Exercise A.3.8: *Take $\mathbb{R}^* = \{-\infty\} \cup \mathbb{R} \cup \{\infty\}$ to be the extended reals. Define $d(x, y) = \left| \frac{x}{1+|x|} - \frac{y}{1+|y|} \right|$ if $x, y \in \mathbb{R}$, define $d(\infty, x) = \left| 1 - \frac{x}{1+|x|} \right|$, $d(-\infty, x) = \left| 1 + \frac{x}{1+|x|} \right|$ for all $x \in \mathbb{R}$, and let $d(\infty, -\infty) = 2$.*
a) *Show that $(\mathbb{R}^*, d)$ is a metric space.*
b) *Suppose $\{x_n\}$ is a sequence of real numbers such that for every $M \in \mathbb{R}$, there exists an N such that $x_n \geq M$ for all $n \geq N$. Show that $\lim x_n = \infty$ in $(\mathbb{R}^*, d)$.*
c) *Show that a sequence of real numbers converges to a real number in $(\mathbb{R}^*, d)$ if and only if it converges in $\mathbb{R}$ with the standard metric.*

Exercise A.3.9: *Let (X, d) be a metric space and $\{x_n\}$ a sequence in X. Prove that $\{x_n\}$ converges to $p \in X$ if and only if every subsequence of $\{x_n\}$ has a subsequence that converges to p.*

A.3.2i Convergence in euclidean space

In $\mathbb{R}^n$, a sequence converges if and only if every component converges:

Proposition A.3.8. *Let $\{x_j\}_{j=1}^{\infty}$ be a sequence in $\mathbb{R}^n$, where $x_j = \left(x_{j,1}, x_{j,2}, \ldots, x_{j,n}\right) \in \mathbb{R}^n$. Then $\{x_j\}_{j=1}^{\infty}$ converges if and only if $\{x_{j,k}\}_{j=1}^{\infty}$ converges for every $k = 1, 2, \ldots, n$, in which case*

$$\lim_{j \to \infty} x_j = \left(\lim_{j \to \infty} x_{j,1}, \lim_{j \to \infty} x_{j,2}, \ldots, \lim_{j \to \infty} x_{j,n}\right).$$

Proof. Suppose the sequence $\{x_j\}_{j=1}^{\infty}$ converges to $y = (y_1, y_2, \ldots, y_n) \in \mathbb{R}^n$. Given $\epsilon > 0$, there exists an M, such that for all $j \geq M$,

$$d(y, x_j) < \epsilon.$$

Fix some $k = 1, 2, \ldots, n$. For $j \geq M$,

$$\left|y_k - x_{j,k}\right| = \sqrt{\left(y_k - x_{j,k}\right)^2} \leq \sqrt{\sum_{\ell=1}^{n} \left(y_\ell - x_{j,\ell}\right)^2} = d(y, x_j) < \epsilon.$$

Hence the sequence $\{x_{j,k}\}_{j=1}^{\infty}$ converges to y_k.

For the other direction, suppose $\{x_{j,k}\}_{j=1}^{\infty}$ converges to y_k for every $k = 1, 2, \ldots, n$. Given $\epsilon > 0$, pick an M, such that if $j \geq M$, then $\left|y_k - x_{j,k}\right| < \epsilon/\sqrt{n}$ for all $k = 1, 2, \ldots, n$. Then

$$d(y, x_j) = \sqrt{\sum_{k=1}^{n} \left(y_k - x_{j,k}\right)^2} < \sqrt{\sum_{k=1}^{n} \left(\frac{\epsilon}{\sqrt{n}}\right)^2} = \sqrt{\sum_{k=1}^{n} \frac{\epsilon^2}{n}} = \epsilon.$$

That is, the sequence $\{x_j\}$ converges to $y = (y_1, y_2, \ldots, y_n) \in \mathbb{R}^n$. □

Example A.3.9: For $\mathbb{C}$, the proposition says that $\{z_j\}_{j=1}^{\infty} = \{x_j + iy_j\}_{j=1}^{\infty}$ converges to $z = x + iy$ if and only if $\{x_j\}$ converges to x and $\{y_j\}$ converges to y.

Exercise A.3.10: *Consider $\mathbb{R}^n$, and let d be the standard euclidean metric. Let $d'(x, y) = \sum_{\ell=1}^{n} |x_\ell - y_\ell|$ and $d''(x, y) = \max\{|x_1 - y_1|, |x_2 - y_2|, \cdots, |x_n - y_n|\}$.*
   a) Use Exercise A.1.3, to show that $(\mathbb{R}^n, d')$ and $(\mathbb{R}^n, d'')$ are metric spaces.
   b) Let $\{x_j\}_{j=1}^{\infty}$ be a sequence in $\mathbb{R}^n$ and $p \in \mathbb{R}^n$. Prove that the following statements are equivalent:
      1) $\{x_j\}$ converges to p in $(\mathbb{R}^n, d)$.
      2) $\{x_j\}$ converges to p in $(\mathbb{R}^n, d')$.
      3) $\{x_j\}$ converges to p in $(\mathbb{R}^n, d'')$.

A.3.3i   Convergence and topology

The topology—the set of open sets of a space—encodes which sequences converge.

Proposition A.3.10. *Let (X, d) be a metric space and $\{x_n\}$ a sequence in X. Then $\{x_n\}$ converges to $x \in X$ if and only if for every open neighborhood U of x, there exists an $M \in \mathbb{N}$ such that for all $n \geq M$ we have $x_n \in U$.*

Proof. Suppose $\{x_n\}$ converges to x. Let U be an open neighborhood of x. There exists an $\epsilon > 0$ such that $B(x, \epsilon) \subset U$. As the sequence converges, find an $M \in \mathbb{N}$ such that for all $n \geq M$, we have $d(x, x_n) < \epsilon$, or in other words $x_n \in B(x, \epsilon) \subset U$.

Let us prove the other direction. Given $\epsilon > 0$, let $U = B(x, \epsilon)$ be the neighborhood of x. Then there is an $M \in \mathbb{N}$ such that for $n \geq M$, we have $x_n \in U = B(x, \epsilon)$, or in other words, $d(x, x_n) < \epsilon$. $\square$

A closed set contains the limits of its convergent sequences.

Proposition A.3.11. *Let (X, d) be a metric space, $E \subset X$ a closed set, and $\{x_n\}$ a sequence in E that converges to some $x \in X$. Then $x \in E$.*

Proof. Let us prove the contrapositive. Suppose $\{x_n\}$ is a sequence in X that converges to $x \in E^c$. As E^c is open, Proposition A.3.10 says that there is an M such that for all $n \geq M$, $x_n \in E^c$. So $\{x_n\}$ is not a sequence in E. $\square$

To take a closure of a set A, we take A, and we throw in points that are limits of sequences in A.

Proposition A.3.12. *Let (X, d) be a metric space and $A \subset X$. Then $x \in \overline{A}$ if and only if there exists a sequence $\{x_n\}$ of elements in A such that $\lim x_n = x$.*

Proof. Let $x \in \overline{A}$. For every $n \in \mathbb{N}$, by Proposition A.2.21 there exists a point $x_n \in B(x, 1/n) \cap A$. As $d(x, x_n) < 1/n$, we have $\lim x_n = x$.

For the other direction, see Exercise A.3.11. $\square$

Exercise A.3.11: *Finish the proof of Proposition A.3.12: Let (X, d) be a metric space and $A \subset X$. Let $x \in X$ be such that there exists a sequence $\{x_n\}$ in A that converges to x. Prove that $x \in \overline{A}$.*

Exercise A.3.12: *Suppose $\{U_n\}_{n=1}^{\infty}$ is a decreasing ($U_{n+1} \subset U_n$ for all n) sequence of open sets in a metric space (X, d) such that $\bigcap_{n=1}^{\infty} U_n = \{p\}$ for some $p \in X$. Suppose $\{x_n\}$ is a sequence of points in X such that $x_n \in U_n$. Does $\{x_n\}$ necessarily converge to p? Prove or construct a counterexample.*

Exercise A.3.13: *Let $E \subset X$ be closed and let $\{x_n\}$ be a sequence in X converging to $p \in X$. Suppose $x_n \in E$ for infinitely many $n \in \mathbb{N}$. Show $p \in E$.*

Exercise A.3.14: *Suppose $\{V_n\}_{n=1}^{\infty}$ is a sequence of open sets in (X, d) such that $V_{n+1} \supset V_n$ for all n. Let $\{x_n\}$ be a sequence such that $x_n \in V_{n+1} \setminus V_n$ and suppose $\{x_n\}$ converges to $p \in X$. Show that $p \in \partial V$ where $V = \bigcup_{n=1}^{\infty} V_n$.*

A.4i \ Completeness and compactness

A.4.1i \ Cauchy sequences and completeness

Definition A.4.1. Let (X, d) be a metric space. A sequence $\{x_n\}$ in X is a *Cauchy sequence* if for every $\epsilon > 0$, there exists an $M \in \mathbb{N}$ such that for all $n \geq M$ and all $k \geq M$, we have

$$d(x_n, x_k) < \epsilon.$$

Proposition A.4.2. *A convergent sequence in a metric space is Cauchy.*

Proof. Suppose $\{x_n\}$ converges to x. Given $\epsilon > 0$, there is an M such that $d(x, x_n) < \epsilon/2$ for all $n \geq M$. Hence, $d(x_n, x_k) \leq d(x_n, x) + d(x, x_k) < \epsilon/2 + \epsilon/2 = \epsilon$ for all $n, k \geq M$. $\square$

Definition A.4.3. Let (X, d) be a metric space. We say X is *complete* or *Cauchy-complete* if every Cauchy sequence $\{x_n\}$ in X converges to an $x \in X$.

Proposition A.4.4. *The space $\mathbb{R}^n$ with the standard metric is a complete metric space.*

We assume the reader has seen the proof of completeness in $\mathbb{R} = \mathbb{R}^1$, and we reduce the completeness in $\mathbb{R}^n$ to the one-dimensional case.

Proof. Let $\{x_j\}_{j=1}^{\infty}$ be a Cauchy sequence in $\mathbb{R}^n$, where $x_j = (x_{j,1}, x_{j,2}, \ldots, x_{j,n}) \in \mathbb{R}^n$. Given $\epsilon > 0$, there exists an M such that $d(x_i, x_j) < \epsilon$ for all $i, j \geq M$.

Fix some $k = 1, 2, \ldots, n$. For $i, j \geq M$,

$$\left| x_{i,k} - x_{j,k} \right| = \sqrt{\left(x_{i,k} - x_{j,k} \right)^2} \leq \sqrt{\sum_{\ell=1}^{n} \left(x_{i,\ell} - x_{j,\ell} \right)^2} = d(x_i, x_j) < \epsilon.$$

Hence the sequence $\{x_{j,k}\}_{j=1}^{\infty}$ is Cauchy. As $\mathbb{R}$ is complete, the sequence converges; there exists a $y_k \in \mathbb{R}$ such that $y_k = \lim_{j \to \infty} x_{j,k}$. Write $y = (y_1, y_2, \ldots, y_n) \in \mathbb{R}^n$. By Proposition A.3.8, $\{x_j\}$ converges to $y \in \mathbb{R}^n$, and hence $\mathbb{R}^n$ is complete. $\square$

A subset of $\mathbb{R}^n$ with the subspace metric need not be complete. For example, $(0, 1]$ with the subspace metric is not complete as $\{1/n\}$ is a Cauchy sequence in $(0, 1]$ with no limit in $(0, 1]$. However, once we have one complete metric space, any closed subspace is also a complete metric space. After all, one way to think of a closed set is that it contains all points that can be reached from the set via a sequence. The proof is again an exercise.

Proposition A.4.5. *Suppose (X, d) is a complete metric space and $E \subset X$ is closed. Then E is a complete metric space with the subspace topology.*

Exercise **A.4.1:** *Prove Proposition A.4.5.*

Example A.4.6: Another very useful example of a complete metric space is the space of continuous functions on a closed interval with the uniform norm, $C([a, b], \mathbb{R})$. See Corollary B.1.8 in the next appendix.

A.4.2i   Compactness

Definition A.4.7. Let (X, d) be a metric space and $K \subset X$. The set K is said to be *compact* if for any collection of open sets $\{U_\lambda\}_{\lambda \in I}$ such that

$$K \subset \bigcup_{\lambda \in I} U_\lambda,$$

there exists a finite subset $\{\lambda_1, \lambda_2, \dots, \lambda_k\} \subset I$ such that

$$K \subset U_{\lambda_1} \cup U_{\lambda_2} \cup \cdots \cup U_{\lambda_k}.$$

A collection of open sets $\{U_\lambda\}_{\lambda \in I}$ as above is said to be an *open cover* of K. A way to say that K is compact is to say that *every open cover of K has a finite subcover*.

Example A.4.8: Let $\mathbb{R}$ be the metric space with the standard metric.

The set $\mathbb{R}$ is not compact. Proof: Take the sets $U_n = (-n, n)$. It is an open cover, but the union of a finite subset of these sets is just $(-n, n)$ for some n.

The set $(0, 1) \subset \mathbb{R}$ is also not compact. Proof: Take the sets $U_n = (1/n, 1 - 1/n)$ for $n = 3, 4, 5, \dots$. As above $(0, 1) = \bigcup_{n=3}^{\infty} U_n$, but the union of finitely many is just U_n again and not all of $(0, 1)$.

The set $\{0\} \subset \mathbb{R}$ is compact. Proof: Given any open cover $\{U_\lambda\}_{\lambda \in I}$, there must exist a λ_0 such that $0 \in U_{\lambda_0}$ as it is a cover, so U_{λ_0} gives a finite subcover.

We will prove below that $[0, 1]$, and in fact any closed and bounded interval $[a, b]$ is compact.

> *Exercise A.4.2:* Let (X, d) be a metric space and A a finite subset of X. Show that A is compact.
>
> *Exercise A.4.3:* Let $A = \{1/n : n \in \mathbb{N}\} \subset \mathbb{R}$.
>   a) Show that A is not compact directly using the definition.
>   b) Show that $A \cup \{0\}$ is compact directly using the definition.
>
> *Exercise A.4.4:*
>   a) Show that the union of finitely many compact sets is a compact set.
>   b) Find an example where the union of infinitely many compact sets is not compact.

Proposition A.4.9. Let (X, d) be a metric space. A compact set $K \subset X$ is closed and bounded.

Proof. Let K be a compact set. Fix $p \in X$. We have the open cover

$$K \subset \bigcup_{n=1}^{\infty} B(p, n) = X.$$

If K is compact, then there exists some set of indices $n_1 < n_2 < \ldots < n_k$ such that

$$K \subset B(p, n_1) \cup B(p, n_2) \cup \cdots \cup B(p, n_k) = B(p, n_k).$$

So K is bounded. See left-hand side of Figure A.9.

Next, we show that a set that is not closed is not compact. Suppose $\overline{K} \neq K$, that is, there is a point $x \in \overline{K} \setminus K$. We have the open cover

$$K \subset \bigcup_{n=1}^{\infty} C(x, 1/n)^c.$$

If we take any finite collection of indices $n_1 < n_2 < \ldots < n_k$, then

$$C(x, 1/n_1)^c \cup C(x, 1/n_2)^c \cup \cdots \cup C(x, 1/n_k)^c = C(x, 1/n_k)^c$$

As x is in the closure of K, then $C(x, 1/n_k) \cap K \neq \emptyset$. So there is no finite subcover and K is not compact. See right-hand side of Figure A.9.   □

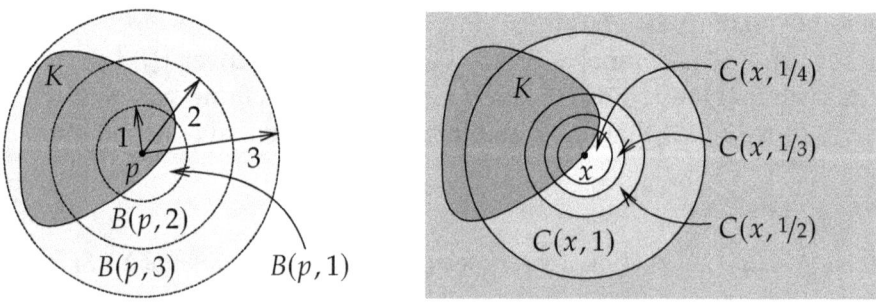

Figure A.9: Proving compact set is bounded (left) and closed (right).

We prove below that in a finite-dimensional euclidean space, every closed bounded set is compact. So closed bounded sets of $\mathbb{R}^n$ are examples of compact sets. It is not true that in every metric space, closed and bounded is equivalent to compact. A simple example is an incomplete metric space such as $(0, 1)$ with the subspace metric from $\mathbb{R}$. There are many complete and very useful metric spaces where closed and bounded is not enough to give compactness: $C([a, b], \mathbb{R})$ is a complete metric space, but the closed unit ball $C(0, 1)$ is not compact, see Exercise A.4.9. However, see also Exercise A.4.11. As this issue is such a common mistake, let me repeat it in italic: *Closed and bounded is not the same as compact.*

A useful property of compact sets in a metric space is that every sequence in the set has a convergent subsequence converging to a point in the set. Such sets are called *sequentially compact*. Let us prove that in the context of metric spaces, a set is compact if and only if it is sequentially compact. First we prove a lemma.

Lemma A.4.10 (Lebesgue covering lemma*). *Let (X, d) be a metric space and $K \subset X$. Suppose every sequence in K has a subsequence convergent in K. Given an open cover $\{U_\lambda\}_{\lambda \in I}$ of K, there exists a $\delta > 0$ such that for every $x \in K$, there exists a $\lambda \in I$ with $B(x, \delta) \subset U_\lambda$.*

Proof. We prove the lemma by contrapositive. If the conclusion is not true, then there is an open cover $\{U_\lambda\}_{\lambda \in I}$ of K with the following property. For every $n \in \mathbb{N}$ there exists an $x_n \in K$ such that $B(x_n, 1/n)$ is not a subset of any U_λ. Take any $x \in K$. There is a $\lambda \in I$ such that $x \in U_\lambda$. As U_λ is open, there is an $\epsilon > 0$ such that $B(x, \epsilon) \subset U_\lambda$. Take M such that $1/M < \epsilon/2$. If $y \in B(x, \epsilon/2)$ and $n \geq M$, then

$$B(y, 1/n) \subset B(y, 1/M) \subset B(y, \epsilon/2) \subset B(x, \epsilon) \subset U_\lambda,$$

where $B(y, \epsilon/2) \subset B(x, \epsilon)$ follows by triangle inequality. See Figure A.10. Thus $y \neq x_n$. In other words, for all $n \geq M$, $x_n \notin B(x, \epsilon/2)$. The sequence cannot have a subsequence converging to x. As $x \in K$ was arbitrary we are done. $\qquad\square$

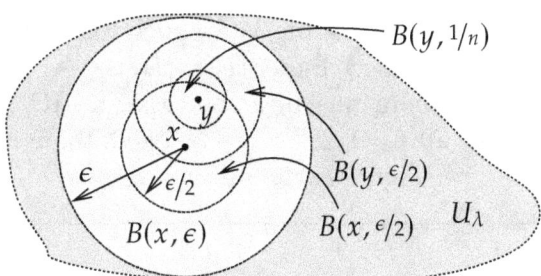

Figure A.10: Proof of Lebesgue covering lemma. Note that $B(y, \epsilon/2) \subset B(x, \epsilon)$ by triangle inequality.

It is important to recognize what the lemma says. It says that if K is sequentially compact, then for any given cover, there is a single $\delta > 0$. The δ depends on the cover, but, of course, it does not depend on x.

For example, let $K = [-10, 10]$ and for $n \in \mathbb{Z}$ let $U_n = (n, n + 2)$ define an open cover. Take $x \in K$. There is an $n \in \mathbb{Z}$, such that $n \leq x < n + 1$. If $n \leq x < n + 1/2$, then $B(x, 1/2) \subset U_{n-1}$. If $n + 1/2 \leq x < n + 1$, then $B(x, 1/2) \subset U_n$. So $\delta = 1/2$. If instead we take the open cover by $U_n' = \left(\frac{n}{2}, \frac{n+2}{2}\right)$, the best δ is $1/4$.

Theorem A.4.11. *Let (X, d) be a metric space. Then $K \subset X$ is compact if and only if every sequence in K has a subsequence converging to a point in K.*

Proof. Claim: *Let $K \subset X$ be a subset of X and $\{x_n\}$ a sequence in K. Suppose that for each $x \in K$, there is a ball $B(x, \alpha_x)$ for some $\alpha_x > 0$ such that $x_n \in B(x, \alpha_x)$ for only finitely many $n \in \mathbb{N}$. Then K is not compact.*

---

*The number δ is sometimes called the Lebesgue number of the cover.

Proof of the claim: Notice

$$K \subset \bigcup_{x \in K} B(x, \alpha_x).$$

Any finite collection of these balls contains at most finitely many elements of $\{x_n\}$, and so there must be an $x_n \in K$ not in their union. Therefore, K is not compact and the claim is proved.

Suppose K is compact and $\{x_n\}$ is a sequence in K. Then there exists an $x \in K$ such that for any $\delta > 0$, $B(x, \delta)$ contains x_k for infinitely many $k \in \mathbb{N}$. The ball $B(x, 1)$ contains some x_k, so let $n_1 = k$. Suppose n_{j-1} is defined. There must exist an $\ell > n_{j-1}$ such that $x_\ell \in B(x, 1/j)$. Define $n_j = \ell$. We now possess a subsequence $\{x_{n_j}\}_{j=1}^{\infty}$. Since $d(x, x_{n_j}) < 1/j$, Proposition A.3.5 says $\lim x_{n_j} = x$.

For the other direction, suppose every sequence in K has a subsequence converging in K. Take an open cover $\{U_\lambda\}_{\lambda \in I}$ of K. Using the Lebesgue covering lemma above, find a $\delta > 0$ such that for every $x \in K$, there is a $\lambda \in I$ with $B(x, \delta) \subset U_\lambda$.

Pick $x_1 \in K$ and find $\lambda_1 \in I$ such that $B(x_1, \delta) \subset U_{\lambda_1}$. If $K \subset U_{\lambda_1}$, we stop as we have found a finite subcover. Otherwise, there must be a point $x_2 \in K \setminus U_{\lambda_1}$. Note that $d(x_2, x_1) \geq \delta$. There must exist some $\lambda_2 \in I$ such that $B(x_2, \delta) \subset U_{\lambda_2}$. We work inductively. Suppose λ_{n-1} is defined. Either $U_{\lambda_1} \cup U_{\lambda_2} \cup \cdots \cup U_{\lambda_{n-1}}$ is a finite cover of K, in which case we stop, or there must be a point $x_n \in K \setminus (U_{\lambda_1} \cup U_{\lambda_2} \cup \cdots \cup U_{\lambda_{n-1}})$. Note that $d(x_n, x_j) \geq \delta$ for all $j = 1, 2, \ldots, n - 1$. Next, there must be some $\lambda_n \in I$ such that $B(x_n, \delta) \subset U_{\lambda_n}$. See Figure A.11.

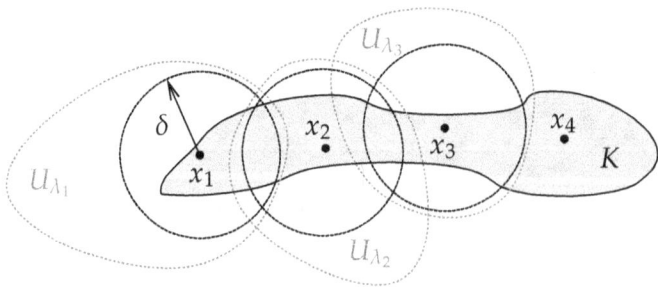

Figure A.11: Covering K by U_λ. The points x_1, x_2, x_3, x_4, the three sets $U_{\lambda_1}, U_{\lambda_2}, U_{\lambda_2}$, and the first three balls of radius δ are drawn.

Either at some point we obtain a finite subcover of K, or we obtain an infinite sequence $\{x_n\}$ as above. For contradiction, suppose that there is no finite subcover and we have the sequence $\{x_n\}$. For all n and k, $n \neq k$, we have $d(x_n, x_k) \geq \delta$, so no subsequence of $\{x_n\}$ can be Cauchy. Hence, no subsequence of $\{x_n\}$ can be convergent, which is a contradiction.   □

Example A.4.12: The Bolzano–Weierstrass theorem for sequences of real numbers says that a bounded sequence in $\mathbb{R}$ has a convergent subsequence. Therefore, any sequence in a closed interval $[a, b] \subset \mathbb{R}$ has a convergent subsequence. The limit must also be in $[a, b]$ as limits preserve non-strict inequalities. Hence a closed bounded interval $[a, b] \subset \mathbb{R}$ is compact.

Proposition A.4.13. *Let (X, d) be a metric space and let $K \subset X$ be compact. If $E \subset K$ is a closed set, then E is compact.*

Proof. Because K is closed, E is closed in K if and only if it is closed in X. See Proposition A.2.11. Let $\{x_n\}$ be a sequence in E. It is also a sequence in K. Therefore, it has a convergent subsequence $\{x_{n_j}\}$ that converges to some $x \in K$. As E is closed, the limit of a sequence in E is also in E and so $x \in E$. Thus E must be compact. $\qquad \square$

Theorem A.4.14 (Heine–Borel). *A closed bounded subset $K \subset \mathbb{R}^n$ is compact.*

So subsets of $\mathbb{R}^n$ are compact if and only if they are closed and bounded, a condition that is much easier to check. Let us reiterate that the Heine–Borel theorem only holds for $\mathbb{R}^n$ and not for metric spaces in general. In general, compact implies closed and bounded, but not vice versa.

Proof. For $\mathbb{R} = \mathbb{R}^1$ if $K \subset \mathbb{R}$ is closed and bounded, then any sequence $\{x_k\}$ in K is bounded, so it has a convergent subsequence by the Bolzano–Weierstrass theorem. As K is closed, the limit of the subsequence must be an element of K. So K is compact.

Let us carry out the proof for $n = 2$ and leave arbitrary n as an exercise. As $K \subset \mathbb{R}^2$ is bounded, there exists a set $B = [a, b] \times [c, d] \subset \mathbb{R}^2$ such that $K \subset B$. We will show that B is compact. Then K, being a closed subset of a compact B, is also compact.

Let $\{(x_k, y_k)\}_{k=1}^{\infty}$ be a sequence in B. That is, $a \leq x_k \leq b$ and $c \leq y_k \leq d$ for all k. A bounded sequence of real numbers has a convergent subsequence, so there is a subsequence $\{x_{k_j}\}_{j=1}^{\infty}$ that is convergent. The subsequence $\{y_{k_j}\}_{j=1}^{\infty}$ is also a bounded sequence so there exists a subsequence $\{y_{k_{j_\ell}}\}_{\ell=1}^{\infty}$ that is convergent. A subsequence of a convergent sequence is still convergent, so $\{x_{k_{j_\ell}}\}_{\ell=1}^{\infty}$ is convergent. Let

$$x = \lim_{\ell \to \infty} x_{k_{j_\ell}} \qquad \text{and} \qquad y = \lim_{\ell \to \infty} y_{k_{j_\ell}}.$$

By Proposition A.3.8, $\{(x_{k_{j_\ell}}, y_{k_{j_\ell}})\}_{\ell=1}^{\infty}$ converges to (x, y). As $a \leq x_k \leq b$ and $c \leq y_k \leq d$ for all k, we know that $(x, y) \in B$. $\qquad \square$

Exercise A.4.5: *Prove Theorem A.4.14 for arbitrary dimension. Hint: The trick is to use the correct notation.*

Proposition A.4.15. *Suppose (X, d) is a metric space and $E_1, E_2, \ldots,$ are nonempty compact subsets of X such that $E_1 \supset E_2 \supset E_3 \supset \cdots$. Then*

$$\bigcap_{k=1}^{\infty} E_k \neq \emptyset.$$

Proof. Suppose $E_1, E_2, \ldots$ are as in the statement except we do not assume they are nonempty. Compact sets are closed, so their complement is open. Consider $U_k = X \setminus E_k$. Suppose that the intersection is empty. Then $\{U_k\}$ is an open cover of E_1, which is compact, and hence there is a finite subcover. As the sets are nested, $U_\ell \subset U_{\ell+1}$ for all ℓ, we have $E_1 \subset U_k$ for some k. Thus E_k is empty.   □

Example A.4.16: Let (X, d) be a metric space with the discrete metric, that is, $d(x, y) = 1$ if $x \neq y$. Suppose X is an infinite set. Then:

   (i) (X, d) is a complete metric space.

   (ii) Any subset $K \subset X$ is closed and bounded.

   (iii) A subset $K \subset X$ is compact if and only if it is a finite set.

   (iv) The conclusion of the Lebesgue covering lemma is always satisfied with any $\delta \in (0, 1)$, even for noncompact $K \subset X$.

The proofs of the statements are either trivial or are relegated to the exercises below.

Remark A.4.17. A subtle point about Cauchy sequences, completeness, compactness, and convergence is that compactness and convergence only depend on the topology, that is, on which sets are the open sets. On the other hand, Cauchy sequences and completeness depend on the actual metric.

Exercise A.4.6: Let (X, d) be a metric space with the discrete metric.
  a) Prove that X is complete.
  b) Prove that X is compact if and only if X is a finite set.

Exercise A.4.7: Show that a compact set K (in any metric space) is itself a complete metric space (using the subspace metric).

Exercise A.4.8: Show that there exists a metric on $\mathbb{R}$ that makes $\mathbb{R}$ into a compact set.

Exercise A.4.9: Let $C([0, 1], \mathbb{R})$ be the metric space of Example A.1.6. Let 0 denote the zero function. Show that the closed ball $C(0, 1)$ is not compact (even though it is closed and bounded). Hint: Construct continuous functions $f_n : [0, 1] \to \mathbb{R}$ such that $d(f_n, 0) = 1$ and $d(f_n, f_k) = 1$ for all $n \neq k$.

Exercise A.4.10: Let $C([0, 1], \mathbb{R})$ be the metric space of Example A.1.6. Let K be the set of $f \in C([0, 1], \mathbb{R})$ such that f is equal to a quadratic polynomial, i.e., $f(x) = a + bx + cx^2$, and such that $|f(x)| \leq 1$ for all $x \in [0, 1]$, that is $f \in C(0, 1)$. Show that K is compact.

Exercise A.4.11: Let (X, d) be a complete metric space. Show that $K \subset X$ is compact if and only if K is closed and such that for every $\epsilon > 0$ there exists a finite set of points $x_1, x_2, \ldots, x_n$ with $K \subset \bigcup_{j=1}^{n} B(x_j, \epsilon)$. Note: Such a set K is said to be totally bounded, so in a complete metric space a set is compact if and only if it is closed and totally bounded.

Exercise **A.4.12:** *Take* $\mathbb{N} \subset \mathbb{R}$ *using the standard metric. Find an open cover of* $\mathbb{N}$ *such that the conclusion of the Lebesgue covering lemma does not hold.*

Exercise **A.4.13:** *Prove the general Bolzano–Weierstrass theorem: Any bounded sequence* $\{x_k\}$ *in* $\mathbb{R}^n$ *has a convergent subsequence.*

Exercise **A.4.14:** *Let* X *be a metric space and* C *the set of nonempty compact subsets of* X. *Using the Hausdorff metric from Exercise A.1.5, show that* (C, d_H) *is a metric space. That is, show that if* L *and* K *are nonempty compact subsets, then* $d_H(L, K) = 0$ *if and only if* $L = K$.

Exercise **A.4.15:** *Let* (X, d) *be an incomplete metric space. Show that there exists a closed and bounded set* $E \subset X$ *that is not compact.*

Exercise **A.4.16:** *Let* (X, d) *be a metric space and* $K \subset X$. *Prove that* K *is compact as a subset of* (X, d) *if and only if* K *is compact as a subset of itself with the subspace metric.*

Exercise **A.4.17:** *Let* (X, d) *be a complete metric space. We say a set* $S \subset X$ *is relatively compact if the closure* $\overline{S}$ *is compact. Prove that* $S \subset X$ *is relatively compact if and only if given any sequence* $\{x_n\}$ *in* S, *there exists a subsequence* $\{x_{n_k}\}$ *that converges (in* X*).*

A.5*i*   Continuous functions

A.5.1*i*   Continuity

Definition A.5.1. Let (X, d_X) and (Y, d_Y) be metric spaces and $c \in X$. Then $f \colon X \to Y$ is *continuous at* c if for every $\epsilon > 0$ there is a $\delta > 0$ such that whenever $x \in X$ and $d_X(x, c) < \delta$, then $d_Y\big(f(x), f(c)\big) < \epsilon$.

    If $f \colon X \to Y$ is continuous at all $c \in X$, then we say that f is a *continuous function*.

Proposition A.5.2. *Let* (X, d_X) *and* (Y, d_Y) *be metric spaces. Then* $f \colon X \to Y$ *is continuous at* $c \in X$ *if and only if for every sequence* $\{x_n\}$ *in* X *converging to* c, *the sequence* $\{f(x_n)\}$ *converges to* $f(c)$.

Proof. Suppose f is continuous at c. Let $\{x_n\}$ be a sequence in X converging to c. Given $\epsilon > 0$, there is a $\delta > 0$ such that $d_X(x, c) < \delta$ implies $d_Y\big(f(x), f(c)\big) < \epsilon$. So take M such that for all $n \geq M$, we have $d_X(x_n, c) < \delta$, and then $d_Y\big(f(x_n), f(c)\big) < \epsilon$. Hence $\{f(x_n)\}$ converges to $f(c)$.

    Now suppose f is not continuous at c. Then there exists an $\epsilon > 0$, such that for every $n \in \mathbb{N}$ there is an $x_n \in X$, with $d_X(x_n, c) < 1/n$ such that $d_Y\big(f(x_n), f(c)\big) \geq \epsilon$. So $\{x_n\}$ converges to c, but $\{f(x_n)\}$ does not converge to $f(c)$. $\qquad\square$

Example A.5.3: Suppose $f: \mathbb{R}^2 \to \mathbb{R}$ is a polynomial. That is,

$$f(x, y) = \sum_{k=0}^{d} \sum_{\ell=0}^{d-k} a_{k\ell} \, x^k y^\ell$$

$$= a_{00} + a_{10} \, x + a_{01} \, y + a_{20} \, x^2 + a_{11} \, xy + a_{02} \, y^2 + \cdots + a_{0d} \, y^d,$$

for some $d \in \mathbb{N}$ (the degree) and $a_{k\ell} \in \mathbb{R}$. Then we claim f is continuous. Let $\{(x_n, y_n)\}_{n=1}^{\infty}$ be a sequence in $\mathbb{R}^2$ that converges to $(x, y) \in \mathbb{R}^2$. Therefore, $\lim x_n = x$ and $\lim y_n = y$. Then

$$\lim_{n \to \infty} f(x_n, y_n) = \lim_{n \to \infty} \sum_{k=0}^{d} \sum_{\ell=0}^{d-k} a_{k\ell} \, x_n^k y_n^\ell = \sum_{k=0}^{d} \sum_{\ell=0}^{d-k} a_{k\ell} \, x^k y^\ell = f(x, y).$$

So f is continuous at (x, y), and as (x, y) was arbitrary, f is continuous everywhere. Similarly, a polynomial in n variables is continuous.

Be careful about taking limits separately. It is not enough that for every y, the function $g(x) = f(x, y)$ is continuous, and for every x, the function $h(y) = f(x, y)$ is continuous. The function $f(x, y)$ could still be discontinuous.

Exercise A.5.1: Let $f: \mathbb{R}^2 \to \mathbb{R}$ be defined by $f(0,0) = 0$, and $f(x, y) = \frac{xy}{x^2+y^2}$ if $(x, y) \neq (0, 0)$. See Figure A.12.
  a) Show that for each fixed x, the function that takes y to $f(x, y)$ is continuous. Similarly for each fixed y, the function that takes x to $f(x, y)$ is continuous.
  b) Show that f is not continuous.

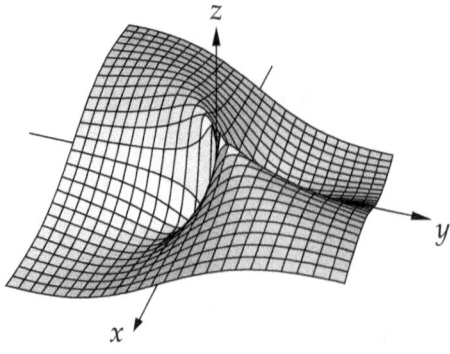

Figure A.12: Graph of $\frac{xy}{x^2+y^2}$.

Example A.5.4: Consider $f : X \to \mathbb{C}$ on a metric space X. Write $f(p) = u(p) + iv(p)$, where $u : X \to \mathbb{R}$ and $v : X \to \mathbb{R}$ are the real and imaginary parts. Then f is continuous at $c \in X$ if and only if its real and imaginary parts are continuous at c. This fact follows because $\{f(p_n) = u(p_n) + iv(p_n)\}_{n=1}^{\infty}$ converges to $f(p) = u(p) + iv(p)$ if and only if $\{u(p_n)\}$ converges to $u(p)$ and $\{v(p_n)\}$ converges to $v(p)$.

Proposition A.5.5. *Let (X, d) be a metric space.*

  (i) If $p \in X$, then $f : X \to \mathbb{R}$ defined by $f(x) = d(x, p)$ is continuous.

  (ii) Given a nonempty set $S \subset X$, the function

$$f(x) = \inf_{p \in S} d(x, p)$$

  is continuous.

Proof. The reverse triangle inequality $|f(x) - f(y)| = |d(x, p) - d(y, p)| \le d(x, y)$ gives part (i).

  For (ii), S being nonempty implies that f is real-valued. Take $x, y \in X$ and assume without loss of generality that $f(x) > f(y)$. For any $\epsilon > 0$, there exists a q such that $\inf_{p \in S} d(y, p) \ge d(y, q) - \epsilon$. Using this inequality together with $\inf_{p \in S} d(x, p) \le d(x, q)$ and the reverse triangle inequality gives

$$f(x) - f(y) = \inf_{p \in S} d(x, p) - \inf_{p \in S} d(y, p) \le d(x, q) - d(y, q) + \epsilon \le d(x, y) + \epsilon.$$

Since the inequality holds for every ϵ, f is continuous. $\qquad\qquad\square$

Exercise A.5.2: *Take the metric space of continuous functions $C([0, 1], \mathbb{R})$. Let $k : [0, 1] \times [0, 1] \to \mathbb{R}$ be a continuous function. Given $f \in C([0, 1], \mathbb{R})$ define*

$$\varphi_f(x) = \int_0^1 k(x, y) f(y) \, dy.$$

  a) Show that $T(f) = \varphi_f$ defines a function $T : C([0, 1], \mathbb{R}) \to C([0, 1], \mathbb{R})$.
  b) Show that T is continuous.

Exercise A.5.3: *Let (X, d) be a metric space. Define a metric on $X \times X$ as in Exercise A.1.3 part b, and show that $g : X \times X \to \mathbb{R}$ defined by $g(x, y) = d(x, y)$ is continuous.*

Exercise A.5.4: *Let $C([a, b], \mathbb{R})$ be the set of continuous functions and $C^1([a, b], \mathbb{R})$ the set of once continuously differentiable functions on $[a, b]$. Define*

$$d_C(f, g) = \|f - g\|_S \qquad and \qquad d_{C^1}(f, g) = \|f - g\|_S + \|f' - g'\|_S,$$

where $\|\cdot\|_S$ is the uniform norm. By Example A.1.6 and Exercise A.1.8, $C([a, b], \mathbb{R})$ with d_C is a metric space and so is $C^1([a, b], \mathbb{R})$ with d_{C^1}.
  a) Prove that the derivative operator $D : C^1([a, b], \mathbb{R}) \to C([a, b], \mathbb{R})$ defined by $D(f) = f'$ is continuous.
  b) On the other hand, if we consider the metric d_C on $C^1([a, b], \mathbb{R})$, then prove the derivative operator is no longer continuous. Hint: Consider $\sin(nx)$.

Exercise **A.5.5:** *Define*

$$f(x, y) = \begin{cases} \frac{2xy}{x^4+y^2} & \text{if } (x, y) \neq (0, 0), \\ 0 & \text{if } (x, y) = (0, 0). \end{cases}$$

a) *Show that for every fixed y the function that takes x to $f(x, y)$ is continuous and hence Riemann integrable.*

b) *For every fixed x, the function that takes y to $f(x, y)$ is continuous.*

c) *Show that f is not continuous at $(0, 0)$.*

d) *Now show that $g(y) = \int_0^1 f(x, y)\, dx$ is not continuous at $y = 0$.*

Note: *Feel free to use what you know about* arctan *from calculus, in particular that* $\frac{d}{ds}[\arctan(s)] = \frac{1}{1+s^2}$.

A.5.2*i*   Compactness and continuity

Continuous maps do not map closed sets to closed sets. For example, $f : (0, 1) \to \mathbb{R}$ defined by $f(x) = x$ takes the set $(0, 1)$, which is closed in $(0, 1)$, to the set $(0, 1)$, which is not closed in $\mathbb{R}$. On the other hand, continuous maps do preserve compact sets.

Lemma A.5.6. *Let (X, d_X) and (Y, d_Y) be metric spaces and $f : X \to Y$ a continuous function. If $K \subset X$ is a compact set, then $f(K)$ is a compact set.*

Proof. Write a sequence in $f(K)$ as $\{f(x_n)\}_{n=1}^\infty$, where $\{x_n\}_{n=1}^\infty$ is a sequence in K. The set K is compact, so there is a subsequence $\{x_{n_\ell}\}_{\ell=1}^\infty$ that converges to some $x \in K$. By continuity,

$$\lim_{\ell \to \infty} f(x_{n_\ell}) = f(x) \in f(K).$$

So every sequence in $f(K)$ has a subsequence convergent to a point in $f(K)$, and $f(K)$ is compact by Theorem A.4.11. $\qquad\square$

As before, $f : X \to \mathbb{R}$ achieves an *absolute minimum* at $c \in X$ if

$$f(x) \geq f(c) \qquad \text{for all } x \in X.$$

On the other hand, f achieves an *absolute maximum* at $c \in X$ if

$$f(x) \leq f(c) \qquad \text{for all } x \in X.$$

Theorem A.5.7. *Let (X, d) be a nonempty compact metric space and $f : X \to \mathbb{R}$ continuous. Then f achieves an absolute minimum and maximum on X. In particular, f is bounded.*

Proof. As X is compact and f is continuous, $f(X) \subset \mathbb{R}$ is compact. Hence $f(X)$ is closed and bounded. In particular, $\sup f(X) \in f(X)$ and $\inf f(X) \in f(X)$, because both the sup and the inf can be achieved by sequences in $f(X)$ and $f(X)$ is closed. Therefore, there is some $x \in X$ such that $f(x) = \sup f(X)$ and some $y \in X$ such that $f(y) = \inf f(X)$. $\qquad\square$

Exercise A.5.6: *Let (X, d) be a metric space. Use Exercise A.5.3 to prove that if K_1 and K_2 are compact subsets of X, then there exists a $p \in K_1$ and $q \in K_2$ such that $d(p, q)$ is minimal, that is, $d(p, q) = \inf\{d(x, y) \colon x \in K_1, y \in K_2\}$.*

Exercise A.5.7: *Let (X, d) be a compact metric space, let $C(X, \mathbb{R})$ be the set of real-valued continuous functions. Define*

$$d(f, g) = \|f - g\|_S = \sup_{x \in X} |f(x) - g(x)|.$$

a) Show that d makes $C(X, \mathbb{R})$ into a metric space.
b) Show that for each $x \in X$, the evaluation function $E_x \colon C(X, \mathbb{R}) \to \mathbb{R}$ defined by $E_x(f) = f(x)$ is a continuous function.

A.5.3*i*   Continuity and topology

Let us see how to define continuity in terms of the topology, that is, the open sets. We have already seen that topology determines which sequences converge, and so it is no wonder that the topology also determines continuity of functions.

Lemma A.5.8. *Let (X, d_X) and (Y, d_Y) be metric spaces. A function $f \colon X \to Y$ is continuous at $c \in X$ if and only if for every open neighborhood U of $f(c)$ in Y, the set $f^{-1}(U)$ contains an open neighborhood of c in X. See Figure A.13.*

In other words, $f^{-1}(U)$ is a not-necessarily-open neighborhood of c.

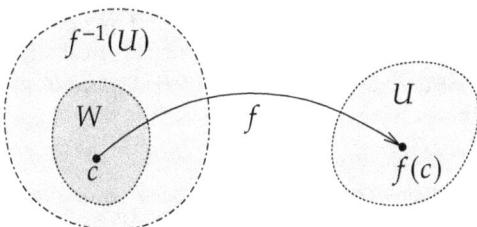

Figure A.13: For every neighborhood U of $f(c)$, the set $f^{-1}(U)$ contains an open neighborhood W of c.

Proof. First suppose that f is continuous at c. Let U be an open neighborhood of $f(c)$ in Y, then $B_Y(f(c), \epsilon) \subset U$ for some $\epsilon > 0$. By continuity of f, there exists a $\delta > 0$ such that whenever x is such that $d_X(x, c) < \delta$, then $d_Y(f(x), f(c)) < \epsilon$. In other words,

$$B_X(c, \delta) \subset f^{-1}(B_Y(f(c), \epsilon)) \subset f^{-1}(U),$$

and $B_X(c, \delta)$ is an open neighborhood of c.

For the other direction, let $\epsilon > 0$ be given. If $f^{-1}\big(B_Y(f(c),\epsilon)\big)$ contains an open neighborhood W of c, it contains a ball. That is, there is some $\delta > 0$ such that

$$B_X(c,\delta) \subset W \subset f^{-1}\big(B_Y(f(c),\epsilon)\big).$$

That means precisely that if $d_X(x,c) < \delta$ then $d_Y\big(f(x),f(c)\big) < \epsilon$, and so f is continuous at c.   □

Theorem A.5.9. *Let (X,d_X) and (Y,d_Y) be metric spaces. A function $f: X \to Y$ is continuous if and only if for every open $U \subset Y$, $f^{-1}(U)$ is open in X.*

The proof follows from Lemma A.5.8 and is left as an exercise.

Exercise A.5.8: Prove Theorem A.5.9. Hint: Use Lemma A.5.8.

Example A.5.10: Let $f: X \to Y$ be a continuous function. Theorem A.5.9 tells us that if $E \subset Y$ is closed, then $f^{-1}(E) = X \setminus f^{-1}(E^c)$ is also closed. Therefore, given a continuous $f: X \to \mathbb{R}$, the *zero set* of f, that is, $f^{-1}(0) = \{x \in X : f(x) = 0\}$, is closed.

The set where f is nonnegative, that is, $f^{-1}([0,\infty)) = \{x \in X : f(x) \geq 0\}$, is closed. On the other hand, the set where f is positive, $f^{-1}\big((0,\infty)\big) = \{x \in X : f(x) > 0\}$, is open.

Exercise A.5.9: Consider $\mathbb{N} \subset \mathbb{R}$ with the standard metric. Let (X,d) be a metric space and $f: X \to \mathbb{N}$ a continuous function.
  a) Prove that if X is connected, then f is constant (the range of f is a single value).
  b) Find an example where X is disconnected and f is not constant.

Exercise A.5.10: Suppose (X,d_X), (Y,d_Y) are metric spaces and $f: X \to Y$ is continuous. Let $A \subset X$.
  a) Show that $f(\overline{A}) \subset \overline{f(A)}$.
  b) Show that the subset can be proper.

Exercise A.5.11: Suppose $f: X \to Y$ is continuous for metric spaces (X,d_X) and (Y,d_Y). Show that if X is connected, then $f(X)$ is connected.

Exercise A.5.12: Prove the following version of the intermediate value theorem. Let (X,d) be a connected metric space and $f: X \to \mathbb{R}$ a continuous function. Suppose that there exist $x_0, x_1 \in X$ and $y \in \mathbb{R}$ such that $f(x_0) < y < f(x_1)$. Then prove that there exists a $z \in X$ such that $f(z) = y$. Hint: See Exercise A.5.11.

Exercise A.5.13: Let (X,d_X) and (Y,d_Y) be metric spaces and $f: X \to Y$ be a one-to-one and onto continuous function. Suppose X is compact. Prove that the inverse $f^{-1}: Y \to X$ is continuous.

A.5.4*i*   Uniform continuity

As for continuous functions on the real line, in the definition of continuity it is sometimes convenient to be able to pick one δ for all points.

Definition A.5.11. Let (X, d_X) and (Y, d_Y) be metric spaces. Then $f \colon X \to Y$ is *uniformly continuous* if for every $\epsilon > 0$ there is a $\delta > 0$ such that whenever $p, q \in X$ and $d_X(p, q) < \delta$, then $d_Y(f(p), f(q)) < \epsilon$.

A uniformly continuous function is continuous, but not necessarily vice versa. It is "vice versa" if X is compact.

Theorem A.5.12. *Let (X, d_X) and (Y, d_Y) be metric spaces. Suppose $f \colon X \to Y$ is continuous and X is compact. Then f is uniformly continuous.*

Proof. Let $\epsilon > 0$ be given. For each $c \in X$, pick $\delta_c > 0$ such that $d_Y(f(x), f(c)) < \epsilon/2$ whenever $x \in B(c, \delta_c)$. The balls $B(c, \delta_c)$ cover X, and the space X is compact. Apply the Lebesgue covering lemma to obtain a $\delta > 0$ such that for every $x \in X$, there is a $c \in X$ for which $B(x, \delta) \subset B(c, \delta_c)$.

If $p, q \in X$ where $d_X(p, q) < \delta$, find a $c \in X$ such that $B(p, \delta) \subset B(c, \delta_c)$. Then $q \in B(c, \delta_c)$. By the triangle inequality and the definition of δ_c,

$$d_Y(f(p), f(q)) \leq d_Y(f(p), f(c)) + d_Y(f(c), f(q)) < \epsilon/2 + \epsilon/2 = \epsilon. \qquad \square$$

Example A.5.13: Useful examples of uniformly continuous functions are the so-called *Lipschitz continuous* functions. That is, if (X, d_X) and (Y, d_Y) are metric spaces, then $f \colon X \to Y$ is called Lipschitz or K-Lipschitz if there exists a $K \in \mathbb{R}$ such that

$$d_Y(f(p), f(q)) \leq K d_X(p, q) \qquad \text{for all } p, q \in X.$$

A Lipschitz function is uniformly continuous: Take $\delta = \epsilon/K$. A function can be uniformly continuous but not Lipschitz: $\sqrt{x}$ on $[0, 1]$ is uniformly continuous but not Lipschitz (exercise).

It is worth mentioning that, if a function is Lipschitz, it tends to be easiest to simply show it is Lipschitz even if we are only interested in knowing continuity (or uniform continuity).

Exercise A.5.14: *Show that $\sqrt{x}$ is uniformly continuous on $[0, 1]$ but not Lipschitz.*

Exercise A.5.15:
   a) *Show that $f \colon (c, \infty) \to \mathbb{R}$ for some $c > 0$ defined by $f(x) = 1/x$ is Lipschitz continuous.*
   b) *Show that $f \colon (0, \infty) \to \mathbb{R}$ defined by $f(x) = 1/x$ is not Lipschitz continuous nor uniformly continuous.*

Exercise A.5.16: *Suppose $f \colon \mathbb{R} \to \mathbb{R}$ is a differentiable function such that f' is a bounded function. Prove f is a Lipschitz continuous function.*

Exercise A.5.17: *Prove that the map T defined in Exercise A.5.2 is Lipschitz continuous.*

Exercise A.5.18: *Let $f : \mathbb{R} \to \mathbb{R}$ be a polynomial of degree $d \geq 2$. Show that f is not Lipschitz continuous.*

A.5.5*i*   Cluster points and continuous limits

Definition A.5.14. Let (X, d) be a metric space and $S \subset X$. A point $p \in X$ is called a *cluster point* of S if for every $\epsilon > 0$, the set $B(p, \epsilon) \cap S \setminus \{p\}$ is not empty.

Definition A.5.15. Let (X, d_X), (Y, d_Y) be metric spaces, $S \subset X$, $p \in X$ a cluster point of S, and $f : S \to Y$ a function. Suppose there exists an $L \in Y$ and for every $\epsilon > 0$, there exists a $\delta > 0$ such that whenever $x \in S \setminus \{p\}$ and $d_X(x, p) < \delta$, then

$$d_Y\big(f(x), L\big) < \epsilon.$$

Then $f(x)$ *converges* to L as x goes to p, and L is the *limit* of $f(x)$ as x goes to p. We write

$$\lim_{x \to p} f(x) \overset{\text{def}}{=} L.$$

If $f(x)$ does not converge as x goes to p, we say f *diverges* at p.

We again used the definite article without showing that the limit is unique. We leave the proof of uniqueness as an exercise.

Proposition A.5.16. *Let (X, d_X) and (Y, d_Y) be metric spaces, $S \subset X$, $p \in X$ a cluster point of S, and let $f : S \to Y$ be a function such that $f(x)$ converges as x goes to p. Then the limit of $f(x)$ as x goes to p is unique.*

Exercise A.5.19: *Prove Proposition A.5.16.*

In a metric space, continuous limits may be replaced by sequential limits. We leave the proof as an exercise. The upshot is that we really only need to prove things for sequential limits.

Lemma A.5.17. *Let (X, d_X) and (Y, d_Y) be metric spaces, $S \subset X$, $p \in X$ a cluster point of S, and let $f : S \to Y$ be a function.*
Then $f(x)$ converges to $L \in Y$ as x goes to p if and only if for every sequence $\{x_n\}$ in $S \setminus \{p\}$ such that $\lim x_n = p$, the sequence $\{f(x_n)\}$ converges to L.

Exercise A.5.20: *Prove Lemma A.5.17.*

By applying Proposition A.5.2 or the definition directly we find (exercise) that for cluster points p of $S \subset X$, the function $f : S \to Y$ is continuous at p if and only if

$$\lim_{x \to p} f(x) = f(p).$$

Exercise A.5.21: *Let (X, d) be a metric space, $S \subset X$, and $p \in X$. Prove that p is a cluster point of S if and only if $p \in \overline{S \setminus \{p\}}$.*

Exercise A.5.22: *Let (X, d_X) and (Y, d_Y) be metric spaces, $S \subset X$, $p \in X$ a cluster point of S, and let $f : S \to Y$ be a function. Prove that $f : S \to Y$ is continuous at p if and only if $\lim_{x \to p} f(x) = f(p)$.*

B*i*   Results From Basic Analysis

I refuse to answer that question on the grounds that I don't know the answer.

—*Douglas Adams*

For this book, we assume as a prerequisite a basic knowledge of analysis on the real line. Let us, however, survey some basic results that the reader might not have seen in such a course, and that are useful in the text. Furthermore, we require some of these results in metric spaces and although their proofs are essentially the same as on the real line, it is worth it to put them down. The text is partly adapted from [L1] and [L2]. See those two texts for more details.

B.1*i*   Sequences of functions

B.1.1*i*   Pointwise and uniform convergence and the uniform norm

In the following, S is any set.

Definition B.1.1. The sequence $\{f_n\}_{n=1}^{\infty}$ of functions $f_n \colon S \to \mathbb{R}$ *converges pointwise* to $f \colon S \to \mathbb{R}$, if for every $x \in S$,
$$f(x) = \lim_{n \to \infty} f_n(x).$$

If we say $f_n \colon S \to \mathbb{R}$ *converges to f on $T \subset S$* we mean that the restrictions of f_n to T converge pointwise to f. As limits of sequences of numbers are unique, the limit function f is unique.

Pointwise convergence does not preserve much structure about f. For example, a pointwise limit of continuous functions is not continuous, see the exercises.

Definition B.1.2. Let $f_n \colon S \to \mathbb{R}$ and $f \colon S \to \mathbb{R}$ be functions. The sequence $\{f_n\}$ *converges uniformly* to f, if for every $\epsilon > 0$ there exists an $N \in \mathbb{N}$ such that for all $n \geq N$,
$$\left| f_n(x) - f(x) \right| < \epsilon \qquad \text{for all } x \in S.$$

In uniform convergence, N cannot depend on x. Given $\epsilon > 0$, we must find an N that works for all $x \in S$. See Figure B.1 for an illustration. It can easily be seen that uniform convergence implies pointwise convergence. The converse does not hold.

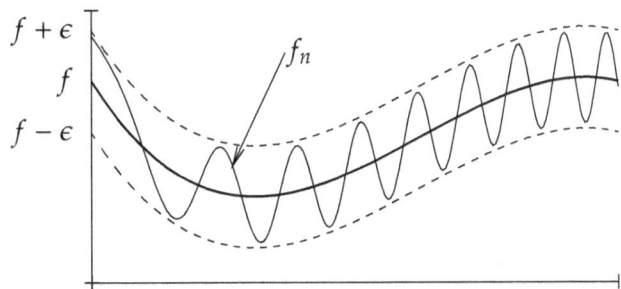

Figure B.1: In uniform convergence, for $n \geq N$, the functions f_n are within a strip of $\pm\epsilon$ from f.

Exercise B.1.1: *Let $f_n(x) = x^n$ be functions on $[0, 1]$.*
 a) *Show that $\{f_n\}$ converges pointwise to a discontinuous function.*
 b) *Prove that $\{f_n\}$ converges pointwise but not uniformly.*

Exercise B.1.2: *Suppose $f_n : S \to \mathbb{R}$ are functions that converge uniformly to $f : S \to \mathbb{R}$. Suppose $A \subset S$. Show that the sequence of restrictions $\{f_n|_A\}$ converges uniformly to $f|_A$.*

Exercise B.1.3:
 a) *Suppose $\{f_n\}$ and $\{g_n\}$ defined on some set A converge to f and g respectively pointwise, and let $a, b \in \mathbb{R}$. Show that $\{a f_n + b g_n\}$ converges pointwise to $a f + b g$.*
 b) *Show the same for uniform convergence.*

Exercise B.1.4: *Find an example of a sequence of functions $\{f_n\}$ and $\{g_n\}$ that converge uniformly to some f and g on some set A, but such that $\{f_n g_n\}$ (the multiple) does not converge uniformly to $f g$ on A.*

Exercise B.1.5: *Suppose there exists a sequence of functions $\{g_n\}$ uniformly converging to 0 on A. Now suppose we have a sequence of functions $\{f_n\}$ and a function f on A such that*

$$\left| f_n(x) - f(x) \right| \leq g_n(x)$$

for all $x \in A$. Show that $\{f_n\}$ converges uniformly to f on A.

Exercise B.1.6: *Let $\{f_n\}$, $\{g_n\}$ and $\{h_n\}$ be sequences of functions on some set S. Suppose $\{f_n\}$ and $\{h_n\}$ converge uniformly to some function $f : S \to \mathbb{R}$ and suppose $f_n(x) \leq g_n(x) \leq h_n(x)$ for all $x \in S$. Show that $\{g_n\}$ converges uniformly to f.*

Exercise B.1.7: *Prove that if a sequence of functions $f_n : S \to \mathbb{R}$ converge uniformly to a bounded function $f : S \to \mathbb{R}$, then there exists an N such that for all $n \geq N$, the f_n are bounded.*

Definition B.1.3. For $f : S \to \mathbb{R}$, define the *uniform norm*,

$$\|f\|_S \overset{\text{def}}{=} \sup\{|f(x)| : x \in S\}.$$

Note that if f is not bounded, then $\|f\|_S = \infty$. Therefore, unless dealing with bounded functions, we treat the norm as an extended real, so it is not what people would call a "norm" unless we restrict to bounded functions.

Proposition B.1.4. *A sequence* $f_n : S \to \mathbb{R}$ *converges uniformly to* $f : S \to \mathbb{R}$ *if and only if*

$$\lim_{n \to \infty} \|f_n - f\|_S = 0.$$

Exercise **B.1.8:** *Prove the proposition.*

We may say $\{f_n\}$ *converges to* f *in uniform norm* instead of *converges uniformly*. The proposition says that the two notions are the same thing. It is generally easiest to think about uniform convergence of functions using metric spaces. A Cauchy sequence of functions in the uniform norm is said to be *Cauchy in the uniform norm* or *uniformly Cauchy*.

Proposition B.1.5. *The set of bounded real-valued functions on* S *is a complete metric space with the metric* $d(f, g) = \|f - g\|_S$. *In particular, if a sequence is uniformly Cauchy, then it is uniformly convergent.*

Exercise **B.1.9:** *Prove the proposition. There are two things to prove. First, prove that the set is a metric space, that is, $d(f, g)$ is a metric. Second, prove that it is complete.*

Remark B.1.6. It is perhaps surprising that on the set of functions $f : S \to \mathbb{R}$ for an uncountable S, there is no metric that gives pointwise convergence. You could even require f to be bounded and/or continuous, and there is still no metric. A metric space (X, d) is so-called *first countable*, that is, at each $x \in X$ there exists a sequence of neighbourhoods U_j such that any neighbourhood U of x contains one of the U_js, a so-called countable neighborhood basis. In a metric space, $B(x, 1/n)$ does the job. But functions on an uncountable set S with pointwise convergence do not have a first countable topology. We do not want to wade too deep into general topology to prove this fact.

B.1.2*i* Continuity of the limit

If we have a sequence $\{f_n\}$ of continuous functions, is the limit continuous? We have seen that for pointwise convergence, it need not be the case, see Exercise B.1.1. If we, however, require the convergence to be uniform, the limits can be interchanged.

Theorem B.1.7. *Let S be a metric space. Let $\{f_n\}$ be a sequence of continuous functions $f_n \colon S \to \mathbb{R}$ converging uniformly to $f \colon S \to \mathbb{R}$. Then f is continuous.*

Proof. Let $x \in S$ be fixed. Let $\{x_n\}$ be a sequence in S converging to x. Let $\epsilon > 0$ be given. As $\{f_k\}$ converges uniformly to f, we find a $k \in \mathbb{N}$ such that

$$\left| f_k(y) - f(y) \right| < \epsilon/3$$

for all $y \in S$. As f_k is continuous at x, we find an $N \in \mathbb{N}$ such that for $m \geq N$ we have

$$\left| f_k(x_m) - f_k(x) \right| < \epsilon/3.$$

Thus for $m \geq N$,

$$
\begin{aligned}
\left| f(x_m) - f(x) \right| &= \left| f(x_m) - f_k(x_m) + f_k(x_m) - f_k(x) + f_k(x) - f(x) \right| \\
&\leq \left| f(x_m) - f_k(x_m) \right| + \left| f_k(x_m) - f_k(x) \right| + \left| f_k(x) - f(x) \right| \\
&< \epsilon/3 + \epsilon/3 + \epsilon/3 = \epsilon.
\end{aligned}
$$

Therefore, $\{f(x_m)\}$ converges to $f(x)$ and hence f is continuous at x. As x was arbitrary, f is continuous everywhere. □

In the language of metric spaces, as uniform limits of continuous functions are continuous, the set of bounded continuous functions is a complete metric space. The proof is left as an exercise. More precisely, let $C_b(S, \mathbb{R})$ denote the set of bounded real-valued continuous functions on S. We use the uniform norm as metric and $C_b(S, \mathbb{R})$ is a metric space for any S. If S is compact, then all continuous functions are bounded and $C(S, \mathbb{R})$ itself is a metric space.

Corollary B.1.8. *Let S be a metric space. Then $C_b(S, \mathbb{R})$ is a complete metric space. If S is compact, then $C(S, \mathbb{R})$ is a complete metric space.*

Exercise B.1.10: Prove Corollary B.1.8.

Definition B.1.9. A sequence of functions $f_n \colon S \to \mathbb{R}$ *converges uniformly on compact subsets* if for every compact $K \subset S$ the sequence $\{f_n\}$ converges uniformly on K.

Corollary B.1.10. *Let $U \subset \mathbb{R}^n$ be open. If $f_n \colon U \to \mathbb{R}$ is a sequence of continuous functions converging uniformly on compact subsets, then the limit is continuous.*

Exercise B.1.11: Prove the corollary.

B.1.3i   Integral of the limit

As with continuity, if we simply require pointwise convergence, then the integral of a limit of a sequence of functions need not be equal to the limit of the integrals.

Example B.1.11: Let χ_T be the characteristic function of a set T, that is, $\chi_T(x) = 1$ if $x \in T$ and $\chi_T(x) = 0$ otherwise. The functions $n\chi_{(0,1/n)}$ all integrate (on the interval $[0,1]$) to 1. Their pointwise limit is 0 (whose integral is 0).

If we require the convergence to be uniform, the limits can be interchanged.

Theorem B.1.12. *Let $\{f_n\}$ be a sequence of Riemann integrable functions $f_n \colon [a,b] \to \mathbb{R}$ converging uniformly to $f \colon [a,b] \to \mathbb{R}$. Then f is Riemann integrable and*

$$\int_a^b f(x)\,dx = \lim_{n\to\infty} \int_a^b f_n(x)\,dx.$$

In the following, let $\overline{\int_a^b} f(x)\,dx$ and $\underline{\int_a^b} f(x)\,dx$ denote the upper and lower Darboux integral. Briefly,

$$\overline{\int_a^b} f(t)\,dt \stackrel{\text{def}}{=} \inf\left\{ \int_a^b s(t)\,dt : s \text{ is a step function and } f(t) \le s(t) \text{ for } t \in [a,b]\right\},$$

$$\underline{\int_a^b} f(t)\,dt \stackrel{\text{def}}{=} \sup\left\{ \int_a^b s(t)\,dt : s \text{ is a step function and } s(t) \le f(t) \text{ for } t \in [a,b]\right\}.$$

The definition of the Riemann integral using Darboux sums and integrals is beyond the scope of this book, but let us just mention that if the upper and lower Darboux integrals are equal, then a function is Riemann integrable, and the common value is the integral. Given this fact, let us prove the theorem.

Proof. Let $\epsilon > 0$ be given. As f_n goes to f uniformly, we find an $M \in \mathbb{N}$ such that for all $n \ge M$, we have $\left| f_n(x) - f(x) \right| < \frac{\epsilon}{2(b-a)}$ for all $x \in [a,b]$. In particular, by reverse triangle inequality $\left| f(x) \right| < \frac{\epsilon}{2(b-a)} + \left| f_n(x) \right|$ for all x, hence f is bounded as f_n is bounded. Note that f_n is integrable and compute

$$\overline{\int_a^b} f(x)\,dx - \underline{\int_a^b} f(x)\,dx$$

$$= \overline{\int_a^b} \left(f(x) - f_n(x) + f_n(x) \right) dx - \underline{\int_a^b} \left(f(x) - f_n(x) + f_n(x) \right) dx$$

$$\le \overline{\int_a^b} \left(f(x) - f_n(x) \right) dx + \overline{\int_a^b} f_n(x)\,dx - \underline{\int_a^b} \left(f(x) - f_n(x) \right) dx - \int_a^b f_n(x)\,dx$$

$$= \overline{\int_a^b} \left(f(x) - f_n(x) \right) dx + \int_a^b f_n(x)\,dx - \underline{\int_a^b} \left(f(x) - f_n(x) \right) dx - \int_a^b f_n(x)\,dx$$

$$= \overline{\int_a^b} (f(x) - f_n(x)) \, dx - \underline{\int_a^b} (f(x) - f_n(x)) \, dx$$

$$\leq \frac{\epsilon}{2(b-a)}(b-a) + \frac{\epsilon}{2(b-a)}(b-a) = \epsilon.$$

The first inequality is due to the upper integral being only subadditive ($\overline{\int}(a+b) \leq \overline{\int}a + \overline{\int}b$) and the lower integral being superadditive. The final inequality follows from the fact that for all $x \in [a,b]$, we have $\frac{-\epsilon}{2(b-a)} < f(x) - f_n(x) < \frac{\epsilon}{2(b-a)}$. As $\epsilon > 0$ was arbitrary, f is Riemann integrable.

We compute $\int_a^b f(x) \, dx$. For $n \geq M$ (M is the same as above),

$$\left| \int_a^b f(x) \, dx - \int_a^b f_n(x) \, dx \right| = \left| \int_a^b (f(x) - f_n(x)) \, dx \right|$$

$$\leq \frac{\epsilon}{2(b-a)}(b-a) = \frac{\epsilon}{2} < \epsilon.$$

Therefore, $\left\{ \int_a^b f_n(x) \, dx \right\}$ converges to $\int_a^b f(x) \, dx$. □

Remark B.1.13. While we do not require the Lebesgue integral in this book, note that for the Lebesgue integral, a much stronger convergence theorem holds. In particular, the dominated convergence theorem implies that if $\{f_n\}$ is a sequence of measurable functions on $[a,b]$, converging pointwise to $f\colon [a,b] \to \mathbb{R}$, and such that $\{f_n\}$ is uniformly bounded (there is a single $B \in \mathbb{R}$ such that $\|f_n\|_{[a,b]} \leq B$ for all n), then

$$\int_a^b f(x) \, dx = \lim_{n \to \infty} \int_a^b f_n(x) \, dx.$$

Here, of course, the integrals must be the Lebesgue integrals, not the Riemann integrals. The pointwise limit of Riemann integrable functions need not even be Riemann integrable.

Exercise B.1.12: *Compute* $\lim\limits_{n \to \infty} \int_1^2 e^{-nx^2} \, dx$.

Exercise B.1.13: *Find a sequence of Riemann integrable functions* $f_n \colon [0,1] \to \mathbb{R}$ *such that* $\{f_n\}$ *converges to zero pointwise, and such that*
  a) $\left\{ \int_0^1 f_n(x) \, dx \right\}_{n=1}^{\infty}$ *increases without bound,*
  b) $\left\{ \int_0^1 f_n(x) \, dx \right\}_{n=1}^{\infty}$ *is the sequence* $-1, 1, -1, 1, -1, 1, \ldots$..

B.1.4*i*   Derivative of the limit

Uniform convergence is enough to swap limits with integrals. It is not, however, enough to swap limits with derivatives, unless the derivatives themselves converge uniformly.

Example B.1.14: The functions $f_n(x) = \frac{\sin(nx)}{n}$ converge uniformly to 0. See Figure B.2. The derivative of the limit is 0. But $f_n'(x) = \cos(nx)$, and that sequence does not converge even pointwise: For example, $f_n'(\pi) = (-1)^n$. Furthermore, $f_n'(0) = 1$ for all n, which does converge, but not to 0.

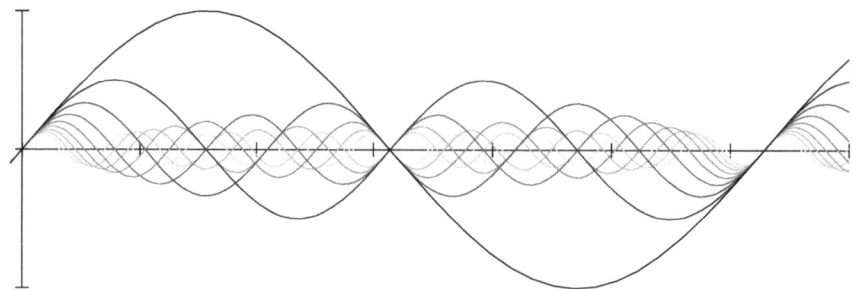

Figure B.2: Graphs of $\frac{\sin(nx)}{n}$ for $n = 1, 2, \dots, 10$, with higher n in lighter gray.

The following theorem is true even if we do not assume continuity of the derivatives, but the proof is more difficult.

Theorem B.1.15. *Let I be a bounded interval and let $f_n : I \to \mathbb{R}$ be continuously differentiable functions. Suppose $\{f_n'\}$ converges uniformly to $g : I \to \mathbb{R}$, and suppose $\{f_n(c)\}_{n=1}^{\infty}$ is a convergent sequence for some $c \in I$. Then $\{f_n\}$ converges uniformly to a continuously differentiable function $f : I \to \mathbb{R}$, and $f' = g$.*

Proof. Define $f(c) = \lim_{n \to \infty} f_n(c)$. As f_n' are continuous and hence Riemann integrable, then via the fundamental theorem of calculus, we find that for $x \in I$,

$$f_n(x) = f_n(c) + \int_c^x f_n'(t)\, dt.$$

As $\{f_n'\}$ converges uniformly on I, it converges uniformly on $[c, x]$ (or $[x, c]$ if $x < c$). Thus, the limit on the right-hand side exists. Define f at the remaining points by

$$f(x) = \lim_{n \to \infty} f_n(c) + \lim_{n \to \infty} \int_c^x f_n'(t)\, dt = f(c) + \int_c^x g(t)\, dt.$$

The function g is continuous, being the uniform limit of continuous functions. Hence, f is differentiable and $f'(x) = g(x)$ for all $x \in I$ by the fundamental theorem of calculus.

It remains to prove uniform convergence. Suppose I has a lower bound a and upper bound b. Let $\epsilon > 0$ be given. Take M such that for $n \geq M$, we have $|f(c) - f_n(c)| < \epsilon/2$ and $|g(x) - f_n'(x)| < \frac{\epsilon}{2(b-a)}$ for all $x \in I$. Then,

$$
\begin{aligned}
|f(x) - f_n(x)| &= \left| f(c) + \int_c^x g - f_n(c) - \int_c^x f_n'(t)\, dt \right| \\
&\leq |f(c) - f_n(c)| + \left| \int_c^x g(t)\, dt - \int_c^x f_n'(t)\, dt \right| \\
&= |f(c) - f_n(c)| + \left| \int_c^x (g(t) - f_n'(t))\, dt \right| \\
&< \frac{\epsilon}{2} + \frac{\epsilon}{2(b-a)}(b-a) = \epsilon. \qquad \square
\end{aligned}
$$

The proof goes through without boundedness of I, except for the uniform convergence of f_n to f. For an example let $I = \mathbb{R}$ and $f_n(x) = x/n$. Then $f_n'(x) = 1/n$, which converges uniformly to 0. However, $\{f_n\}$ converges to 0 only pointwise.

Example B.1.16: In Exercise A.1.8, you proved that the set of once continuously differentiable functions on $[a, b]$, that is, $C^1([a, b], \mathbb{R})$, is a metric space with the so-called C^1 metric (or C^1 norm)

$$
d(f, g) = \|f - g\|_{C^1([a,b],\mathbb{R})} \overset{\text{def}}{=} \|f - g\|_{[a,b]} + \|f' - g'\|_{[a,b]}.
$$

The theorem says that $C^1([a, b], \mathbb{R})$ is a complete metric space.

> **Exercise B.1.14:** *Find an explicit example of a sequence of differentiable functions on $[-1, 1]$ that converge uniformly to a function f such that f is not differentiable. Hint: Perhaps $\sqrt{x^2 + (1/n)^2}$?*
>
> **Exercise B.1.15:** *Let $f_n(x) = \frac{x^n}{n}$. Show that $\{f_n\}$ converges uniformly to a differentiable function f on $[0, 1]$ (find f). However, show that $f'(1) \neq \lim_{n \to \infty} f_n'(1)$.*

B.2i Continuity, Fubini, derivatives under the integral

B.2.1i Continuity

Let $f(x, y)$ be a function of two variables and define

$$
g(y) = \int_a^b f(x, y)\, dx.
$$

Question is: Is g continuous? We are really asking when do two limits commute, which is not always possible, so some extra hypothesis is necessary. A sufficient (but not necessary) condition is that f is continuous on a closed rectangle.

Proposition B.2.1. *If $f : [a,b] \times [c,d] \to \mathbb{R}$ is a continuous function, then $g : [c,d] \to \mathbb{R}$ defined by*

$$g(y) = \int_a^b f(x,y)\,dx \qquad \text{is continuous.}$$

Proof. Fix $y \in [c,d]$, and let $\{y_n\}$ be a sequence in $[c,d]$ converging to y. Let $\epsilon > 0$ be given. As f is continuous on the compact set $[a,b] \times [c,d]$, f is uniformly continuous. Thus, there exists a $\delta > 0$ such that whenever $\tilde{y} \in [c,d]$ and $|\tilde{y} - y| < \delta$, we have $|f(x,\tilde{y}) - f(x,y)| < \frac{\epsilon}{b-a}$ for all $x \in [a,b]$. Suppose $|\tilde{y} - y| < \delta$. Then

$$\left| g(\tilde{y}) - g(y) \right| = \left| \int_a^b f(x,\tilde{y})\,dx - \int_a^b f(x,y)\,dx \right|$$

$$= \left| \int_a^b \bigl(f(x,\tilde{y}) - f(x,y) \bigr)\,dx \right| \le (b-a) \frac{\epsilon}{b-a} = \epsilon. \qquad \square$$

In applications, if we are interested in continuity at y_0, we just need to apply the proposition in $[a,b] \times [y_0 - \epsilon, y_0 + \epsilon]$ for some small $\epsilon > 0$. For example, if f is continuous in $[a,b] \times \mathbb{R}$, then g is continuous on $\mathbb{R}$.

Exercise **B.2.1:** *Prove a stronger version of Proposition B.2.1: If $f : (a,b) \times (c,d) \to \mathbb{R}$ is a bounded continuous function, then $g : (c,d) \to \mathbb{R}$ defined by*

$$g(y) = \int_a^b f(x,y)\,dx \qquad \text{is continuous.}$$

Hint: First integrate over $[a + 1/n, b - 1/n]$.

B.2.2i Fubini's theorem

Fubini's theorem says that under some mild conditions one can generally swap the order of integrals in an iterated integral. We prove the following simple case of Fubini that is generally enough for the purposes of this book.

Theorem B.2.2 (Fubini). *Suppose $f : [a,b] \times [c,d] \to \mathbb{R}$ is continuous. Then*

$$\int_c^d \int_a^b f(x,y)\,dx\,dy = \int_a^b \int_c^d f(x,y)\,dy\,dx.$$

One of the tricky bits about Fubini for the Riemann integral is that the integrand of the outer integral, for example $\int_a^b f(x,y)\,dx$ as a function of y, is not necessarily Riemann integrable even if $f(x,y)$ is Riemann integrable as a function of two variables. However, by the previous subsection, if f is continuous, then $\int_a^b f(x,y)\,dx$ is a continuous function of y and hence integrable. So for continuous functions we sidestep the integrability issues.*

---

*For more complicated scenarios, the reader is encouraged to just learn the Lebesgue integral.

Proof. As $[a, b] \times [c, d]$ is compact, f is uniformly continuous. So for any $\epsilon > 0$, there is a $\delta > 0$ such that if $|x - x'| < \delta$ and $|y - y'| < \delta$, then $|f(x, y) - f(x', y')| < \epsilon$. Let

$$g_n(x, y) = \begin{cases} f(a, y) & \text{if } x = a, \\ f\left(a + \frac{k(b-a)}{n}, y\right) & \text{if } a + \frac{(k-1)(b-a)}{n} < x \leq a + \frac{k(b-a)}{n}, \ k = 1, \ldots, n. \end{cases}$$

This g_n is the "right-hand rule" step function with subinterval length $\frac{b-a}{n}$: Integrating g_n with respect to x is the right-hand rule for integrating f. Let n be large enough so that $\frac{b-a}{n} < \delta$. Then, via the uniform continuity estimate, we find that $|f(x, y) - g_n(x, y)| < \epsilon$ for all x and y. So

$$\left| \int_a^b f(x, y) \, dx - \int_a^b g_n(x, y) \, dx \right| \leq \int_a^b |f(x, y) - g_n(x, y)| \leq (b - a)\epsilon.$$

So $\int_a^b g_n(x, y) \, dx$ converges uniformly as a function of y to $\int_a^b f(x, y) \, dx$. Finally, the integral of g_n is just the right-hand rule. Putting it all together,

$$\int_c^d \int_a^b f(x, y) \, dx \, dy = \int_c^d \left(\lim_{n \to \infty} \int_a^b g_n(x, y) \, dx \right) dy$$

$$= \lim_{n \to \infty} \int_c^d \left(\int_a^b g_n(x, y) \, dx \right) dy$$

$$= \lim_{n \to \infty} \int_c^d \left(\sum_{k=1}^n \frac{b - a}{n} f\left(a + \frac{k(b - a)}{n}, y\right) \right) dy$$

$$= \lim_{n \to \infty} \sum_{k=1}^n \frac{b - a}{n} \int_c^d f\left(a + \frac{k(b - a)}{n}, y\right) dy$$

$$= \int_a^b \int_c^d f(x, y) \, dy \, dx.$$

The final equation is simply the realization that $\int_c^d f(x, y) \, dy$ is a continuous function of x, hence integrable, and what we have is just the right-hand rule for the integral of this function over $[a, b]$. $\square$

Exercise B.2.2: *Suppose $f(x, y) = 1$ if $x \in \mathbb{Q}$ and $y = 1/2$ and 0 otherwise. Using the Riemann integral, prove that one of $\int_0^1 \int_0^1 f(x, y) \, dx \, dy$ and $\int_0^1 \int_0^1 f(x, y) \, dy \, dx$ exists and the other does not (the integrand is not a well-defined function).*

Exercise B.2.3: *Compute*

$$\int_0^1 \int_0^1 \frac{x^2 - y^2}{(x^2 + y^2)^2} \, dx \, dy \qquad \text{and} \qquad \int_0^1 \int_0^1 \frac{x^2 - y^2}{(x^2 + y^2)^2} \, dy \, dx.$$

You will need to interpret the integrals as improper, that is, the limit of $\int_\epsilon^1$ as $\epsilon \downarrow 0$.

B.2.3i Differentiation under the integral

Let $f(x, y)$ be a function of two variables and

$$g(y) = \int_a^b f(x, y)\, dx.$$

If f is continuous on $[a, b] \times [c, d]$, then Proposition B.2.1 says that g is continuous on $[c, d]$. Suppose f is differentiable in y. Can we "differentiate under the integral?" Differentiation is a limit and we are again asking when do two limits, namely integration and differentiation, commute. The first question we face is the integrability of $\frac{\partial f}{\partial y}$, but the formula can fail even if $\frac{\partial f}{\partial y}$ is integrable as a function of x for every fixed y. We prove a simple, but perhaps the most useful, version of the theorem.

Theorem B.2.3 (Leibniz integral rule). *Suppose* $f \colon [a, b] \times [c, d] \to \mathbb{R}$ *is a continuous function, such that* $\frac{\partial f}{\partial y}$ *exists for all* $(x, y) \in [a, b] \times [c, d]$ *and is continuous. Define*

$$g(y) = \int_a^b f(x, y)\, dx.$$

Then $g \colon [c, d] \to \mathbb{R}$ *is continuously differentiable and*

$$g'(y) = \int_a^b \frac{\partial f}{\partial y}(x, y)\, dx.$$

The hypotheses on f and $\frac{\partial f}{\partial y}$ can be weakened to a degree. See, e.g., Exercise B.2.10. The proof below requires that $\frac{\partial f}{\partial y}$ exists and is continuous as a function of two variables, and the x interval must be the entire closed interval $[a, b]$. The y interval $[c, d]$ can be replaced by a small (open or closed) interval if needed, and in applications, we often make $[c, d]$ be a small interval around the point where we need to differentiate.

Proof. Fix $y \in [c, d]$ and let $\epsilon > 0$ be given. As $\frac{\partial f}{\partial y}$ is continuous on $[a, b] \times [c, d]$, it is uniformly continuous. In particular, there exists $\delta > 0$ such that whenever $y_1 \in [c, d]$ with $|y_1 - y| < \delta$ and all $x \in [a, b]$, we have

$$\left| \frac{\partial f}{\partial y}(x, y_1) - \frac{\partial f}{\partial y}(x, y) \right| < \epsilon.$$

Suppose h is such that $y + h \in [c, d]$ and $|h| < \delta$. Fix x for a moment and apply the mean value theorem to find a y_1 between y and $y + h$ such that

$$\frac{f(x, y + h) - f(x, y)}{h} = \frac{\partial f}{\partial y}(x, y_1).$$

As $|y_1 - y| \leq |h| < \delta$,

$$\left| \frac{f(x, y + h) - f(x, y)}{h} - \frac{\partial f}{\partial y}(x, y) \right| = \left| \frac{\partial f}{\partial y}(x, y_1) - \frac{\partial f}{\partial y}(x, y) \right| < \epsilon.$$

This argument worked for every $x \in [a, b]$. Therefore, as a function of x

$$x \mapsto \frac{f(x, y + h) - f(x, y)}{h} \quad \text{converges uniformly to} \quad x \mapsto \frac{\partial f}{\partial y}(x, y) \quad \text{as } h \to 0.$$

We defined uniform convergence for sequences although the idea is the same. You may replace h with a sequence of nonzero numbers $\{h_n\}$ converging to 0 such that $y + h_n \in [c, d]$ and let $n \to \infty$.

Consider the difference quotient of g,

$$\frac{g(y + h) - g(y)}{h} = \frac{\int_a^b f(x, y + h) \, dx - \int_a^b f(x, y) \, dx}{h} = \int_a^b \frac{f(x, y + h) - f(x, y)}{h} \, dx.$$

Uniform convergence implies the limit can be taken underneath the integral. So

$$\lim_{h \to 0} \frac{g(y + h) - g(y)}{h} = \int_a^b \lim_{h \to 0} \frac{f(x, y + h) - f(x, y)}{h} \, dx = \int_a^b \frac{\partial f}{\partial y}(x, y) \, dx.$$

Then g' is continuous on $[c, d]$ by Proposition B.2.1. $\qquad \square$

Example B.2.4: Consider

$$\int_0^1 \frac{x - 1}{\ln(x)} \, dx.$$

The integral exists as the function under the integral extends continuously to $[0, 1]$, see Exercise B.2.4. Trouble is finding it. Introduce a parameter y and define a function:

$$g(y) = \int_0^1 \frac{x^y - 1}{\ln(x)} \, dx.$$

The function $\frac{x^y - 1}{\ln(x)}$ also extends to a continuous function of x and y for $(x, y) \in [0, 1] \times [0, 1]$ (also in the exercise). See Figure B.3.

Therefore, g is a continuous function of on $[0, 1]$, and $g(0) = 0$. For $0 < \epsilon < 1$, the y derivative of the integrand, x^y, is continuous on $[0, 1] \times [\epsilon, 1]$. Therefore, for $y > 0$ we may differentiate under the integral sign

$$g'(y) = \int_0^1 \frac{\ln(x) x^y}{\ln(x)} \, dx = \int_0^1 x^y \, dx = \frac{1}{y + 1}.$$

We know g is continuous on $[0, 1]$, $g(0) = 0$, and for $y \in (0, 1)$, g is differentiable and $g'(y) = \frac{1}{y+1}$. So $g(1) = \int_0^1 g'(y) \, dy = \ln(2)$. In other words,

$$\int_0^1 \frac{x - 1}{\ln(x)} \, dx = \ln(2).$$

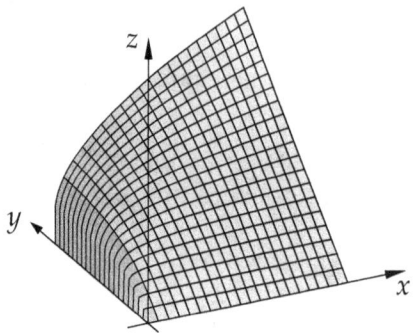

Figure B.3: The graph $z = \frac{x^y - 1}{\ln(x)}$ on $[0,1] \times [0,1]$.

Exercise B.2.4: *Prove the two statements that were asserted in Example B.2.4:*

a) *Prove $\frac{x-1}{\ln(x)}$ extends to a continuous function of $[0,1]$. That is, there exists a continuous function on $[0,1]$ that equals $\frac{x-1}{\ln(x)}$ on $(0,1)$.*

b) *Prove $\frac{x^y - 1}{\ln(x)}$ extends to a continuous function on $[0,1] \times [0,1]$.*

Exercise B.2.5: *Suppose $h \colon \mathbb{R} \to \mathbb{R}$ is continuous and $g \colon \mathbb{R} \to \mathbb{R}$ is continuously differentiable and compactly supported (g is zero outside a compact interval). Define*

$$f(x) = \int_{-\infty}^{\infty} h(y)g(x - y)\, dy.$$

Show that f is differentiable.

Exercise B.2.6: *Suppose $f \colon \mathbb{R} \to \mathbb{R}$ is an infinitely differentiable function (all derivatives exist) such that $f(0) = 0$.*

a) *Show that there exists an infinitely differentiable function $g \colon \mathbb{R} \to \mathbb{R}$ such that $f(x) = x\, g(x)$.*

b) *Show that if $f'(0) \neq 0$, then $g(0) \neq 0$.*

Hint: First write $f(x) = \int_0^x f'(s)\, ds$ and then rewrite the integral to go from 0 to 1.

Exercise B.2.7: *Let $U \subset \mathbb{R}^n$ be open and suppose $f(x, y_1, y_2, \ldots, y_n)$ is a continuous function defined on $[0,1] \times U \subset \mathbb{R}^{n+1}$. Suppose $\frac{\partial f}{\partial y_1}, \frac{\partial f}{\partial y_2}, \ldots, \frac{\partial f}{\partial y_n}$ exist and are continuous on $[0,1] \times U$. Prove that $F \colon U \to \mathbb{R}$ defined by*

$$F(y_1, y_2, \ldots, y_n) = \int_0^1 f(x, y_1, y_2, \ldots, y_n)\, dx$$

is continuously differentiable (the partial derivatives exist and are continuous).

Exercise B.2.8: *Work out the following counterexample: Let*

$$f(x, y) = \begin{cases} \frac{xy^3}{(x^2+y^2)^2} & \text{if } x \neq 0 \text{ or } y \neq 0, \\ 0 & \text{if } x = 0 \text{ and } y = 0. \end{cases}$$

a) *Prove that for each fixed y the function $x \mapsto f(x, y)$ is Riemann integrable on $[0, 1]$ and*

$$g(y) = \int_0^1 f(x, y)\, dx = \frac{y}{2y^2 + 2}.$$

Therefore $g'(y)$ exists and we get the continuous function

$$g'(y) = \frac{1 - y^2}{2(y^2 + 1)^2}.$$

b) *Prove $\frac{\partial f}{\partial y}$ exists at all x and y and compute it.*

c) *Show that for all y*

$$\int_0^1 \frac{\partial f}{\partial y}(x, y)\, dx \qquad \text{exists, but} \qquad g'(0) \neq \int_0^1 \frac{\partial f}{\partial y}(x, 0)\, dx.$$

Exercise B.2.9: *Work out the following counterexample: Let*

$$f(x, y) = \begin{cases} x \sin\left(\frac{y}{x^2+y^2}\right) & \text{if } (x, y) \neq (0, 0), \\ 0 & \text{if } (x, y) = (0, 0). \end{cases}$$

a) *Prove f is continuous on all of $\mathbb{R}^2$. Therefore the following function is well-defined for every $y \in \mathbb{R}$:*

$$g(y) = \int_0^1 f(x, y)\, dx.$$

b) *Prove $\frac{\partial f}{\partial y}$ exists for all (x, y), but is not continuous at $(0, 0)$.*

c) *Show that $\int_0^1 \frac{\partial f}{\partial y}(x, 0)\, dx$ does not exist even if we take improper integrals, that is, that the limit $\lim_{h \downarrow 0} \int_h^1 \frac{\partial f}{\partial y}(x, 0)\, dx$ does not exist.*

Note: Feel free to use what you know about sine and cosine from calculus.

Exercise B.2.10: *Strengthen the Leibniz integral rule in the following way. Suppose $f: (a, b) \times (c, d) \to \mathbb{R}$ is a bounded continuous function, such that $\frac{\partial f}{\partial y}$ exists for all $(x, y) \in (a, b) \times (c, d)$ and is continuous and bounded. Define*

$$g(y) = \int_a^b f(x, y)\, dx.$$

Then $g: (c, d) \to \mathbb{R}$ is continuously differentiable and

$$g'(y) = \int_a^b \frac{\partial f}{\partial y}(x, y)\, dx.$$

Hint: See also Exercise B.2.1 and Theorem B.1.15.

B.3*i*   The derivative in several real variables

B.3.1*i*   The derivative

In the following, the norm $\|\cdot\|$ of a vector in $\mathbb{R}^n$ is the euclidean norm $\|x\| = \sqrt{x_1^2 + \cdots + x_n^2}$. When applied to a linear mapping (a matrix) it is the *operator norm*:

$$\|A\| \overset{\text{def}}{=} \sup_{\|x\|=1} \|Ax\|.$$

The following exercise collects some key facts about the operator norm for the reader who has not seen this norm yet.

Exercise **B.3.1:**
- a) Prove that if A is a linear mapping between finite-dimensional vector spaces, then $\|A\| < \infty$.
- b) Prove that if A is a linear mapping of vector spaces, then $\|Ax\| \leq \|A\|\|x\|$.
- c) Find an explicit 2×2 matrix A and a vector $x \in \mathbb{R}^2$ such that $\|Ax\| < \|A\|\|x\|$.
- d) If A is a $1 \times n$ or $n \times 1$ matrix, then the operator norm $\|A\|$ is the same as the euclidean norm of the entries of A.

The derivative of $f \colon \mathbb{R} \to \mathbb{R}$ at $x \in \mathbb{R}$ exists if there is a number a (the derivative of f at x) such that

$$\lim_{h \to 0} \left| \frac{f(x+h) - f(x)}{h} - a \right| = \lim_{h \to 0} \frac{|f(x+h) - f(x) - ah|}{|h|} = 0.$$

Multiplying by a is a linear map in one dimension: $h \mapsto ah$. So the derivative is a linear map. Let us extend this idea to more variables.

Definition B.3.1. Let $U \subset \mathbb{R}^n$ be open. We say $f \colon U \to \mathbb{R}^m$ is *(real) differentiable* at $x \in U$ if there exists a linear $A \colon \mathbb{R}^n \to \mathbb{R}^m$ such that

$$\lim_{\substack{h \to 0 \\ h \in \mathbb{R}^n}} \frac{\|f(x+h) - f(x) - Ah\|}{\|h\|} = 0.$$

We write $Df|_x = A$ and we say A is the *(real) derivative* of f at x. When f is (real) differentiable at every $x \in U$, we say that f is *(real) differentiable*. See Figure B.4.

Intuitively, f is differentiable at x if f "infinitesimally close" to a linear map near x. We cheated a bit and said that A is *the* derivative, let us prove that we were justified.

Proposition B.3.2. *Let $U \subset \mathbb{R}^n$ be open, $f \colon U \to \mathbb{R}^m$ a function, $x \in U$, and $A, B \colon \mathbb{R}^n \to \mathbb{R}^m$ are linear such that*

$$\lim_{h \to 0} \frac{\|f(x+h) - f(x) - Ah\|}{\|h\|} = 0 \quad \text{and} \quad \lim_{h \to 0} \frac{\|f(x+h) - f(x) - Bh\|}{\|h\|} = 0.$$

Then $A = B$.

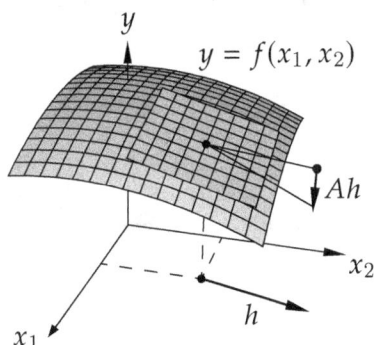

Figure B.4: Illustration of a derivative for a function $f \colon \mathbb{R}^2 \to \mathbb{R}$. The vector h is shown in the $x_1 x_2$-plane based at (x_1, x_2), and the vector $Ah \in \mathbb{R}^1$ is shown along the y direction.

Proof. Suppose $h \in \mathbb{R}^n$, $h \neq 0$. Compute

$$\frac{\|(A - B)h\|}{\|h\|} = \frac{\|-\big(f(x+h) - f(x) - Ah\big) + f(x+h) - f(x) - Bh\|}{\|h\|}$$

$$\leq \frac{\|f(x+h) - f(x) - Ah\|}{\|h\|} + \frac{\|f(x+h) - f(x) - Bh\|}{\|h\|}.$$

So $\frac{\|(A-B)h\|}{\|h\|} = \left\|(A - B)\frac{h}{\|h\|}\right\| \to 0$ as $h \to 0$. Any point on the unit sphere can be written as $\frac{h}{\|h\|}$ for an arbitrarily small h, and a linear mapping vanishing on the unit sphere is zero everywhere. $\square$

Example B.3.3: If $f(x) = Ax$ for a linear mapping A, then $Df|_x = A$:

$$\frac{\|f(x+h) - f(x) - Ah\|}{\|h\|} = \frac{\|A(x+h) - Ax - Ah\|}{\|h\|} = \frac{0}{\|h\|} = 0.$$

Example B.3.4: Let $f \colon \mathbb{R}^2 \to \mathbb{R}^2$ be defined by

$$f(x, y) = \big(f_1(x, y), f_2(x, y)\big) = (1 + x + 2y + x^2, 2x + 3y + xy).$$

Let us show that f is differentiable at the origin and compute the derivative, directly using the definition. If the derivative exists, it can be represented by a 2-by-2 matrix $\left[\begin{smallmatrix} a & b \\ c & d \end{smallmatrix}\right]$. Suppose $h = (h_1, h_2)$. We need the following expression to go to zero.

$$\frac{\|f(h_1, h_2) - f(0, 0) - (ah_1 + bh_2, ch_1 + dh_2)\|}{\|(h_1, h_2)\|} =$$

$$\frac{\sqrt{\big((1-a)h_1 + (2-b)h_2 + h_1^2\big)^2 + \big((2-c)h_1 + (3-d)h_2 + h_1 h_2\big)^2}}{\sqrt{h_1^2 + h_2^2}}.$$

APPENDIX B. RESULTS FROM BASIC ANALYSIS

If we choose $a = 1, b = 2, c = 2, d = 3$, the expression becomes

$$\frac{\sqrt{h_1^4 + h_1^2 h_2^2}}{\sqrt{h_1^2 + h_2^2}} = |h_1| \frac{\sqrt{h_1^2 + h_2^2}}{\sqrt{h_1^2 + h_2^2}} = |h_1|.$$

And this expression does indeed go to zero as $h \to 0$. The function f is differentiable at the origin and the derivative $Df|_0$ is represented by the matrix $\begin{bmatrix} 1 & 2 \\ 2 & 3 \end{bmatrix}$.

Proposition B.3.5. *Let $U \subset \mathbb{R}^n$ be open and $f : U \to \mathbb{R}^m$ be differentiable at $p \in U$. Then f is continuous at p.*

Proof. Another way to write the differentiability of f at p is to consider

$$r(h) = f(p + h) - f(p) - Df|_p h.$$

The function f is differentiable at p if $\frac{\|r(h)\|}{\|h\|}$ goes to zero as $h \to 0$, so $r(h)$ goes to zero. The derivative $Df|_p$ is a linear map from $\mathbb{R}^n$ to $\mathbb{R}^m$ and hence $Df|_p h \to 0$ as $h \to 0$. Thus, $f(p + h)$ goes to $f(p)$ as $h \to 0$. □

The derivative is itself a linear operator on the space of differentiable functions.

Proposition B.3.6. *Suppose $U \subset \mathbb{R}^n$ is open, $f : U \to \mathbb{R}^m$ and $g : U \to \mathbb{R}^m$ are differentiable at p, and $\alpha \in \mathbb{R}$. Then the functions $f + g$ and αf are differentiable at p, and*

$$D(f + g)|_p = Df|_p + Dg|_p \qquad \text{and} \qquad D(\alpha f)|_p = \alpha Df|_p.$$

Proof. Let $h \in \mathbb{R}^n$, $h \neq 0$. The proposition follows from the following estimates:

$$\frac{\left\| f(p + h) + g(p + h) - (f(p) + g(p)) - (Df|_p + Dg|_p)h \right\|}{\|h\|}$$

$$\leq \frac{\left\| f(p + h) - f(p) - Df|_p h \right\|}{\|h\|} + \frac{\left\| g(p + h) - g(p) - Dg|_p h \right\|}{\|h\|},$$

and

$$\frac{\left\| \alpha f(p + h) - \alpha f(p) - \alpha Df|_p h \right\|}{\|h\|} = |\alpha| \frac{\left\| f(p + h)) - f(p) - Df|_p h \right\|}{\|h\|}. \qquad \Box$$

If $A : \mathbb{R}^n \to \mathbb{R}^m$ and $B : \mathbb{R}^m \to \mathbb{R}^k$ are linear, then they are their own derivative. The composition BA, a linear map from $\mathbb{R}^n$ to $\mathbb{R}^k$, is also its own derivative, and so the derivative of the composition is the composition of the derivatives. As differentiable maps are "infinitesimally close" to linear maps, they have the same property:

Theorem B.3.7 (Chain rule)**.** *Let $U \subset \mathbb{R}^n$ be open and let $f : U \to \mathbb{R}^m$ be differentiable at $p \in U$. Let $V \subset \mathbb{R}^m$ be open, $f(U) \subset V$ and let $g : V \to \mathbb{R}^\ell$ be differentiable at $f(p)$. Then*

$$F(x) = g(f(x))$$

is differentiable at p and

$$DF|_p = Dg|_{f(p)} Df|_p.$$

Proof. Let $A = Df|_p$ and $B = Dg|_{f(p)}$. Take a nonzero $h \in \mathbb{R}^n$, write $q = f(p)$, $k = f(p+h) - f(p)$, and let

$$r(h) = f(p+h) - f(p) - Ah.$$

Then $r(h) = k - Ah$ or $Ah = k - r(h)$, and $f(p+h) = q + k$. Then

$$\frac{\|F(p+h) - F(p) - BAh\|}{\|h\|} = \frac{\|g(f(p+h)) - g(f(p)) - BAh\|}{\|h\|}$$

$$= \frac{\|g(q+k) - g(q) - B(k - r(h))\|}{\|h\|}$$

$$\leq \frac{\|g(q+k) - g(q) - Bk\|}{\|h\|} + \|B\| \frac{\|r(h)\|}{\|h\|}$$

First, $\|B\|$ is constant and f is differentiable at p, so $\|B\| \frac{\|r(h)\|}{\|h\|} \to 0$ as $h \to 0$. Next, if $k = 0$, then $\frac{\|g(q+k) - g(q) - Bk\|}{\|h\|} = 0$ and so assume $k \neq 0$. We have

$$\frac{\|g(q+k) - g(q) - Bk\|}{\|h\|} = \frac{\|g(q+k) - g(q) - Bk\|}{\|k\|} \frac{\|f(p+h) - f(p)\|}{\|h\|}.$$

As f is continuous at p, we have that $k \to 0$ as $h \to 0$. Therefore, $\frac{\|g(q+k) - g(q) - Bk\|}{\|k\|} \to 0$ because g is differentiable at q. Finally,

$$\frac{\|f(p+h) - f(p)\|}{\|h\|} \leq \frac{\|f(p+h) - f(p) - Ah\|}{\|h\|} + \|A\|.$$

As f is differentiable at p, the right-hand side stays bounded as $h \to 0$. So $\frac{\|g(q+k) - g(q) - Bk\|}{\|h\|} \to 0$ as $h \to 0$. Hence, $\frac{\|F(p+h) - F(p) - BAh\|}{\|h\|} \to 0$, and $DF|_p = BA$. $\square$

Let us prove a "mean value theorem" for vector-valued functions. For a function $\varphi \colon [a, b] \to \mathbb{R}^n$, we think of the derivative $D\varphi|_{t_0}$ as a vector in $\mathbb{R}^n$, and as a vector we will write it as $\varphi'(t_0)$. It is not hard to check that the entries of the matrix $D\varphi|_{t_0}$ (a linear map from $\mathbb{R}$ to $\mathbb{R}^n$) are just the derivatives of the components of φ, and $D\varphi|_{t_0} h = h\varphi'(t_0)$. Then $\|\varphi'(t_0)\|$ is the euclidean norm in $\mathbb{R}^n$. In fact, in this setting it is the same as the operator norm of $D\varphi|_{t_0}$.

Lemma B.3.8. *If $\varphi \colon [a, b] \to \mathbb{R}^n$ is differentiable on (a, b) and continuous on $[a, b]$, then there exists a $t_0 \in (a, b)$ such that*

$$\|\varphi(b) - \varphi(a)\| \leq (b - a)\|\varphi'(t_0)\|.$$

Proof. By the mean value theorem on the scalar-valued function $t \mapsto (\varphi(b) - \varphi(a)) \cdot \varphi(t)$, where the dot is the dot product, we obtain that there is a $t_0 \in (a, b)$ such that

$$\|\varphi(b) - \varphi(a)\|^2 = (\varphi(b) - \varphi(a)) \cdot (\varphi(b) - \varphi(a))$$

$$= (\varphi(b) - \varphi(a)) \cdot \varphi(b) - (\varphi(b) - \varphi(a)) \cdot \varphi(a)$$

$$= (b - a)(\varphi(b) - \varphi(a)) \cdot \varphi'(t_0).$$

By the Cauchy–Schwarz inequality

$$\|\varphi(b) - \varphi(a)\|^2 = (b - a)\big(\varphi(b) - \varphi(a)\big) \cdot \varphi'(t_0) \leq (b - a)\|\varphi(b) - \varphi(a)\|\,\|\varphi'(t_0)\|. \quad \square$$

Recall that a set U is convex if whenever $x, y \in U$, the line segment from x to y lies in U.

Proposition B.3.9. *Let $U \subset \mathbb{R}^n$ be a convex open set, $f : U \to \mathbb{R}^m$ a differentiable function, and M be such that*

$$\|Df|_x\| \leq M \qquad \text{for all } x \in U.$$

Then f is Lipschitz with constant M, that is,

$$\|f(x) - f(y)\| \leq M\|x - y\| \qquad \text{for all } x, y \in U.$$

Proof. Fix $x, y \in U$. By convexity, $(1 - t)x + ty \in U$ for all $t \in [0, 1]$. Next

$$\frac{d}{dt}\Big[f\big((1 - t)x + ty\big)\Big] = Df|_{((1-t)x+ty)}(y - x).$$

By the mean value theorem above, for some $t_0 \in (0, 1)$,

$$\|f(x) - f(y)\| \leq \left\|\frac{d}{dt}\Big|_{t=t_0}\Big[f\big((1 - t)x + ty\big)\Big]\right\|$$

$$\leq \big\|Df|_{((1-t_0)x+t_0y)}\big\|\,\|y - x\| \leq M\|y - x\|. \qquad \square$$

Let us solve the differential equation $Df = 0$.

Corollary B.3.10. *If $U \subset \mathbb{R}^n$ is open and connected, $f : U \to \mathbb{R}^m$ is differentiable, and $Df|_x = 0$ for all $x \in U$, then f is constant.*

Proof. For any $x \in U$, there is an open ball $B(x, \delta) \subset U$. The ball $B(x, \delta)$ is convex. Since $\|Df|_y\| \leq 0$ for all $y \in B(x, \delta)$, then $\|f(x) - f(y)\| \leq 0\|x - y\| = 0$. Thus $f^{-1}(c)$ is open for any $c \in \mathbb{R}^m$. Suppose $f^{-1}(c)$ is nonempty. The two sets

$$U' = f^{-1}(c), \qquad U'' = f^{-1}\big(\mathbb{R}^m \setminus \{c\}\big)$$

are open and disjoint, and further $U = U' \cup U''$. As U' is nonempty and U is connected, then $U'' = \emptyset$. So $f(x) = c$ for all $x \in U$. $\qquad \square$

Exercise B.3.2: Using only the definition of the derivative, show that the following $f : \mathbb{R}^2 \to \mathbb{R}^2$ are differentiable at the origin and find their derivative.
   a) $f(x, y) = (1 + x + xy, x)$,
   b) $f(x, y) = \big(y - y^{10}, x\big)$,
   c) $f(x, y) = \big((x + y + 1)^2, (x - y + 2)^2\big)$.

Exercise B.3.3: Define $f : \mathbb{R}^2 \to \mathbb{R}^2$ by $f(x, y) = \big(x, y + \varphi(x)\big)$ for some differentiable function φ of one variable. Show f is differentiable and find Df.

Exercise B.3.4: Suppose $f : \mathbb{R}^n \to \mathbb{R}$ and $h : \mathbb{R}^n \to \mathbb{R}$ are differentiable and $Df|_x = Dh|_x$ for all $x \in \mathbb{R}^n$. Prove that if $f(0) = h(0)$, then $f(x) = h(x)$ for all $x \in \mathbb{R}^n$.

B.3.2i   The derivative in terms of partial derivatives

Partial derivatives are easier to compute with all the machinery of calculus, and they provide a way to compute the derivative of a function.

Proposition B.3.11. *Let $U \subset \mathbb{R}^n$ be open and let $f: U \to \mathbb{R}^m$ be differentiable at $p \in U$. Then all the partial derivatives at p exist and, in terms of the standard bases of $\mathbb{R}^n$ and $\mathbb{R}^m$, $Df|_p$ is represented by the matrix*

$$
\begin{bmatrix}
\frac{\partial f_1}{\partial x_1}\big|_p & \frac{\partial f_1}{\partial x_2}\big|_p & \cdots & \frac{\partial f_1}{\partial x_n}\big|_p \\
\frac{\partial f_2}{\partial x_1}\big|_p & \frac{\partial f_2}{\partial x_2}\big|_p & \cdots & \frac{\partial f_2}{\partial x_n}\big|_p \\
\vdots & \vdots & \ddots & \vdots \\
\frac{\partial f_m}{\partial x_1}\big|_p & \frac{\partial f_m}{\partial x_2}\big|_p & \cdots & \frac{\partial f_m}{\partial x_n}\big|_p
\end{bmatrix}.
$$

In other words,

$$
Df|_p\, e_j = \sum_{k=1}^{m} \frac{\partial f_k}{\partial x_j}\bigg|_p e_k,
$$

where e_j denote the vectors of the standard basis in the appropriate space. Recall that the standard basis element e_j is the vector with all zeros except a 1 at the jth entry.

Proof. Fix a j and note that for nonzero h,

$$
\left\| \frac{f(p + he_j) - f(p)}{h} - Df|_p\, e_j \right\| = \left\| \frac{f(p + he_j) - f(p) - Df|_p\, he_j}{h} \right\|
$$
$$
= \frac{\| f(p + he_j) - f(p) - Df|_p\, he_j \|}{\| he_j \|}.
$$

As h goes to 0, the right-hand side goes to zero by differentiability of f, and hence

$$
\lim_{h \to 0} \frac{f(p + he_j) - f(p)}{h} = Df|_p\, e_j.
$$

Let us represent f by components $f = (f_1, f_2, \ldots, f_m)$, since it is vector-valued. Taking a limit in $\mathbb{R}^m$ is the same as taking the limit in each component separately. For any k,

$$
\frac{\partial f_k}{\partial x_j}\bigg|_p = \lim_{h \to 0} \frac{f_k(p + he_j) - f_k(p)}{h}
$$

exists and is equal to the kth component of $Df|_p\, e_j$, and we are done. $\square$

The converse of the proposition is not true. Just because the partial derivatives exist, does not mean that the function is differentiable. However, when the partial derivatives are continuous, the converse holds.

Definition B.3.12. Let $U \subset \mathbb{R}^n$ be open. We say $f \colon U \to \mathbb{R}^m$ is *continuously differentiable*, if all partial derivatives $\frac{\partial f_j}{\partial x_k}$ exist and are continuous.*

Proposition B.3.13. *Let $U \subset \mathbb{R}^n$ be open. If $f \colon U \to \mathbb{R}^m$ is continuously differentiable, then f is differentiable.*

Proof. Fix $p \in U$. We do induction on dimension. The case $n = 1$ is left as an exercise. Suppose the conclusion is true for $\mathbb{R}^{n-1}$, that is, if we restrict to the first $n-1$ variables, the function is differentiable. The first $n-1$ partial derivatives of f restricted to the set where the last coordinate is fixed are the same as those for f. In the following, by a slight abuse of notation, we think of $\mathbb{R}^{n-1}$ as a subset of $\mathbb{R}^n$, that is, the set in $\mathbb{R}^n$ where $x_n = 0$. In other words, we identify the vectors $(x_1, x_2, \ldots, x_{n-1})$ and $(x_1, x_2, \ldots, x_{n-1}, 0)$. Let

$$
A = \begin{bmatrix} \frac{\partial f_1}{\partial x_1}\big|_p & \cdots & \frac{\partial f_1}{\partial x_n}\big|_p \\ \vdots & \ddots & \vdots \\ \frac{\partial f_m}{\partial x_1}\big|_p & \cdots & \frac{\partial f_m}{\partial x_n}\big|_p \end{bmatrix}, \quad A' = \begin{bmatrix} \frac{\partial f_1}{\partial x_1}\big|_p & \cdots & \frac{\partial f_1}{\partial x_{n-1}}\big|_p \\ \vdots & \ddots & \vdots \\ \frac{\partial f_m}{\partial x_1}\big|_p & \cdots & \frac{\partial f_m}{\partial x_{n-1}}\big|_p \end{bmatrix}, \quad v = \begin{bmatrix} \frac{\partial f_1}{\partial x_n}\big|_p \\ \vdots \\ \frac{\partial f_m}{\partial x_n}\big|_p \end{bmatrix}.
$$

Let $\epsilon > 0$ be given. By the induction hypothesis, there is a $\delta > 0$ such that for any $h' \in \mathbb{R}^{n-1}$ with $\|h'\| < \delta$,

$$
\frac{\|f(p + h') - f(p) - A'h'\|}{\|h'\|} < \epsilon.
$$

By continuity of the partial derivatives, suppose δ is small enough so that

$$
\left| \frac{\partial f_k}{\partial x_n}\Big|_{p+h} - \frac{\partial f_k}{\partial x_n}\Big|_p \right| < \epsilon,
$$

for all k and all $h \in \mathbb{R}^n$ with $\|h\| < \delta$.

　　Suppose $h = h' + t e_n$ is a vector in $\mathbb{R}^n$, where $h' \in \mathbb{R}^{n-1}$, $t \in \mathbb{R}$, such that $\|h\| < \delta$. Then $\|h'\| \leq \|h\| < \delta$. Note that $Ah = A'h' + tv$.

$$
\begin{aligned}
\|f(p + h) - f(p) - Ah\| &= \|f(p + h' + t e_n) - f(p + h') - tv + f(p + h') - f(p) - A'h'\| \\
&\leq \|f(p + h' + t e_n) - f(p + h') - tv\| + \|f(p + h') - f(p) - A'h'\| \\
&\leq \|f(p + h' + t e_n) - f(p + h') - tv\| + \epsilon \|h'\|.
\end{aligned}
$$

As all the partial derivatives exist, by the mean value theorem, for each k there is a θ_k, where $|\theta_k| \leq |t|$, such that

$$
f_k(p + h' + t e_n) - f_k(p + h') = t \frac{\partial f_k}{\partial x_n}\Big|_{(p + h' + \theta_k e_n)}.
$$

---

*Alternatively, people define f being continuously differentiable if $Df|_x$ is a continuous function taking $x \in U$ to the space of linear operators. The propositions in this section say the two definitions are equivalent.

Note that $\|h' + \theta_k e_n\| \le \|h\| < \delta$. To finish,

$$\|f(p+h) - f(p) - Ah\| \le \|f(p + h' + te_n) - f(p+h') - tv\| + \epsilon\|h'\|$$

$$\le \sqrt{\sum_{k=1}^{m} \left(t\frac{\partial f_k}{\partial x_n}\Big|_{(p+h'+\theta_k e_n)} - t\frac{\partial f_k}{\partial x_n}\Big|_p \right)^2} + \epsilon\|h'\|$$

$$\le \sqrt{m}\,\epsilon|t| + \epsilon\|h'\|$$

$$\le (\sqrt{m}+1)\epsilon\|h\|. \qquad\qquad \square$$

Exercise B.3.5: *Prove the base case in Proposition B.3.13: If $n = 1$ and "the partials exist and are continuous," then the function is differentiable. Note that f is vector-valued.*

Exercise B.3.6: *Define a function $f: \mathbb{R}^2 \to \mathbb{R}$ by (see Figure B.5)*

$$f(x,y) = \begin{cases} \frac{xy}{x^2+y^2} & \text{if } (x,y) \ne (0,0), \\ 0 & \text{if } (x,y) = (0,0). \end{cases}$$

a) *Show that partial derivatives $\frac{\partial f}{\partial x}$ and $\frac{\partial f}{\partial y}$ exist at all points (including the origin).*
b) *Show that f is not continuous at the origin (and hence not differentiable).*
c) *Show that the partial derivatives are not continuous.*

Exercise B.3.7: *Define $f: \mathbb{R}^2 \to \mathbb{R}$ as*

$$f(x,y) = \begin{cases} (x^2+y^2)\sin\big((x^2+y^2)^{-1}\big) & \text{if } (x,y) \ne (0,0), \\ 0 & \text{if } (x,y) = (0,0). \end{cases}$$

Show that f is differentiable at the origin, but that it is not continuously differentiable.

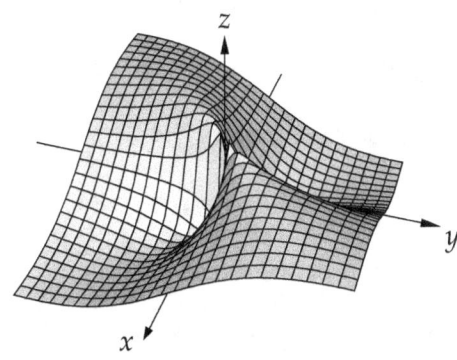

Figure B.5: Graph of $\frac{xy}{x^2+y^2}$.

B.3.3*i*   Fixed point theorem

Before we prove the inverse function theorem we must take a detour to prove a fixed point theorem for metric spaces.

Definition B.3.14. Let (X, d_X) and (Y, d_Y) be metric spaces. A mapping $f: X \to Y$ is said to be a *contraction* (or a contractive map) if it is a k-Lipschitz map for some $k < 1$, i.e., if there exists a $k < 1$ such that

$$d_Y\big(f(p), f(q)\big) \leq k\, d_X(p, q) \qquad \text{for all } p, q \in X.$$

If $f: X \to X$ is a map, $x \in X$ is called a *fixed point* if $f(x) = x$.

Theorem B.3.15 (Contraction mapping principle or Banach fixed point theorem). *Let (X, d) be a nonempty complete metric space and $f: X \to X$ a contraction. Then f has a unique fixed point.*

Proof. Pick any $x_0 \in X$. Define a sequence $\{x_n\}$ by $x_{n+1} = f(x_n)$.

$$d(x_{n+1}, x_n) = d\big(f(x_n), f(x_{n-1})\big) \leq kd(x_n, x_{n-1}) \leq \cdots \leq k^n d(x_1, x_0).$$

Suppose $m > n$, then

$$d(x_m, x_n) \leq \sum_{\ell=n}^{m-1} d(x_{\ell+1}, x_\ell) \leq \sum_{\ell=n}^{m-1} k^\ell d(x_1, x_0) = k^n d(x_1, x_0) \sum_{\ell=0}^{m-n-1} k^\ell$$

$$\leq k^n d(x_1, x_0) \sum_{\ell=0}^{\infty} k^\ell = k^n d(x_1, x_0) \frac{1}{1-k}.$$

So the sequence is Cauchy. Since X is complete, let $x = \lim x_n$. We claim that x is our unique fixed point.

Fixed point? The function f is a contraction, so it is Lipschitz continuous:

$$f(x) = f(\lim x_n) = \lim f(x_n) = \lim x_{n+1} = x.$$

Unique? Let x and y both be fixed points.

$$d(x, y) = d\big(f(x), f(y)\big) \leq k\, d(x, y).$$

As $k < 1$, we have $d(x, y) = 0$ and hence $x = y$. The theorem is proved. $\square$

The proof is constructive. Not only do we know a unique fixed point exists. We also know how to find it. Start with any $x_0 \in X$ and iterate $f(x_0), f(f(x_0)), f(f(f(x_0)))$, etc.

Exercise **B.3.8:**
  a) *Find an example of a contraction $f: X \to X$ of a non-complete metric space X with no fixed point.*
  b) *Find a 1-Lipschitz map $f: X \to X$ of a complete metric space X with no fixed point.*

Exercise **B.3.9:** *Let $f(x) = x - \frac{x^2-2}{2x}$ (you may recognize Newton's method for $\sqrt{2}$).*
  a) *Prove $f([1,\infty)) \subset [1,\infty)$.*
  b) *Prove that $f: [1,\infty) \to [1,\infty)$ is a contraction.*
  c) *Apply the fixed point theorem to find an $x \geq 1$ such that $f(x) = x$, and show that $x = \sqrt{2}$.*

B.3.4*i* Inverse function theorem

Intuitively, we again consider that if a function is continuously differentiable, then it locally "behaves like" the derivative (a linear function). The idea of the inverse function theorem is that if a function is continuously differentiable and the derivative is invertible, the function is (locally) invertible.

Theorem B.3.16 (Inverse function theorem). *Suppose $U \subset \mathbb{R}^n$ is open, $f: U \to \mathbb{R}^n$ is continuously differentiable, $p \in U$, and $Df|_p$ is invertible (that is, $\det Df|_p \neq 0$). Then there exist open sets $V, W \subset \mathbb{R}^n$ such that $p \in V \subset U$, $f(V) = W$, the restriction $f|_V$ is injective (one-to-one), and hence a $g: W \to V$ exists such that $g(y) = (f|_V)^{-1}(y)$ for all $y \in W$. See Figure B.6. Furthermore, g is continuously differentiable and*

$$Dg|_y = (Df|_x)^{-1}, \qquad \text{for all } x \in V, y = f(x).$$

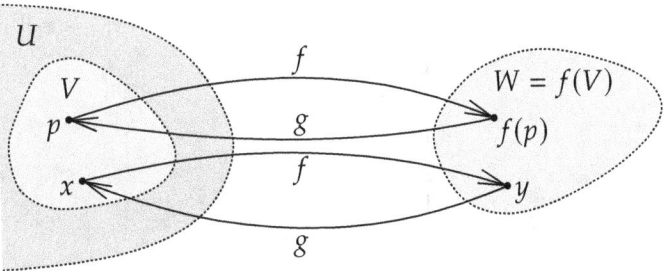

Figure B.6: Setup of the inverse function theorem in $\mathbb{R}^n$.

Proof. Write $A = Df|_p$. As Df is continuous, there exists an open ball V around p such that

$$\|A - Df|_x\| < \frac{1}{2\|A^{-1}\|} \qquad \text{for all } x \in V.$$

The inequality implies that $Df|_x$ is invertible for all $x \in V$ (see exercise below).

Given $y \in \mathbb{R}^n$, define $\varphi_y \colon C \to \mathbb{R}^n$ by

$$\varphi_y(x) = x + A^{-1}\big(y - f(x)\big).$$

As A^{-1} is one-to-one, $\varphi_y(x) = x$ (x is a fixed point) if only if $y - f(x) = 0$, or in other words $f(x) = y$. Using the chain rule we obtain

$$D\varphi_y|_x = I - A^{-1}Df|_x = A^{-1}\big(A - Df|_x\big).$$

So for $x \in V$,

$$\|D\varphi_y|_x\| \leq \|A^{-1}\|\,\|A - Df|_x\| < \tfrac{1}{2}.$$

As V is a ball, it is convex. Hence,

$$\|\varphi_y(x_1) - \varphi_y(x_2)\| \leq \frac{1}{2}\|x_1 - x_2\| \qquad \text{for all } x_1, x_2 \in V.$$

In other words, φ_y is a contraction defined on V, though we so far do not know what is the range of φ_y. We cannot yet apply the fixed point theorem, but we can say that φ_y has at most one fixed point in V: If $\varphi_y(x_1) = x_1$ and $\varphi_y(x_2) = x_2$, then $\|x_1 - x_2\| = \|\varphi_y(x_1) - \varphi_y(x_2)\| \leq \frac{1}{2}\|x_1 - x_2\|$, so $x_1 = x_2$. That is, there exists at most one $x \in V$ such that $f(x) = y$, and so $f|_V$ is one-to-one.

Let $W = f(V)$ and let $g \colon W \to V$ be the inverse of $f|_V$. We need to show that W is open. Take a $y_0 \in W$. There is a unique $x_0 \in V$ such that $f(x_0) = y_0$. Let $r > 0$ be small enough such that the closed ball $C(x_0, r) \subset V$ (such $r > 0$ exists as V is open). Suppose y is such that

$$\|y - y_0\| < \frac{r}{2\|A^{-1}\|}.$$

If we show that $y \in W$, then we have shown that W is open. If $x_1 \in C(x_0, r)$, then

$$\begin{aligned}
\|\varphi_y(x_1) - x_0\| &\leq \|\varphi_y(x_1) - \varphi_y(x_0)\| + \|\varphi_y(x_0) - x_0\| \\
&\leq \frac{1}{2}\|x_1 - x_0\| + \|A^{-1}(y - y_0)\| \\
&\leq \frac{1}{2}r + \|A^{-1}\|\,\|y - y_0\| \\
&< \frac{1}{2}r + \|A^{-1}\|\frac{r}{2\|A^{-1}\|} = r.
\end{aligned}$$

So φ_y takes $C(x_0, r)$ into $B(x_0, r) \subset C(x_0, r)$. It is a contraction on $C(x_0, r)$ and $C(x_0, r)$ is complete (closed subset of $\mathbb{R}^n$ is complete). Apply the contraction mapping principle (Theorem B.3.15) to obtain a fixed point x, i.e., $\varphi_y(x) = x$. That is, $f(x) = y$, and $y \in f\big(C(x_0, r)\big) \subset f(V) = W$. Therefore, W is open.

Next we need to show that g is continuously differentiable and compute its derivative. First, let us show that it is differentiable. Consider $y \in W$ and $k \in \mathbb{R}^n$, $k \neq 0$, such that $y + k \in W$. Because $f|_V$ is a one-to-one and onto mapping of V onto

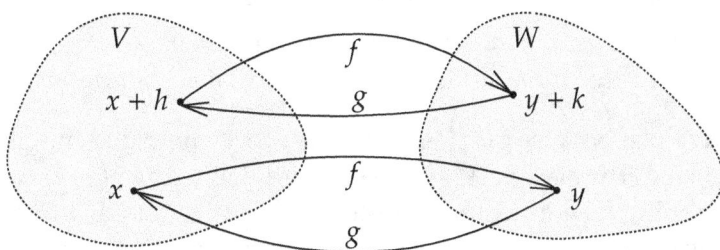

Figure B.7: Proving that g is differentiable.

W, there are unique $x \in V$ and $h \in \mathbb{R}^n$, where $h \neq 0$ and $x + h \in V$, such that $f(x) = y$ and $f(x + h) = y + k$. In other words, $g(y) = x$ and $g(y + k) = x + h$. See Figure B.7.

We can still squeeze some information from the fact that φ_y is a contraction.

$$\varphi_y(x + h) - \varphi_y(x) = h + A^{-1}\big(f(x) - f(x + h)\big) = h - A^{-1}k.$$

So

$$\|h - A^{-1}k\| = \|\varphi_y(x + h) - \varphi_y(x)\| \leq \frac{1}{2}\|x + h - x\| = \frac{\|h\|}{2}.$$

By the inverse triangle inequality, $\|h\| - \|A^{-1}k\| \leq \frac{1}{2}\|h\|$. Hence,

$$\|h\| \leq 2\|A^{-1}k\| \leq 2\|A^{-1}\|\,\|k\|.$$

In particular, as k goes to 0, so does h.

As $x \in V$, then $Df|_x$ is invertible. Let $B = \big(Df|_x\big)^{-1}$, which is what we think the derivative of g at y is. Then

$$
\begin{aligned}
\frac{\|g(y + k) - g(y) - Bk\|}{\|k\|} &= \frac{\|h - Bk\|}{\|k\|} \\
&= \frac{\|h - B\big(f(x + h) - f(x)\big)\|}{\|k\|} \\
&= \frac{\|B\big(f(x + h) - f(x) - Df|_x h\big)\|}{\|k\|} \\
&\leq \|B\| \frac{\|h\|}{\|k\|} \frac{\|f(x + h) - f(x) - Df|_x h\|}{\|h\|} \\
&\leq 2\|B\|\,\|A^{-1}\| \frac{\|f(x + h) - f(x) - Df|_x h\|}{\|h\|}.
\end{aligned}
$$

As k goes to 0, so does h, and, as f is differentiable, the right-hand side goes to zero. So g is differentiable, and B is precisely what we claimed $Dg|_y$ to be.

Let us show g is continuously differentiable. The function $g\colon W \to V$ is continuous (it is differentiable), Df is a continuous function from V to the space of $n \times n$ matrices (each entry is continuous), and the inverse of a matrix M is $M^{-1} = \frac{1}{\det M} \operatorname{adj} M$, so $M \mapsto M^{-1}$ is continuous outside the set where $\det M = 0$. As $Dg|_y = (Df|_{g(y)})^{-1}$ is the composition of these three continuous functions, it is continuous.   □

Example B.3.17: Just because $Df|_x$ is invertible everywhere does not mean that f is one-to-one globally. The map $f : \mathbb{R}^2 \setminus \{0\} \to \mathbb{R}^2 \setminus \{0\}$ defined by $f(x, y) = (x^2 - y^2, 2xy)$, that is, the mapping $z \mapsto z^2$ in the complex plane. It is not hard to check that the derivative is invertible on $\mathbb{R}^2 \setminus \{0\}$. On the other hand, the mapping is 2-to-1 globally (except at the origin). For every $(a, b) \neq (0, 0)$, there are exactly two solutions to $x^2 - y^2 = a$ and $2xy = b$.

The invertibility of the derivative is not a necessary condition, just sufficient, for having a continuous inverse and being an open mapping. For example, the function $f(x) = x^3$ is an open mapping from $\mathbb{R}$ to $\mathbb{R}$ and is globally one-to-one with a continuous inverse, although the inverse is not differentiable at $x = 0$.

Remark B.3.18. As a side note, there is a famous related, and as yet unsolved, problem called the *Jacobian conjecture*. If $F : \mathbb{R}^n \to \mathbb{R}^n$ (or more famously $F : \mathbb{C}^n \to \mathbb{C}^n$) is polynomial (each component is a polynomial) and $\det DF$ is a nonzero constant, does F have a polynomial inverse? The inverse function theorem gives a local C^1 inverse, but can one always find a global polynomial inverse is the question.

Exercise B.3.10:
   a) *Suppose A is a linear operator on $\mathbb{R}^n$ such that $\|I - A\| < 1$ (I is the identity). Prove that A is invertible.*
   b) *For two linear operators A and B on $\mathbb{R}^n$ where A is invertible, prove that $\|A - B\| < \frac{1}{\|A^{-1}\|}$ implies that B is invertible.*

Exercise B.3.11: *Define $f : \mathbb{R}^2 \to \mathbb{R}^2$ by $f(x, y) = (x, y + h(x))$ for some continuously differentiable function h of one variable.*
   a) *Show that f is one-to-one and onto.*
   b) *Compute Df.*
   c) *Show that Df is invertible at all points, and compute its inverse.*

Exercise B.3.12: *Define $f : \mathbb{R}^2 \to \mathbb{R}^2$*

$$f(x, y) = \begin{cases} (x^2 \sin(1/x) + x/2, y) & \text{if } x \neq 0, \\ (0, y) & \text{if } x = 0. \end{cases}$$

   a) *Show that f is differentiable everywhere.*
   b) *Show that $Df|_{(0,0)}$ is invertible.*
   c) *Show that f is not one-to-one in every neighborhood of the origin (it is not locally invertible, that is, the inverse function theorem does not work).*
   d) *Show that f is not continuously differentiable.*

Exercise **B.3.13** (Polar coordinates)*: Define a mapping $F(r, \theta) = \big(r\cos(\theta), r\sin(\theta)\big)$.*

    a) *Show that F is continuously differentiable (for all $(r, \theta) \in \mathbb{R}^2$).*

    b) *Compute $DF|_{(0,\theta)}$ for all θ.*

    c) *Show that if $r \neq 0$, then $DF|_{(r,\theta)}$ is invertible, and so an inverse of F exists locally as long as $r \neq 0$.*

    d) *Show that $F \colon \mathbb{R}^2 \to \mathbb{R}^2$ is onto, and for each point $(x, y) \in \mathbb{R}^2$, the set $F^{-1}(x, y)$ is infinite.*

    e) *Show that $F|_{(0,\infty)\times[0,2\pi)}$ is one-to-one and onto $\mathbb{R}^2 \setminus \{(0,0)\}$.*

C*i* Basic Notation and Terminology

Let us quickly review some basic notation used. We use $\mathbb{C}$, $\mathbb{R}$ for complex and real numbers (i for imaginary unit), $\mathbb{N} = \{1, 2, 3, \ldots\}$ for the natural numbers, $\mathbb{Z}$ for all integers, and $\mathbb{Q}$ for rational real numbers.

We denote the set subtraction by $Y \setminus X$ (all elements of Y that are not in X). We write the complement of a set as X^c, in which case the ambient set should be clear. The topological closure of a set X is denoted by $\overline{X}$ and its boundary by ∂X. By ∂X we may also mean the path that gives the topological boundary traversed counterclockwise. We write the interior of X as X°.

The notation $f: X \to Y$ is a function with domain X and codomain Y. By $f(S)$ we mean the direct image of S by f. By f^{-1} we mean the inverse image of sets and single points, and if f is bijective (one-to-one and onto), we use it for the inverse mapping. To define a function without necessarily giving it a name, we use

$$x \mapsto F(x),$$

where $F(x)$ would generally be some formula giving the output. The notation $f|_S$ means the restriction of f to S: a function $f|_S: S \to Y$ such that $f|_S(x) = f(x)$ for all $x \in S$. For derivatives, vertical bar means evaluation, $\frac{\partial f}{\partial x}\big|_p$ means $\frac{\partial f}{\partial x}$ evaluated at p. To say that two functions f and g are identically equal, that is that $f(x) = g(x)$ for all x in the domain, we write

$$f \equiv g.$$

The notation $f \circ g$ denotes the composition defined by $x \mapsto f(g(x))$.

For one-sided limits we use

$$\lim_{t \uparrow a} f(t) \quad \left(= \lim_{\substack{t \to a \\ t < a}} f(t) \right) \quad \text{and} \quad \lim_{t \downarrow a} f(t) \quad \left(= \lim_{\substack{t \to a \\ t > a}} f(t) \right),$$

as these seemed the clearer option in some of the situations in this book. We may write $\{x_n\}$ for a sequence $\{x_n\}_{n=1}^{\infty}$ and similarly $\lim x_n$ instead of $\lim_{n \to \infty} x_n$ when it is clear that n is the index of the sequence.

To define X to be Y rather than just show equality, we write

$$X \stackrel{\text{def}}{=} Y.$$

Further Reading

[B] Ralph P. Boas, *Invitation to complex analysis*, 2nd ed., MAA Textbooks, Mathematical Association of America, Washington, DC, 2010. Revised by Harold P. Boas. MR2674618

[C1] John B. Conway, *Functions of one complex variable*, 2nd ed., Graduate Texts in Mathematics, vol. 11, Springer-Verlag, New York-Berlin, 1978. MR503901

[C2] _____, *Functions of one complex variable. II*, Graduate Texts in Mathematics, vol. 159, Springer-Verlag, New York, 1995. MR1344449

[L1] Jiří Lebl, *Basic analysis I: Introduction to real analysis, Volume I.* https://www.jirka.org/ra/.

[L2] _____, *Basic analysis II: Introduction to real analysis, Volume II.* https://www.jirka.org/ra/.

[L3] _____, *Tasty bits of several complex variables.* https://www.jirka.org/scv/.

[R1] Walter Rudin, *Principles of mathematical analysis*, 3rd ed., McGraw-Hill Book Co., New York-Auckland-Düsseldorf, 1976. International Series in Pure and Applied Mathematics. MR0385023

[R2] _____, *Real and complex analysis*, 3rd ed., McGraw-Hill Book Co., New York, 1987. MR924157

[U] David C. Ullrich, *Complex made simple*, Graduate Studies in Mathematics, vol. 97, American Mathematical Society, Providence, RI, 2008. MR2450873

Index

List of Notation

Notation	Description	Page
$\mathbb{N}$	natural numbers $\{1, 2, 3, \ldots\}$	8
$\mathbb{Z}$	integers	8
$\mathbb{Q}$	rational numbers	8
$\mathbb{R}$	real numbers	8
$\mathbb{C}$	complex numbers	9
i	$\sqrt{-1}$	9
$\bar{z}$	complex conjugate	9
$\operatorname{Re} z$	real part	9
$\operatorname{Im} z$	imaginary part	9
$|z|$	modulus	10
$\Delta_r(a)$	disc	11
$\mathbb{D}$	unit disc	11
$\mathbb{H}$	upper half-plane $\{z \in \mathbb{C} : \operatorname{Im} z > 0\}$	11
$\exp(z)$, e^z	the exponential	14
$\arg z$	argument of z	16
$\operatorname{Arg} z$	the principal branch of the $\arg z$	17
$\mathbb{C}_\infty$	the Riemann sphere	18
∞	the Riemann sphere infinity	18
$\mathbb{CP}^1$	one-dimensional projective space	22
$[z : w]$	point in $\mathbb{CP}^1$	23
$(z_1, z_2; z_3, z_4)$	cross ratio	25
$o(|h|)$	little-oh notation	26
$f'(z)$	complex derivative	27

Notation	Description	Page
$\dfrac{df}{dz}$	complex derivative	27
$\dfrac{\partial}{\partial z}, \dfrac{\partial}{\partial \bar{z}}$	Wirtinger operators	32
$\displaystyle\sum_{n=0}^{\infty} c_n(z-p)^n$	power series (at p)	41
$C^1, C^1(X,Y)$	continuously differentiable functions	52, 236
$\displaystyle\int_{\gamma} f(z)\,dz$	path integral	53
dz	$dz = dx + i\,dy$	55
$d\bar{z}$	$d\bar{z} = dx - i\,dy$	55
$\displaystyle\int_{\partial U} f(z)\,dz$	integral over boundary	56
$\displaystyle\int_{\gamma} f(z)\,|dz|$	arclength integral	57
$\displaystyle\int_{\Gamma} f(z)\,dz, \int_{a_1\gamma_1+\cdots+a_n\gamma_n} f(z)\,dz$	integral over a cycle	58
$\Gamma_1 = \Gamma_2$	cycle equivalence	58
$-\Gamma$	cycle equivalence	59
$[z,w]$	line segment from z to w	60
dA	area form, $dA = dx\,dy = r\,dr\,d\theta$	70
$\|f\|_K$	supremum or uniform norm of f	79, 268
$Cf, C[f]$	Cauchy transform	82
φ_a	automorphism of $\mathbb{D}$, $\frac{z-a}{1-\bar{a}z}$	87
$\operatorname{Log} z$	principal branch of the log	91
$\log z, \log|z|$	the logarithm	92
$n(\gamma;p)$	the winding number of γ around p	94, 111
$d(p,X)$	distnace from p to set X	96
$\operatorname{ann}(p;r_1,r_2)$	annulus	103
$\displaystyle\sum_{n=-\infty}^{\infty} c_n(z-p)^n$	Laurent series (at p)	104
$\operatorname{Res}(f;p)$	residue of f at p	131

Notation	Description	Page
∇^2	Laplacian	165
$P_r(\theta)$	Poisson kernel for the unit disc	171
$Pf, P[f]$	Poisson integral of f	172, 174
$\prod\limits_{n=1}^{\infty}(1 + a_n)$	infinite product	195
$E_m(z)$	elementary factor	199
$\widehat{K}$	polynomial hull	214
$d(x, y)$	metric/distance	231
(X, d)	metric space	231
$C(X, Y)$	continuous functions $f : X \to Y$	234
$\mathrm{diam}(S)$	diameter of S	235
$B(p, \delta), B_X(p, \delta)$	open ball in a metric space	237
$C(p, \delta), C_X(p, \delta)$	closed ball in a metric space	237
$\overline{X}$	topological closure	243, 294
X°	interior of X	244, 294
∂X	boundary of X	244, 294
$\{x_n\}, \{x_n\}_{n=1}^{\infty}$	sequence	245
$\{x_{n_j}\}, \{x_{n_j}\}_{j=1}^{\infty}$	subsequence	246
$\lim x_n, \lim\limits_{n \to \infty} x_n$	limit of a sequence	246
$\lim\limits_{x \to c} f(x)$	limit of a function	264
$\|x\|$	euclidean norm for $x \in \mathbb{R}^n$	280
$\|A\|$	operator norm for a linear operator A	280
$Df, Df\|_p$	real derivative of a mapping (at p)	280
$Y \setminus X$	set subraction	294
X^c	complement	294
$f : X \to Y$	a function from X to Y	294
$x \mapsto F(x)$	a function of x	294
$f\|_S$	restriction of f to S	294
$\equiv$	identically equal, equal at all points	294

Notation	Description	Page
$f \circ g$	composition, $p \mapsto f(g(p))$	294
$\lim_{t \uparrow c} f(t)$	one-sided limit (from below)	294
$\lim_{t \downarrow c} f(t)$	one-sided limit (from above)	294
$X \overset{\text{def}}{=} Y$	define X to be Y	294